# Mechanobiology

From Molecular Sensing to Disease

# Mechanobiology
## From Molecular Sensing to Disease

*Edited by*

GLEN L. NIEBUR, PHD
Director
Bioengineering Graduate Program
University of Notre Dame, Notre Dame
Indiana
United States

Professor
Aerospace and Mechanical Engineering
University of Notre Dame

ELSEVIER

---

**Notices**

Practitioners and researchers must always rely on their own experience and knowledge in evaluating and using any information, methods, compounds or experiments described herein. Because of rapid advances in the medical sciences, in particular, independent verification of diagnoses and drug dosages should be made. To the fullest extent of the law, no responsibility is assumed by Elsevier, authors, editors or contributors for any injury and/or damage to persons or property as a matter of products liability, negligence or otherwise, or from any use or operation of any methods, products, instructions, or ideas contained in the material herein.

---

*Publisher:* Oliver Walter
*Acquisition Editor:* Priscilla Braglia
*Editorial Project Manager:* Anna Dubnow
*Production Project Manager:* Sreejith Viswanathan
*Cover Designer:* Alan Studholme

Working together
to grow libraries in
developing countries

www.elsevier.com • www.bookaid.org

*To Marcia and Max*

# List of Contributors

**Daniel P. Ahern, MCh, MRCSI, PhD**
Candidate
Trinity Centre for Biomedical Engineering
Trinity Biomedical Sciences Institute
Trinity College
Dublin, Ireland

Department of Mechanical and Manufacturing
    Engineering
School of Engineering
Trinity College Dublin
Dublin, Ireland

Department of Surgery
School of Medicine
Trinity College Dublin
Dublin, Ireland

**Mark Alber**
Department of Mathematics
University of California
Riverside, CA, United States

Center for Quantitative Modeling in Biology
University of California
Riverside, CA, United States

School of Medicine
University of California
Riverside, CA, United States

Department of Bioengineering
University of California
Riverside, CA, United States

**Johana Barrientos**
Department of Biological Sciences
Wright State University
Dayton, OH, United States

**Michael T.K. Bramson**
Biomedical Engineering Department
Rensselaer Polytechnic Institute
Troy, NY, United States

**Joseph S. Butler, PhD, FRCS**
Clinical Associate Professor
School of Medicine
Trinity College Dublin
Consultant Surgeon
National Spinal Injuries Unit
Department of Trauma & Orthopaedic Surgery
Mater Misericordiae University Hospital
Dublin, Ireland

**Weitao Chen**
Department of Mathematics
University of California
Riverside, CA, United States

Center for Quantitative Modeling in Biology
University of California
Riverside, CA, United States

**David T. Corr**
Biomedical Engineering Department
Rensselaer Polytechnic Institute
Troy, NY, United States

**Matthew E. Dolack**
Department of Mechanical and Nuclear Engineering
The Pennsylvania State University
University Park, PA, United States

**Henry J. Donahue, PhD**
Alice T. and William H. Goodwin Jr.
    Professor and Distinguished Chair
Department of Biomedical Engineering
Virginia Commonwealth University
Richmond, VA, United States

**Michael P. Duffy, MSc, PhD**
Candidate
Department of Biomedical Engineering
Columbia University
New York, NY, United States

**Michael A. Friendman, PhD**
Department of Biomedical Engineering
Virginia Commonwealth University
Richmond, VA, United States

**Diego A. Garzón-Alvarado**
Biomimetics Laboratory
Instituto de Biotecnología
Universidad Nacional de Colombia
Bogotá, Colombia

Numerical Methods and Modeling Research Group
Universidad Nacional de Colombia
Bogotá, Colombia

**Damian C. Genetos, PhD**
Associate Professor
Anatomy, Physiology, and Cell Biology
UC Davis
Davis, CA, United States

School of Veterinary Medicine
Baton Rouge, LA, United States

**Matthew Goelzer, MS**
Mechanical and Biomedical Engineering
Boise State University
Boise, ID, United States

**David A. Hoey, PhD**
Associate Professor
Biomedical Engineering
Trinity Centre for Biomedical Engineering
Trinity Biomedical Sciences Institute
Trinity College
Dublin, Ireland

Department of Mechanical and Manufacturing
    Engineering
School of Engineering
Trinity College Dublin
Dublin, Ireland

Advanced Materials and Bioengineering Research
    Centre
Trinity College Dublin & RCSI
Dublin, Ireland

**Minyi Hu, PhD**
Stony Brook University
Stony Brook, NY, United States

**Ethylin Wang Jabs**
Department of Genetics and Genomic Sciences
Icahn School of Medicine at Mount Sinai
New York, NY, United States

**Reuben H. Kraft**
Department of Mechanical and Nuclear Engineering
The Pennsylvania State University
University Park, PA, United States

Department of Biomedical Engineering
The Pennsylvania State University
University Park, PA, United States

**Chanyoung Lee**
Coulter Department of Biomedical Engineering
Georgia Institute of Technology/Emory University
Atlanta, GA, United States

**Jiun Liou, Jr., PhD**
Department of Bioengineering
Swanson School of Engineering
University of Pittsburgh
Pittsburgh, PA, United States

**Jing Liu, PhD**
Department of Physics
Indiana University-Purdue University Indianapolis
Indianapolis, IN, United States

**Maureen E. Lynch, PhD**
University of Colorado Boulder
Boulder, CO, United States

**Arsalan Marghoub**
Department of Mechanical Engineering
University College London
London, United Kingdom

**Kaitlin P. McCreery, BS**
University of Colorado Boulder
Boulder, CO, United States

**Megan R. Mc Fie, MSc, PhD**
Candidate
School of Engineering and Materials Science
Queen Mary University of London
London, United Kingdom

**Mehran Moazen**
Department of Mechanical Engineering
University College London
London, United Kingdom

**Corey P. Neu, PhD**
University of Colorado Boulder
Boulder, CO, United States

**Glen L. Niebur, PhD**
Professor
Tissue Mechanics Laboratory
Bioengineering Graduate Program and Department of
   Aerospace and Mechanical Engineering
University of Notre Dame
Notre Dame, IN, United States

**Yi-Xian Qin, PhD**
Stony Brook University
Stony Brook, NY, United States

**Joan T. Richtsmeier**
Department of Anthropology
The Pennsylvania State University
University Park, PA, United States

**Ying Ru**
Department of Genetics and Genomic Sciences
Icahn School of Medicine at Mount Sinai
New York, NY, United States

**Benjamin Seelbinder, MS**
University of Colorado Boulder
Boulder, CO, United States

**Jason A. Shar, BS**
Department of Mechanical and Materials Engineering
Wright State University
Dayton, OH, United States

**Jason E. Shoemaker, PhD**
Department of Chemical and Petroleum Engineering
Swanson School of Engineering
Department of Computational and Systems Biology
School of Medicine
McGowan Institute for Regenerative Medicine
University of Pittsburgh
Pittsburgh, PA, United States

**Philippe Sucosky, PhD**
Department of Mechanical and Materials Engineering
Wright State University
Dayton, OH, United States

**Clare L. Thompson, PhD**
Postdoctoral Researcher
School of Engineering and Materials Science
Queen Mary University of London
London, United Kingdom

**William R. Thompson, DPT, PhD**
Assistant Professor
Department of Physical Therapy
School of Health & Human Sciences
Indiana University
Indianapolis, IN, United States

**Gunes Uzer, PhD**
Assistant Professor
Mechanical and Biomedical Engineering
Boise State University
Boise, ID, United States

**Sarah K. Van Houten**
Biomedical Engineering Department
Rensselaer Polytechnic Institute
Troy, NY, United States

**Jonathan P. Vande Geest, PhD**
Department of Bioengineering
Swanson School of Engineering
McGowan Institute for Regenerative Medicine
Department of Ophthalmology
School of Medicine
Louis J. Fox Center for Vision Restoration
University of Pittsburgh
Pittsburgh, PA, United States

**Vijay Velagala**
Department of Chemical and Biomolecular
   Engineering
University of Notre Dame
Notre Dame, IN, United States
Bioengineering Graduate Program
University of Notre Dame
Notre Dame, IN, United States

**Jeremiah J. Zartman**
Department of Chemical and Biomolecular
   Engineering
University of Notre Dame
Notre Dame, IN, United States
Bioengineering Graduate Program
University of Notre Dame
Notre Dame, IN, United States

# Preface: *Mechanobiology*, why not?

## HISTORY OF MECHANOBIOLOGY

Mechanobiology is, in some ways, a new field. The first reference to the term *mechanobiology* in the US government's PubMed database was by Marjolein van der Meulen in 1993.[1] Indeed, the term mechanobiology was coined specifically for this paper at a meeting of the Stanford-Palo Alto Bone Remodeling club. The group had realized that biomechanical neither was the correct description for the phenomenon they were describing nor was it sufficient to use the adjective "mechanical" to modify biological terms.[2] They noted

*The term mechanobiological is used to emphasize that the mechanical effects we are modeling are also dependent upon a biological response.*[1]

Within a few years, a number of papers from laboratories at Stanford and the Palo Alto VA adopted the term. It came into more widespread use throughout the 1990s. One of the seminal definitions of the term appeared in the article *"Why Mechanobiology?"* published in the *Journal of Biomechanics* in 2001.[3] In this article the authors laid out the central precept of mechanobiology for the skeleton:

*The premise of mechanobiology is that these biological processes are regulated by signals to cells generated by mechanical loading, a concept dating back to Roux.[4] The relevant questions include how external and muscle loads are transferred to the tissues, how the cells sense these loads, and how the signals are translated into the cascade of biochemical reactions to produce cell expression or differentiation. Ultimately, we want to predict growth and differentiation in quantitative terms, based on a given force exerted on a given tissue matrix populated by cells.*[3]

This definition has been readily adapted to other tissues by changing "muscle loading" to a variety of force generators and "tissue matrix populated by cells" with any number of tissues or even cells themselves. At the same time that a new field was being defined, the authors noted that it was in some ways a very old field. The underlying principles of mechanobiology were outlined a century earlier in the work by Julius Wolff and Wilhelm Roux. Wolff published his monograph "Das Gesetz der Transformation der Knochen" ("The Law of Bone Remodeling")[5,6] in 1892, and Roux published "Der Kampf Der Theile im Organismus" ("The Challenges to the Parts in the Organism") in which he proposed the idea of functional adaptation to external stimuli in 1881.[4] In 1885, Roux published a second treatise on developmental mechanics ("Die Entwicklungsmechanik; ein neuer Zweig der biologischen Wissenschaft" or "Developmental mechanics: a new branch of biological science").[7,8]

Before the emergence of the term mechanobiology, a significant amount of research was described by the ubiquitously applied misnomer "Wolff's law for …", which was used to describe the concept of tissue adaptation or healing under the influence of mechanical loads in tendons, ligaments,[9–11] and skin,[12] among other tissues.[13]

"Wolff's law" has certainly seen its share of criticism for its simplicity, conflation of differing processes, over-reliance on mechanics relative to hormonal and nutritional factors, and incorrect understanding of mechanics.[14,15] Similarly, Roux's writings intermixed concepts of heredity and adaptation, using examples of ducks and cows. However, one must remember that the existence of cells had only been known for a few decades at the time, Charles Darwin was a contemporary, and the discovery of DNA was far in the future. In this context, these researchers established concepts and asked questions that remain at the heart of mechanobiology today, even if their initial hypotheses have been superseded and augmented with newly discovered biological structures and phenomena over time.

Much of the early work in mechanobiology focused on the musculoskeletal system, formalizing the ideas and incorporating growing knowledge of cellular processes. Advanced mechanical analysis, made possible by the finite element method, provided a means to calculate the spatial distribution of the mechanical signal, which was correlated to locations of bone formation or resorption or to changes in the local density of the bone.[16–18] Similarly, cartilage[19] and tendon[20–22] adaptation and remodeling were simulated using algorithms that incorporated mechanical cues. Cardiovascular mechanobiology research also grew, leveraging similar computational methods but applying differing theories of cellular response and remodeling.[23] As computational power has grown, cell and molecular level knowledge has been incorporated into simulations to include the local cell density, the density of local signaling molecules, and nutrient diffusion.[24,25]

## MECHANOBIOLOGY ACROSS LENGTH SCALES

At the largest length scales, mechanobiology has been used to explain how activities of daily living and environment affect the composition, geometry, or healing of the tissue. For example, multiple mechanobiological models have attempted to predict the shape and scaling of bones with respect to mechanical loading[26,27] or to investigate rehabilitation regimens that can best reestablish normal function.[28]

At the tissue scale the focus is often on the changes in extracellular matrix (ECM) properties. In these models the cell populations can often aggregate as concentrations. The interactions of cells with the matrix are quantified by the local strain or fluid flow. The choice of constitutive model has proven to be important in many instances, with fiber-reinforced or poroelastic descriptions improving the fidelity of calculations of the mechanical behavior. The mechanobiological processes are hypothesized to alter the mechanical properties by defining rate laws for any or all of the constitutive parameters that depend on the number and the state of the various cells. Nutrients or signaling molecules can also be described by concentrations. Their diffusion and convection are typically governed by classical laws, but additional production and consumption laws must also be postulated based on the cell populations present.

In parallel with these macroscopic phenomenological models, novel methods were being developed to determine whether cells actually sense and respond to mechanical cues. Flow chambers, stretched surfaces, patterned surfaces, and materials of varying mechanical properties were all used to determine whether cells altered gene or protein expression in response to mechanical cues. Lower and upper bounds of shearing stresses, stretches, and pressures have since been established for mechanobiological response of a range of cell phenotypes. In a seminal paper, Engler and colleagues[29] showed that culturing stem cells on materials with differing constitutive properties alters their differentiation fate.

Mechanobiology is now studied as an inherently multiscale phenomenon. Starting from the cell scale, the role of mechanotransduction in sensing the presence or localized motions of surrounding cells or ECM is the first stage in the mechanobiological cascade. At this level, it is essential to understand how forces are transmitted to the physiologic structures that connect the cells to the ECM or to other cells—the integrins and the adherens, respectively—as well as the cytoskeleton and cell membrane, and how they are transduced into biochemical actions within the cell. Forces on these structures can induce gene translation, post-translational modification of proteins, or upregulation of secondary messengers that drive protein translation and cell differentiation through both autocrine and paracrine signaling. This signaling may alter local cell phenotypes, cell signaling, further ECM production and degradation, or the organization of ECM molecules.

## WHY NOT MECHANOBIOLOGY?

Most cells contain the machinery necessary to sense and respond to loads, even if other factors may dominate their function. While mechanical loading is obviously an essential part of the function of musculoskeletal and cardiovascular tissues, almost all tissues are subjected to some mechanical forces and deformation. Both epithelial and endothelial surfaces can be subjected to fluid flows and stretching. Even baseline respiration in mammals causes cyclic motion and pressure changes in the thoracic and abdominal cavities, which would induce mechanical stress in the organs. For example, mechanobiology plays a role in liver regeneration[30] and normal kidney function.[31,32] Interestingly, molecules involved in mechanosensing in the kidney—yes-associated protein (YAP) and transcriptional coactivator with PDZ-binding motif (TAZ)—also play a role in bone mechanobiology[33,34] and many other cell lineages.[35] Identification of this and other molecular targets may provide a link to mechanobiological functions in other tissues and organ systems.

Over the past 25 years, mechanobiology has grown into a distinct field with research centers, journals, and conferences dedicated to its study. It is a uniquely interdisciplinary field that rests at an interface between engineering, biology, materials science, and physics. It relies on many other new and growing fields such as bioinformatics, gene editing, and high-resolution imaging to achieve its aims. Mechanobiology is now firmly established in the biology and bioengineering lexicon, and in 2018, 308 articles indexed in the PubMed database specifically used one of the terms mechanobiology or mechanobiological in the abstract or title. There is an excellent textbook on mechanobiology,[36] and courses are offered at many universities. A multisite National Science Foundation is highly active, and there are other institutes and centers at universities throughout the world.

New technologies, including animal models, organ-/tissue-on-a-chip, and engineered tissues, coupled with gene editing and cellular level probes, have opened endless possibilities for mechanobiology research. It is now much easier to probe the mechanical response of systems and to understand how they are coupled to both health and disease. As such, we have entered the era of "why not mechanobiology."

## OVERVIEW

In the first five chapters of this book, mechanobiological effects in different tissues are reviewed and discussed. As mechanobiology has taken root in various fields, the theories and methods have evolved independently. It is often valuable to look across fields to understand new potential pathways and molecules that may play similar or differing roles in other systems. Across these chapters, we can see the recurring themes of understanding the mechanical environment of the tissue and how that is translated to the cells. Many themes are repeated across these tissues, especially the role of cell-matrix and cell-cell adhesions via integrins, cadherens, and connexins. We can also see that similar pathways and cellular structures are involved in all tissues, but with varying effects.

Some key aspects of mechanosensing are addressed in the next two chapters. Cellular structures that can detect deformations of the matrix or the cell membrane, which can in turn be translated to gene and protein expression, cell proliferation, motility, and other physical actions that alter tissues, consume energy, or create waste products. Knowledge of how cells can act as sensors of force or matrix deformation is essential to study mechanobiology and understand the biological pathways that can potentially be activated by mechanics.

The experimental chapters describe unique approaches to study mechanobiology. Animal models, especially gene knockout and knockin models, provide a unique means to study mechanobiology. However, in vitro experiments in the context of tissue engineering and cell culture are essential complements to the in vivo models because they allow the mechanics of the system to be tightly controlled. In the context of tissue engineering, bioreactors can simultaneously serve as both developmental platforms for regenerative medicine and powerful experimental platforms for mechanobiology, especially when coupled with genetically engineered cell sources[37] or small molecule activators or inhibitors of pathways of interest. Underlying biological mechanisms can be further probed by cell culture using reporter cells to quantify the mechanics of individual cells.

The final two chapters consider modeling of mechanobiology in development and tissue morphogenesis. Computational modeling can provide the best approach to develop and test hypotheses in mechanobiology. Multiphysics modeling software allows researchers to explore the interactions of mechanics, transport, and growth and to explore scenarios that lead to experimentally testable hypotheses. Development provides a unique model with which to study mechanobiology because the tissue and organ morphology can change rapidly, and the effects of interventions are easily observed. Researchers have studied the mechanobiology of development using animal[38,39] and even human models.[40,41]

**Glen L. Niebur**

*Tissue Mechanics Laboratory, Bioengineering Graduate Program, Department of Aerospace and Mechanical Engineering, University of Notre Dame, IN, USA*

## REFERENCES

1. van der Meulen MC, Beaupre GS, Carter DR. Mechanobiologic influences in long bone cross-sectional growth. *Bone*. 1993;14:635−642.
2. van der Meulen MC. Origins of the term mechanobiology. *Personal Commun*. 2019. June, 2019.
3. van der Meulen MCH, Huiskes R. Why mechanobiology?; a survey article. *J Biomech*. 2002;35:401−414.
4. Roux W. *Der kampf der teile im organismus*. Leipzig: Engelmann; 1881.
5. Wolff J. *Das gesetz der transformation der knochen*. Berlin: Hirschwald; 1892.
6. Wolff J. *The law of bone remodelling*. Berlin; New York: Springer-Verlag; 1986.
7. Roux W. *Die entwicklungsmechanik; ein neuer zweig der biologischen wissenschaft*. Leipzig: Wilhelm Engelmann; 1905a.
8. Roux W. *Die entwicklungsmechanik; ein neuer zweig der biologischen wissenschaft*. Leipzig: Wilhelm Engelmann; 1905b.
9. McGaw WT. The effect of tension on collagen remodelling by fibroblasts: A stereological ultrastructural study. *Connect Tissue Res*. 1986;14:229−235.
10. Urschel JD, Scott PG, Williams HT. The effect of mechanical stress on soft and hard tissue repair; a review. *Br J Plast Surg*. 1988;41:182−186.

11. Driessen NJ, Peters GW, Huyghe JM, Bouten CV, Baaijens FP. Remodelling of continuously distributed collagen fibres in soft connective tissues. *J Biomech.* 2003;36:1151–1158.

12. Forrester JC, Zederfeldt BH, Hayes TL, Hunt TK. Wolff's law in relation to the healing skin wound. *J Trauma.* 1970;10:770–779.

13. Arem AJ, Madden JW. Is there a Wolff's law for connective tissue? *Surg Forum.* 1974;25:512–514.

14. Bertram JE, Swartz SM. The 'law of bone transformation': A case of crying wolff? *Biol Rev Camb Philos Soc.* 1991;66:245–273.

15. Cowin SC. The false premise of wolff's law. *Forma.* 1997;12:247–262.

16. Fyhrie DP, Carter DR. Femoral head apparent density distribution predicted from bone stresses. *J Biomech.* 1990;23:1–10.

17. Van Rietbergen B, Huiskes R, Weinans H, Sumner DR, Turner TM, Galante JO. ESB research award 1992. The mechanism of bone remodeling and resorption around press-fitted THA stems. *J Biomech.* 1993;26:369–382.

18. Carter DR, Van Der Meulen MC, Beaupre GS. Mechanical factors in bone growth and development. *Bone.* 1996;18:5S–10S.

19. Carter DR, Wong M. Modelling cartilage mechanobiology. *Philos Trans R Soc Lond B Biol Sci.* 2003;358:1461–1471.

20. Malaviya P, Butler DL, Korvick DL, Proch FS. In vivo tendon forces correlate with activity level and remain bounded: Evidence in a rabbit flexor tendon model. *J Biomech.* 1998;31:1043–1049.

21. Malaviya P, Butler DL, Boivin GP, Smith FN, Barry FP, Murphy JM, Vogel KG. An in vivo model for load-modulated remodeling in the rabbit flexor tendon. *J Orthop Res.* 2000;18:116–125.

22. Wren TA, Beaupre GS, Carter DR. Mechanobiology of tendon adaptation to compressive loading through fibrocartilaginous metaplasia. *J Rehabil Res Dev.* 2000;37:135–143.

23. Taber LA, Humphrey JD. Stress-modulated growth, residual stress, and vascular heterogeneity. *J Biomech Eng.* 2001;123:528–535.

24. Checa S, Prendergast PJ. A mechanobiological model for tissue differentiation that includes angiogenesis: A lattice-based modeling approach. *Ann Biomed Eng.* 2009;37:129–145.

25. Khayyeri H, Checa S, Tagil M, O'Brien FJ, Prendergast PJ. Tissue differentiation in an in vivo bioreactor: In silico investigations of scaffold stiffness. *J Mater Sci Mater Med.* 2010;21:2331–2336.

26. Fischer KJ, Jacobs CR, Carter DR. Computational method for determination of bone and joint loads using bone density distributions. *J Biomech.* 1995;28:1127–1135.

27. Fischer KJ, Jacobs CR, Levenston ME, Carter DR. Different loads can produce similar bone density distributions. *Bone.* 1996;19:127–135.

28. Lacroix D, Prendergast PJ. A mechano-regulation model for tissue differentiation during fracture healing: Analysis of gap size and loading. *J Biomech.* 2002;35:1163–1171.

29. Engler AJ, Sen S, Sweeney HL, Discher DE. Matrix elasticity directs stem cell lineage specification. *Cell.* 2006;126:677–689.

30. Song Z, Gupta K, Ng IC, Xing J, Yang YA, Yu H. Mechanosensing in liver regeneration. *Semin Cell Dev Biol.* 2017;71:153–167.

31. Neal CR, Crook H, Bell E, Harper SJ, Bates DO. Three-dimensional reconstruction of glomeruli by electron microscopy reveals a distinct restrictive urinary subpodocyte space. *J Am Soc Nephrol.* 2005;16:1223–1235.

32. Endlich K, Kliewe F, Endlich N. Stressed podocytes-mechanical forces, sensors, signaling and response. *Pflugers Arch.* 2017;469:937–949.

33. Kegelman CD, Mason DE, Dawahare JH, Horan DJ, Vigil GD, Howard SS, Robling AG, Bellido TM, Boerckel JD. Skeletal cell YAP and TAZ combinatorially promote bone development. *FASEB J.* 2018;32:2706–2721.

34. McDermott AM, Herberg S, Mason DE, Collins JM, Pearson HB, Dawahare JH, Tang R, Patwa AN, Grinstaff MW, Kelly DJ, Alsberg E, Boerckel JD. Recapitulating bone development through engineered mesenchymal condensations and mechanical cues for tissue regeneration. *Sci Transl Med.* 2019;11.

35. Dupont S, Morsut L, Aragona M, Enzo E, Giulitti S, Cordenonsi M, Zanconato F, Le Digabel J, Forcato M, Bicciato S, Elvassore N, Piccolo S. Role of yap/taz in mechanotransduction. *Nature.* 2011;474:179–183.

36. Jacobs CR, Huang H, Kwon RY. *Introduction to Cell Mechanics and Mechanobiology.* Garland Science; 2012.

37. Adkar SS, Wu CL, Willard VP, Dicks A, Ettyreddy A, Steward N, Bhutani N, Gersbach CA, Guilak F. Step-wise chondrogenesis of human induced pluripotent stem cells and purification via a reporter allele generated by crispr-cas9 genome editing. *Stem Cells.* 2019;37:65–76.

38. Nowlan NC, Murphy P, Prendergast PJ. A dynamic pattern of mechanical stimulation promotes ossification in avian embryonic long bones. *J Biomech.* 2007.

39. Nowlan NC, Murphy P, Prendergast PJ. Mechanobiology of embryonic limb development. *Ann N Y Acad Sci.* 2007b;1101:389–411.

40. Verbruggen SW, Loo JH, Hayat TT, Hajnal JV, Rutherford MA, Phillips AT, Nowlan NC. Modeling the biomechanics of fetal movements. *Biomech Model Mechanobiol*. 2016;15:995—1004.

41. Verbruggen SW, Kainz B, Shelmerdine SC, Arthurs OJ, Hajnal JV, Rutherford MA, Phillips ATM, Nowlan NC. Altered biomechanical stimulation of the developing hip joint in presence of hip dysplasia risk factors. *J Biomech*. 2018;78:1—9.

# Contents

**CHAPTER 1.1**

# Osteocyte Mechanobiology in Aging and Disease

HENRY J. DONAHUE, PHD • MICHAEL J. FRIENDMAN, PHD • DAMIAN GENETOS, PHD

## 1. INTRODUCTION

Advances in healthcare enable longer lives, although the proportion of quality living years has not kept pace.[1,2] Thus a longer lifespan frequently engenders adverse effects, including loss of muscle mass and function (i.e., sarcopenia), idiopathic or senile osteoporosis, reductions in joint mobility, and osteoarthritis. These disorders promote physical inactivity, which reduces bone mass, microarchitecture, and strength, to increase fracture risk. Although fractures may be successfully repaired, the adverse effects, such as decreased mobility and loss of independence, increase postfracture morbidity and mortality. In the United States, musculoskeletal diseases affect half of the persons aged 18 years or older and three of four people over the age of 65 years.[3] Because women are at a greater risk of osteoporotic fracture, osteoporosis diagnosis and treatment often focus on women; however, mortality rates due to fractures are greater in men within 1 year post fracture.[4] Moreover, the incidence of osteoporosis and increases in skeletal fragility are exacerbated by lifestyle choices, including type 2 diabetes, smoking, and low physical activity. The current frequency of osteoporotic fractures, projected future fracture rate, and associated socioeconomic burden[5] demand a means to reduce, if not eliminate, osteoporosis. However, this is unlikely without a thorough understanding of the cellular processes that contribute to, and are dysregulated by, aging and disease (Fig. 1.1.1).

## 2. MECHANICAL LOADING EFFECTS ON BONE: MECHANOTRANSDUCTION

Historically the skeleton was considered a mineral reservoir, storing calcium and phosphorus until necessary to serve the functions of other organs.[6] As such, postnatal changes in bone mass, diameter, length, etc. were attributed to variations in hormonal milieu or serum ion concentration, rather than a response to the mechanical loading environment.[7] Motivated by serendipitous observations on the relationship between trabecular alignment in the femoral neck and the estimated principal stress directions therein,[8] skeletal adaptation as a consequence of mechanical loading gained support and acceptance. Skeletal adaptation to the mechanical environment may be most easily observed in conditions of disuse: decreasing externally applied loads—through casting, limb immobilization, or microgravity—reduces bone mass via periosteal and endosteal resorption. Conversely, dynamically applied external loads increase bone mass through concerted reductions in osteoclastogenesis, conversion of bone-lining cells to osteoblasts, and osteoblast formation and activation.[9]

Many bones are naturally curved, owing to the combined influences of chondral growth and bone modeling.[8] Applied loads transiently exacerbate their inherent curvature, generating compressive and tensile stress gradients perpendicular to the bone surface. The differential stresses promote site-specific bone resorption (at areas of tension) and formation (at compressive sites) to reduce the magnitude of the applied strain. Yet any compression or tension does not, nor should it, elicit a mechanoadaptive response. Rather than simply respond to any nonzero strain, Frost proposed that time-averaged mechanical strains within or on a bone elicit a skeletal response if they occur above or below a specific strain threshold.[10] Analogous to using a thermostat to establish the temperature of a house, the mechanostat is a

Mechanobiology. https://doi.org/10.1016/B978-0-12-817931-4.00001-7

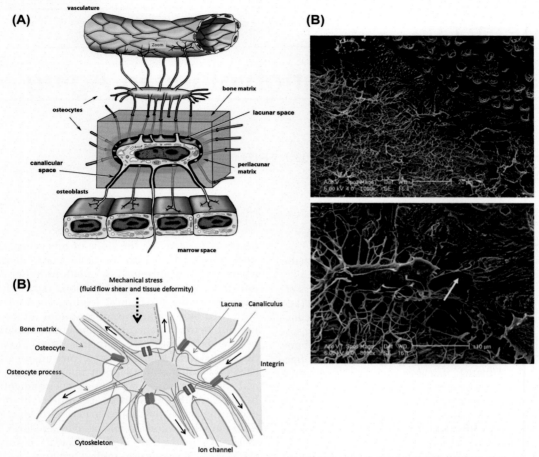

FIG. 1.1.1 **(A)** Depiction of osteocyte morphology, lacunocanalicular organization, and interaction with other osteocytes, surface osteoblasts, bone marrow, and the vasculature. *From S.L. Dallas, M. Prideaux, L.F. Bonewald, The osteocyte: an endocrine cell ... and more, Endocr. Rev., 34 (2013), pp. 658–690.* **(B)** Scanning electron microscopic image of an acid-etched resin embedded murine osteocyte demonstrating its numerous and tortuous canalicular network. *From Lynda F. Bonewald, Osteocyte Biology In Robert Marcus, David Feldman,... Jane A. Cauley Eds. Osteoporosis (Fourth Edition), (2013), pp. 209–234.* **(C)** Mechanisms whereby osteocytes regulate remodeling within and throughout a bone. *From Sakhr, A. Murshid, The role of osteocytes during experimental orthodontic tooth movement: A review, Archives of Oral Biology, 73, (2017), pp. 25–33, 2017.*

theoretic set point that must be surpassed or unmet before initiating structurally appropriate alterations to bone mass and architecture. In this model, strains below a set point promote osteoclast formation and bone resorption in order to reduce bone mass, thereby minimizing the metabolic cost of unnecessary mass, and strains above the set point increase bone mass and the strength to prevent pathologic fractures. Feedback—in the form of increased bone cross section that reduces bone strain—thereby limits the adaptive response.[11] Beyond strain magnitude, other critical determinants of

skeletal response to externally applied load include the nature of the strain (dynamic vs. static) and the time over which it is applied.[9]

Mechanotransduction refers to the complex interactions whereby tissue-level strains are converted to localized biophysical signals that ultimately promote skeletal adaptation; per Duncan and Turner, it consists of four unique stages[12]:

(1) Mechanocoupling: Conversion of tissue-level loads into localized mechanical signals perceived by mechanosensitive cells.

(2) Biochemical coupling: Transduction of localized mechanical signals into biochemical responses in mechanosensitive cells.

(3) Signal transmission: From mechanosensory cell to effector cell.

(4) Effector cell response: Initiation of tissue-level response.

Within this model, the strain induced by bone bending during physical activity or exercise, approximately 400–3000 microstrain ($\mu\varepsilon$), induces a plethora of biophysical signals within the bone tissue, consisting of interstitial fluid flow, direct mechanical strain, hydrostatic pressure, and electrokinetic effects on bone cells.[13,14] Rapid responses (0 s–1 min) to biophysical forces include the generation or liberation of second messengers such as $Ca^{2+}$, cyclic AMP, diacylglycerol, and inositol triphosphate. Over the course of minutes to hours, such messengers subsequently promote the synthesis and secretion of autocrine/paracrine factors (e.g., NO, prostaglandin [PG] $E_2$, and ATP), kinase activation, cytoskeletal rearrangement, transcription factor (nuclear factor [NF]-$\kappa$B, $\beta$-catenin) activity, and gene transcription and translation. Concomitantly, gap junctional intercellular communication (GJIC) and juxtacrine signaling amplifies the local signal among effector cells for initiation of appropriate tissue-level responses (Fig. 1.1.2).[15–17]

Because of their frequency and localization throughout bone, osteocytes are currently considered the primary mechanosensory cell within a bone. Based on in vitro studies of osteoblast mechanosensitivity, osteoblasts (and, by inference, osteocytes) require >5000 $\mu\varepsilon$ in order to elicit second messenger activation and gene transcription,[18,19] the magnitude of which induces pathologic fracture. Therefore it has been proposed that osteocytes possess an ability to amplify the applied tissue-level strain into a localized strain sufficient to elicit osteocyte activation.[20] In this model, canalicular tethering elements, such as integrins and the hyaluronan-rich glycocalyx, connect the osteocyte cell membrane to the extracellular matrix,[21–24] which generates drag forces across the canalicular cell process to amplify the tissue-level strain to a level sufficient to induce an osteocytic response.

Tremendous efforts have identified potential and functional biochemical coupling mechanisms in bone. Thus there are several mechanisms by which osteocytes may detect mechanical signals, and it is likely that most, if not all, of these mechanisms contribute to mechanotransduction in osteocytes.

a) *Integrins.* Integrins are heteromeric membrane-spanning proteins composed of $\alpha$- and $\beta$-chains. Integrins bind focal adhesion kinase (FAK) and transmit force to ERK, Src, and RhoA, leading to stress fiber formation.[25] Fluid-flow-induced shear

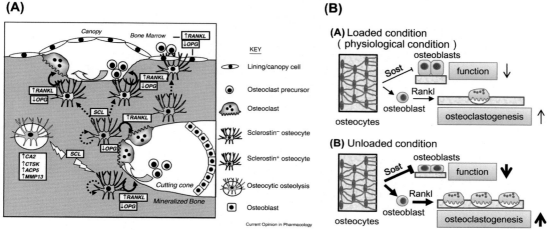

**FIG. 1.1.2** **(A)** Mechanistic model for osteocyte mechanocoupling, wherein fluid shear stress and/or tissue deformation promotes integrin engagement, ion channel activation and opening, and cytoskeletal alignment. *From M. Prideaux, D.M. Findlay, G.J. Atkins, Osteocytes: The master cells in bone remodeling, Curr Opin Pharmacol 28 (2016), pp. 24-30.* **(B)** Proposed influence of mechanical loading or disuse influences bone modeling via sclerostin, osteoprotegerin, and Rankl. OPG, osteoprotegerin. *T. Moriishi, R. Fukuyama, M. Ito, M. Myazaki, Y. Kawai, H. Komori, T. Komori, Osteocyte Network: a Negative Regulatory System for Bone Mass Augmented by the Induction of Rankl in Osteoblasts and Sost in Osteocytes at Unloading, PLoS ONE (2012).*

stress causes conformational changes in integrins that likely activate downstream signaling.[26] Furthermore, recent evidence suggests that pannexin 1, which is implicated as an ATP-releasing channel; the ATP-gated purinergic receptor P2RX7; and the low-voltage transiently opened T-type calcium channel $Ca_v3.2$ co-localize with $\beta_3$ integrin attachment foci on osteocyte processes, suggesting a specialized mechanotransduction complex at these sites.[27] Thus integrins are well positioned to contribute to osteocyte mechanotransduction not only through mechanocoupling of direct substrate strain into an intracellular response but also as a node that integrates multiple mechanocoupling mechanisms.

b) *Cilia*. Over the past several years, the role of cilia in osteocyte mechanotransduction has begun to emerge. Cilia are long antennalike structures that have been implicated in mechanotransduction in several cell types including osteocytes. It has been demonstrated that lengthening primary cilia enhances cellular mechanosensitivity. Osteocytic cells designed to have longer cilia displayed greater increased expression of COX-2 and osteopontin messenger RNA in response to fluid flow than did cells with normal length cilia.[28] Furthermore, Lee *et al.* demonstrated that the primary cilium function as a mechanical and calcium signaling nexus in osteocytes. Additionally, removal of polycystins (Pkd1 and 2 in osteoblasts and osteocytes) or ciliary proteins (Kif3a) impairs mechanotransduction.[29-31] However, while abundant data suggest a role for cilia in osteocyte mechanotransduction, the biochemical coupling linking cilia movement to osteoanabolism remains elusive. Furthermore, cilia are located on cell bodies rather than on dendritic processes, the more ideal location of mechanosensors in osteocytes.

c) *Membrane Channels*. Osteocytic cells express gadolinium-sensitive stretch-activated channels, transient receptor potential (TRP) channels, and voltage-sensitive calcium channels.[32-36] Brown et al. elegantly demonstrated that $Ca_v3.2$ T-type voltage-sensitive calcium channels mediate shear-stress-induced cytosolic calcium in osteocytes through a mechanism involving endoplasmic reticulum calcium dynamics,[37] and Thompson et al. found that the auxiliary $\alpha_2\delta_1$ subunit of the $Ca_v3.2$ channel complex is involved in osteocytic stretch-activated release of ATP. Regarding gadolinium-sensitive stretch-activated channels, there is considerable evidence for their role in osteoblast

mechanotransduction, but there has only been one study describing their role in osteocyte mechanotransduction: Miyauchi[38] demonstrated that gadolinium chloride blocked hypoosmotic stretch-induced increases in intracellular $Ca^{2+}$ ($Ca^{2+}_i$) in rat osteocytes, as well as inhibiting expression of the pore-forming a1c subunit of the $Ca_v1.2$ calcium channel.

Similar to the case with gadolinium-sensitive stretch-activated channels, there are limited studies on the role of TRP channels in osteocyte mechanotransduction. However, in a comprehensive study Lyons et al. demonstrated that TRPV4 was a critical component of the mechanism by which fluid-flow-induced shear stress increases cytosolic $Ca^{2+}$ levels, which then reduces sclerostin expression.[39] As sclerostin inhibits bone formation, these results suggest that osteocytic TRPV is involved in the bone anabolic effects of mechanical signals.

d) *Gap junctions*. Abundant in vitro data suggest that GJIC, largely through gap junction composed of Cx43, plays a critical role in mechanotransduction in bone. In vitro experiments, largely with osteoblasts, demonstrate that gap-junction-deficient cell ensembles are less responsive to mechanical signals[40] and that mechanically induced signals travel from bone cell to bone cell via gap junctions.[41] This suggests that gap junctions sensitize bone to mechanical signals. However, in vivo studies do not support the concept. For instance, bone from mice specifically deficient in osteoblast and osteocyte Cx43 (*Gja1*) is actually more responsive to the anabolic effects of mechanical loading[42,43] and less responsive to the catabolic effects of unloading.[44,45] The mechanism underlying these rather counterintuitive results is not known. However, a noncanonical function of Cx43, that of a molecule that binds $\beta$-catenin, has been proposed.[46,47]

Gap junction hemichannels may also play a role in bone mechanotransduction. Mechanical signals, including fluid-flow-induced shear stress, increase the release of ATP[48] and $PGE_2$[49] via Cx43 hemichannels and, in the case of $PGE_2$ release, this may involve activated AKT kinase.[50] However, one study has demonstrated that, at least in osteoblasts, fluid-flow-induced $PGE_2$ release occurs in Cx43-deficient osteoblasts and is dependent on the expression of pannexin 1.[51] Similarly, Genetos et al.[48] found that osteocytes transfected with Cx43 small interfering RNA were still capable of $PGE_2$ release in response to purinoceptor activation, supporting a mechanism wherein ATP is released via Cx43 or pannexins, which subsequently binds to purinoceptors to induce $PGE_2$ secretion.

Initial consideration of bone mechanobiology focused on the effects of diverse biophysical forces on osteoblast or osteoclast function, as these are the primary effectors of bone adaptation. Furthermore, osteocyte isolation procedures were time-consuming, and the ignorance or unavailability of osteocyte markers prevented procurement of pure populations. The development of fluorescent reporter mice and improved primary osteocyte enrichment approaches have overcome this burden, enabling investigators to directly assay the effects of biophysical forces on osteocytes. Although osteocytes recapitulate many of the biophysical load-induced responses as observed in osteoblasts, for example, rapid and transient increases in $Ca^{2+}{}_i$; release of PGs, nitric oxide, and ATP; and OPG/RANKL, there exist fundamental differences between osteoblastic and osteocytic responses to in vitro mechanical loading. For example, fewer primary osteocytes than primary osteoblasts responded to shear stresses of 1.2–2.4 Pa by mobilizing $Ca^{2+}{}_i$.[52] Similarly, the molecular mechanisms whereby localized biophysical signals are transduced into intracellular responses can alter as a function of osteolineage state, with osteoblasts reliant on $Ca_v1.2$ calcium channels, and osteocytes reliant on $Ca_v3.2$ calcium channels, to mobilize $Ca^{2+}{}_i$.[53] Further, osteocyte-specific responses to mechanical load are observed in the context of osteocyte-specific or -enriched proteins, which are mechanoregulated in osteocytes, but not mechanoregulated in osteoblasts.

## 3. EVIDENCE FOR OSTEOCYTE-DIRECTED SKELETAL RESPONSES

Exercise has many long-lasting benefits to bone, including increasing bone volume, bone strength, and bone tissue quality.[54–57] Bones respond to dynamic, repetitive loading with increased periosteal bone formation, bone mineralization, and tissue quality. This response strengthens the areas of bone tissue exposed to the highest amounts of loading.[58] Periosteal bone formation increases linearly with increasing applied strain,[59] loading frequency, and daily loading cycles.[9] However, bone becomes desensitized to prolonged loading, after approximately 40 high-magnitude loading cycles per day.[60] Effects of increasing loading on bone are also dependent on age and skeletal development. Exercising when young and still undergoing skeletal development offers the greatest benefits to bone mass and bone strength.[61] Benefits to bone mass from exercise decrease throughout adulthood such that, in the elderly, exercise only helps to attenuate bone loss from aging.[62]

Unloading of bones because of disuse, weight loss, or exposure to microgravity during spaceflight causes relatively rapid decreases in bone mass and mineralization. Humans in space suffer bone loss of 0.5%–1.5% of bone mineral density (BMD) per month from exposure to microgravity, even when using a light exercise program of 3–4 days per week.[63,64] Unloading from immobilization also causes similar amounts of bone loss in human and animal studies on Earth.[56,65–67] Bones exposed to extended periods of unloading have increased numbers of empty lacunae and lacunae size, increased porosity, and increased endosteal resorption. These changes make bone more susceptible to fracture and less sensitive to changes in mechanical loading.

Noting variations in lacunar shape, Marotti[68] proposed that osteocytes secrete an inhibitory factor that limits bone deposition by nearby osteoblasts; this hypothesis was refined by Martin,[69] whose model predicted tonic suppression bone-lining cell activation by the said inhibitory factor. Furthermore, the fundamental necessity of osteocytes for mechanosensing is observed in their absence: targeted deletion of osteocytes increases porosity and microdamage and decreases bone formation, strength, and mechanosensitivity.[70] Sclerostin (product of the *Sost* gene), an inhibitor of Wnt signaling expressed primarily in osteocytes, represents such an osteocyte-expressed inhibitor of bone formation. Nonsense mutations in *Sost* produce the sclerosing bone dysplasia sclerosteosis,[71] and a related syndrome, van Buchem disease, results from the deletion of a noncoding distal enhancer.[72–74] Murine models of *Sost* or *ECR5* deletion phenocopy the high bone mass phenotype observed in humans,[75,76] enabling thorough evaluation of the molecular mechanisms of *Sost* transcriptional regulation. The mechanical environment reciprocally influences sclerostin expression, which decreases in limbs of loaded animals and increases following disuse.[77,78] Load-induced changes in *Sost* expression are required for skeletal adaptation: transgenic mice that increase *Sost* expression in response to loading, rather than the natural downregulation that occurs after load, are incapable of an osteoadaptive response,[79] and pharmacologic or genetic inhibition of *Sost* function protects mice from disuse-induced bone loss.[80,81] Mechanoregulated *Sost* expression appears independent of the *ECR5* enhancer, as, despite being mechanosensitive in vitro, *ECR5*$^{-/-}$ mice demonstrate equivalent bone formation or loss compared to wild-type mice in response to loading or unloading, respectively.[82] In contrast, the inhibitory effect of parathyroid hormone (PTH) (1-34) requires *ECR5*.[83]

Osteocytes also have the ability to regulate bone resorption by secreting receptor activator of NF-κB ligand (RANKL). RANKL is also secreted by chondrocytes and osteoblastic cells and is required for differentiation of osteoclast progenitor cells into osteoclasts.[84] Osteocytes also secrete osteoprotegerin (OPG) that works as a decoy receptor,[85,86] binding to RANKL and preventing its osteoclastogenic effects. RANKL and OPG are closely regulated, and the RANKL/OPG ratio is used as an indicator of osteoclastogenic activity.[87] Osteocyte apoptosis induces RANKL and decreases OPG expression in neighboring osteocytes, thereby promoting pro-resorptive adaptation.[88] Thus reciprocal modulation of OPG and RANKL by osteocytes provides a means to target bone, damaged from overuse, for replacement and subsequent reduction of fracture risk.[89,90]

In addition to regulating bone tissue remodeling, osteocytes possess the capacity to remodel the perilacunar area in which they reside. Perilacunar remodeling allows for more rapid mobilization or storage of minerals from bone, as the osteocytes' perilacunar networks contain far more surface area of mineralized tissue than the periosteal and endosteal bone surfaces. Enlarged lacunae, indicative of perilacunar bone resorption, are observed in patients with mineral metabolism disorders,[91] in rats exposed to unloading during spaceflight,[92] in hibernating animals,[93] and in lactating animals.[94,95] Perilacunar remodeling is activated through tumor growth factor (TGF)-β[96] or PTH[95,97] signaling and, thus, is sensitive to both changes in both the humoral milieu (PTH) and factors released from bone matrix during resorption (TGF-β, PTH-related protein [PTHrP]). Furthermore, changes in lacuna size from perilacunar remodeling affect fluid-flow-induced shear stress forces sensed by osteocytes.[98] Thus conditions that change the lacuna size can also change the mechanosensitivity of the bone.

To limit osteocytic influence on bone remodeling to a few scant proteins—sclerostin, OPG, and RANKL—or *via* perilacunar remodeling minimizes the tremendous influence that osteocytes have on skeletal development and repair. Dickkopf-1 (Dkk1) functions similar to sclerostin to inhibit bone formation by inhibiting activation of the Wnt pathway. However, the absence of Dkk1 does not influence bone mass and strength to the extent observed in *Sost*KO mice: neither osteocytic *Dkk1* cKO mice nor mice in which Dkk1 is systemically neutralized by an antibody reveal a robust skeletal response to these interventions, in contrast to the dramatic skeletal phenotype in animals deprived of sclerostin expression or function. Instead, absence of Dkk1 expression or function increases sclerostin expression, indicating a compensatory mechanism between the two Wnt antagonists.[99] Osteoanabolic effects of Wnt signaling appear to require

the transcription factor β-catenin. Mice lacking β-catenin in osteocytes do not fully develop bones and die prematurely,[86] and heterozygous β-catenin osteocyte cKO mice reveal decreased trabecular bone volume and fail to lose cortical bone in response to unloading.[100] Furthermore, constitutively active β-catenin in osteocytes has an osteoanabolic effect resulting from the increasing bone formation and osteoblast differentiation.[101] Yet, other factors that exert osteotropic effects, beyond Wnt signaling, are capable of activating β-catenin signaling and eliciting osteoanabolic effects: load-induced β-catenin signaling in osteocytes is also driven through autocrine $PGE_2$ function[102] and PTH independent of Lrp5,[103-105] as well as fibroblast growth factor signaling.[106]

Osteocytes release many factors that work together to regulate bone remodeling. Communication between neighboring osteocytes and the extracellular environment is crucial for osteocytes to release and propagate these signals in response to changes in mechanical loading. For example, skeletal response to increases in mechanical loading is independent of osteocyte density in areas of maximum loading, supporting the concept that signal propagation from a modest few mechanoactivated osteocytes is necessary to promote skeletal adaptation. Membrane-bound connexins help osteocytes to propagate signals released in response to mechanical loading. Mechanical loading increases expression and activation of connexin 43, allowing for greater release of signals such as $PGE_2$ that affect bone metabolism. Deletion of connexin 43 in osteocytes causes early onset osteopenia, weak bones, and high porosity, properties usually seen in older bones or following increased osteocyte apoptosis.[107,108] This change in phenotype is not seen in osteocytes with normal connexin 43 hemichannels and defective connexin 43 gap junctions,[109] suggesting hemichannels play a predominant role in the response to changes in mechanical loading. Connexin 43-deficient osteocytes are more responsive to increases in mechanical loading, increasing production of $PGE_2$ and decreasing *Sost* expression to a greater extent than in wild-type osteocytes.[110] Connexin 43-deficient osteocytes are less responsive to unloading, although these bones already have signs of bone loss seen after wild-type bones are exposed to microgravity.[44,45] Yet, while cell-cell communication, or release of paracrine factors, is attributed to Cx43 function in bones, evidence indicates that the Cx43 carboxy terminus, independent of gap junction signaling, is obligate for normal skeletal development.[47]

Alternately, direct cell-cell interactions, rather than secretion of soluble factors, reveal tremendous influence on skeletal form and function; this is observed, for example, in the context of Notch signaling, which is activated by direct interactions between a Notch ligand and a Notch receptor in neighboring cells. Notch signaling imparts osteolineage-specific effects on the skeleton: in osteoprogenitors or immature osteoblasts, activation of Notch signaling through overexpression of the Notch1 intracellular domain (NICD) reduces osteoblast differentiation, whereas in osteocytes the same model increases cancellous bone volume through increased bone formation and reduced bone resorption.[111] Notch activation in osteocytes promotes Wnt signaling through reductions in expression of the Wnt antagonists *Sost* and *Dkk1*; mice expressing constitutively active β-catenin show increased Notch receptor expression,[101] and osteocytic Notch activation promotes Wnt signaling, creating a positive feedback between Notch and Wnt signaling.[112]

## 4. OSTEOCYTES, MECHANOTRANSDUCTION, AND AGING

The existence of skeletal involution with increasing chronologic age is without dispute, but attributing this tissue-level response to deficits in specific cells, or specific signaling mechanisms, is nigh impossible owing to the rich diversity of changes in, for example, the hormonal milieu, attenuation of cell function, and increases in the inflammatory environment. Murine models of inbred mice can identify quantitative trait loci to map genes that influence BMD or other parameters[113] or adaptation to mechanical loading,[114] either confirming candidate genes or identifying novel genes for susceptibility to osteoporosis. Similarly, human genetic diseases can reveal novel genes involved in causing osteoporosis (e.g., *LRP5* or *CYP17*), but such variants that have a large effect are rare, whereas common allelic variants have a small effect.[115] Provided the evidence for skeletal adaptation to a changing mechanical environment, the question whether age influences the capacity of an osteoanabolic response to age, seemingly a simple question to resolve with murine models, has inconsistent results. For example, Rubin et al.[116] reported that greater strain was required for the initiation of an osteoanabolic response in the ulna of aged (19-month-old) turkeys versus younger (9-month-old) turkeys, whereas Järvinen et al.[117] found no overt effect of age on the influence of treadmill-running on the femoral neck, and Brodt and Silva[118] reported enhanced endocortical response to axial tibial compression in aged mice. Stark differences in animal models, age ranges, experiment duration, and method of mechanical stimulus prevent direct comparison among

many studies designed to interrogate the effect of chronologic age on osteoanabolic response to loading. Yet, rather than suggest failures in various models, the inability to answer such a facile question may underlie the complexity of organism aging viz. redundancy in mechanoresponsive or mechanoadaptive mechanisms. Indeed, rather than posit a specific defective signaling pathway in osteocytes as the cause for primary osteoporosis, it is necessary to consider osteoporosis as the result of changes in the interaction of osteocytes with their matrix, osteocyte abundance, the hormonal milieu, and cell-autonomous deficits in osteocyte function (Fig. 1.1.3).

### 4.1. Osteocyte Number

The ability to perceive biophysical signals and transduce them to tissue-level responses is predicated upon the presence of an osteocyte, such that alterations in osteocyte number may contribute to osteoporosis. Indeed, reductions in osteocyte number and lacuna are observed in human specimens and murine models, although this highly depends on the anatomic location and type of bone (cortical vs. trabecular). For example, osteocyte lacunar density decreases exponentially with advancing chronologic age in human mid-diaphyseal cortical bone in specimens from both men and women,[119] whereas lacunar number in iliac trabecular bone decreases linearly[120]; yet it was reported that in the femoral head trabecular bone, there is no decrease in osteocyte density until 70 years of age, after which a sharp decline in osteocyte lacunar density appears.[121]

Osteocyte density is also a function of distance from the bone/haversian surfaces—that is, as a function of bone age—with superficial osteocytes (<25 μm from surfaces) resistant to age-related decreases in osteocyte density, whereas osteocyte density in deep bones (>45 μm from surfaces) decreases with age in both men and women.[122] Osteocyte density in both superficial and deep bones is reduced in vertebral iliac trabecular bone from individuals with osteoporotic fractures,[119,123] and the endosteal cortex appears more sensitive to age-dependent osteocyte loss than the periosteal cortex.[124] As osteocytes are mechanosensory cells, their reduction secondary to aging removes a means through which skeletal integrity may be interrogated and improved.

### 4.2. Changes in Osteocyte Microenvironment

The mechanosensory role of osteocytes is grounded in localized biophysical signals that are engendered during locomotion or exercise. As described earlier, such signals

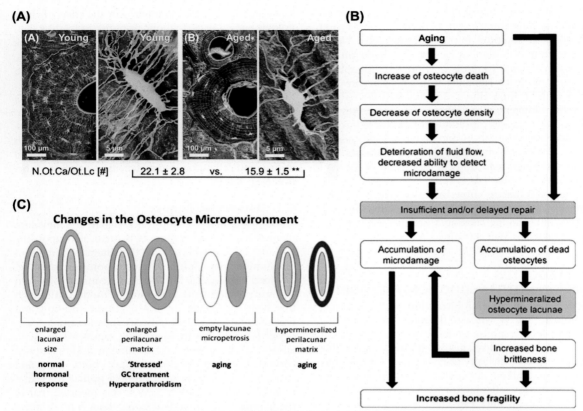

N.Ot.Ca/Ot.Lc [#] | 22.1 ± 2.8 | vs. | 15.9 ± 1.5 **

**FIG. 1.1.3 Age-related changes in osteocyte networks.** **(A)** Scanning electron microscopic images of acid-etched resin embedded human bone sections, revealing decreased osteonal width, osteocyte frequency, and canaliculi in aged bone. *From L.M. Tiede-Lewis, S.L. Dallas, Changes in the osteocyte lacunocanalicular network with aging, Bone 122 (2019), pp. 101-113.* **(B)** A model for osteocyte-mediated effects of age on bone quality and fragility. *From B. Busse, D. Djonic, P. Milanovic, M. Hahn, K. Püschel, R.O. Ritchie, M. Djuric, M. Amling, Decrease in the osteocyte lacunar density accompanied by hypermineralized lacunar occlusion reveals failure and delay of remodeling in aged human bone, Aging Cell 9 (2010), pp. 1065-1075From B. Busse, D. Djonic, P. Milanovic, M. Hahn, K. Püschel, R.O. Ritchie, M. Djuric, M. Amling, Decrease in the osteocyte lacunar density accompanied by hypermineralized lacunar occlusion reveals failure and delay of remodeling in aged human bone, Aging Cell 9 (2010), pp. 1065-1075.* **(C)** Alterations in osteocyte microenvironment via perilacunar remodeling associated with pharmacotherapy, disease, and age. *From L.F. Bonewald. (2011) Osteocyte Mechanosensation and Transduction. In: Noda M. (eds) Mechanosensing Biology. Springer, Tokyo, 2011, pp. 141-155.*

include the shear stresses of interstitial fluid flow across the plasma membrane of osteocyte dendrites and cell body, microperturbations in osteocyte plasma membrane integrity, and tethering of osteocyte dendritic processes to the canalicular wall.[21,23] Thus it is axiomatic that age-related alterations to the microenvironment in which an osteocyte resides would subsequently alter the biophysical signals experienced by an osteocyte.

The integrity of the lacunocanalicular network (LCN) is diminished by microdamage, severing canaliculi and thereby generating a "wall," a "sink," or a "reservoir" to mitigate interstitial fluid flow and mass transport within bone.[125] Thus microdamage effectively decreases the concentration of osteotropic agents, from the vasculature or osteocytes within the LCN to osteocytes distal from the microdamage. Furthermore, compromised transport and exchange of nutrients in response to microdamage would decrease osteocyte viability, as is observed in vivo.[126] Occlusion of the LCN can also occur with aging, a process termed *micropetrosis* by Frost, wherein osteocyte death promotes occlusion of lacunae with hypermineralized calcium phosphate, which could further cause deteriorations in canalicular fluid flow.[124] Even in the absence of lacunar occlusion, microstructural changes in the LCN negatively impact fluid flow in response to loading. Strain amplification within osteocyte lacunae is influenced by pericellular lacunar tissue modulus, with predicted maximum perilacunar strain increasing as the modulus of the perilacunar matrix decreases; conversely, maximum perilacunar strain decreased as perilacunar matrix modulus increased.[127] Furthermore, microstructural changes, such as increases in vascular porosity, reductions in trabecular number, and increases in trabecular separation, in response to ovariectomy fundamentally reduce load-induced interstitial fluid flow,[128] indicating that changes in micro- and nanostructural properties of bone can negatively influence osteocyte mechanosensation. Estrogen deficiency influences strain distribution across osteocytes more than osteoblasts, exerting acute increases yet sustained decreases on applied strain[129] resulting from changes in tissue stiffness and mineral content.[130] Thus localized changes in mineral density, tissue composition, and even LCN tortuosity immediately surrounding the osteocyte[130,131] may serve as a means to sensitize or diminish osteocyte mechanoresponsiveness via osteocytic perilacunar remodeling. Indeed, Hesse et al.[132] used synchrotron radiation phase-contrast nano–computed tomography to reveal a gradient of bone tissue mass density that decreases from the pore-matrix interface into the tissue. Hemmatian et al.[133] used virgin or lactating mice to evaluate the influence of perilacunar remodeling on the activation of β-catenin, as a marker of mechanotransduction, and found that load-induced stabilization of β-catenin was greater in the lactating mouse than in the virgin mouse, suggesting that increased lacunar volume enhances the response to mechanical loading. Although lacunar volume was interrogated as a marker of perilacunar remodeling, other aspects of the lacuna-matrix interface, such as mineral-matrix ratio and carbon-phosphate ratio, were not studied. Conversely, exercise induces perilacunar remodeling; Gardiner et al.[134] reported that perilacunar remodeling occurs during exercise, with alterations in mineral-matrix or carbonate-phosphate ratio observed within 0–5 μm from osteocyte lacunae yet disappearing with greater radial distance from lacunae. Thus there appears to be a finely regulated interaction between lacunocanalicular properties as both a predictor of osteocyte mechanosensation and the result of osteocyte mechanoresponsiveness.

The osteocyte morphology appears to depend on the anatomic location, age, hormonal status, and mechanical loading environment. In humans and rodents, osteocyte morphology changes from relatively flattened to more spherical.[135,136] Similarly, osteocytes within rodents demonstrate an elongated morphology, whereas calvarial osteocytes are more spherical.[137] Differences in osteocyte morphology may result from the mechanical loading environment: the long axis of fibular osteocytes align parallel to the principal loading direction, whereas calvarial osteocytes do not appear to align with any loading direction. In human osteonal bone, osteocyte lacunae are aligned with collagen fiber orientation,[138–140] suggesting that osteocyte orientation within a lacuna occurs during the osteoblast-to-osteocyte transition.[141] Hormonal cues also influence osteocyte morphology: for instance, compared to virgin mice, the lacunar volume of lactating mice is increased, indicating perilacunar remodeling,[133] and exercise-induced changes in PTH secretion promote changes in lacunar volume.[134]

To date, that osteocyte morphology influences mechanosensation and downstream effects on remodeling, bone strength, etc. is inferred from changes in bone material and structural properties. Direct demonstration of altered osteocyte morphology to a given mechanical load is light, although Bacabac et al.[142] demonstrated that round, partially adherent MLO-Y4 osteocytes were more mechanosensitive than flat, adherent osteocytes. Nonetheless, with the possible exception of hormonal influence on osteocyte morphology, whether changes in osteocyte morphology are the cause, or the consequence, of changes in bone health requires more careful and considerate investigation.

## 4.3. Cell-Autonomous Alteration of Osteocyte Function

The reductions in bone mass, density, and microarchitecture that are observed as a result of aging are proposed to result, in part, from decreased mechanosensing. Although there is sufficient evidence (detailed earlier) that applied strains or interstitial fluid flows can be reduced secondary to increases in vascular porosity or lacunocanalicular tortuosity, there is little direct evidence regarding whether osteoblasts or osteocytes per se are less capable of perceiving a given load. Donahue et al.[143] observed a modest decrease in the percent of aged rat osteoblastic cells that responded to oscillatory fluid flow with an increase in cytosolic $Ca^{2+}$ levels. Sterck et al.[144] evaluated the influence of donor cell age on mechanically induced responses, finding that $PGE_2$ and $PGI_2$ release increased with donor cell age, and Bakker et al.[145] reported that pulsatile fluid flow is able to induce the synthesis and release of second messengers (nitric oxide, $PGE_2$) from osteoporotic or osteoarthritic donors. Similarly, relative increases in cytosolic $Ca^{2+}$ levels and *cfos* transcription in fluid flow across osteoblasts are no different in response to loading in osteoblasts from young versus senescent mice, whereas absolute changes are affected as a function of age.[146]

Thus rather than the decreased capacity of osteocytes to perceive a given stimulus, it is perhaps more likely that other age-related changes in cell function contribute to loss of bone with age. For example, there is a loss of osteoprogenitors with age as mesenchymal stem cells shift toward adipogenesis, increased osteoblast apoptosis reduces the duration of bone-forming capacity, and increased marrow adiposity enhances suppression of bone formation through reductions in osteoblastogenesis and mineralization.[147]

Alternately, senescence of osteoblasts or osteocytes can contribute to tissue dysfunction.[148] Markers of cell senescence, including gene transcripts ($p16^{Ink4}$, $Cdkn1a/p21$, and $p53$), telomere shortening, and senescence-associated secretory phenotype, are observed in osteoblasts and osteocytes from older (24 months) mice relative to young (6 months) mice, with greater age-related increases observed in osteocytes relative to osteoblasts.[149] Senescence is no longer an unavoidable consequence of aging, as genetic[150] and pharmacologic[151,152] approaches to reduce or eliminate senescent cell populations revert age-related pathologic conditions. Indeed, genetic or pharmacologic ablation of senescent cells in old mice increases bone mass and strength resulting from decreased endocortical bone resorption and increased endocortical bone formation.[148] That bone resorption was reduced after senescent cell ablation without concomitantly reducing bone formation is in contrast to current antiresorptive therapies (bisphosphonates, anti-RANKL antibody, or estrogen), which reduce both bone formation and resorption. Because the currently identified senolytic agents are already FDA-approved, and a monthly dosing of senolytics was sufficient to reverse changes due to age, such approaches portend their use in the treatment of age-related bone loss.

## 5. OSTEOCYTES AND DISEASE
### 5.1. Cancer
The skeleton is a common site for metastasis of other cancers: multiple myeloma (MM) also begins in the

bone is not derived from bone cells, instead it results from the clonal plasma cell malignant neoplasm, which subsequently causes osteolytic bone lesions in >80% of patients with myeloma.[153] Osteolytic bone lesions are also observed in breast and lung cancers that have metastasized to the skeleton, and mixed osteoblastic/osteolytic lesions are most commonly observed in metastatic prostate cancer (PCa). Complications of skeletal metastasis, termed skeletal related events (SREs), include pathologic fracture, hypercalcemia, spinal cord compression, and chronic pain. Beyond the debilitating impact on quality of life, the incidence of SREs negatively predicts survival: 5-year survival rates for PCa are 56% after skeletal metastasis, but the survival rate plummets to <1% with metastasis and an SRE.[154] The events involved in cancer metastasis to the skeleton involve disseminated tumor cells (1) colonizing the bone microenvironment and (2) adapting to the new environment, avoiding immune surveillance, and entering a state of dormancy; (3) reactivating in response to as-yet-unknown cues, thereafter proliferating and forming micrometastases; and (4) growing uncontrollably, independent of the microenvironment (reviewed in Ref. [155]) (Fig. 1.1.4).

Provided that osteocytes are sentinels against remodeling, and the proclivity and propensity for cancer to subvert normal homeostatic processes, it is logical to suggest that tumor colonization, dormancy, and activation involve, if not require, altered osteocyte function. The most direct way to alter osteocyte function is to promote their death, and osteocyte death is implicated in disseminated tumor cell colonization and adaptation: increased frequency of apoptotic osteocytes is observed in individuals with MM[156] and selective depletion of osteocytes supports MM colonization of bone and tumor growth.[157] Whether the opposite—maintaining osteocyte viability pharmacologically or through other means—reduces tumor burden is under active investigation. Bisphosphonates reduce osteoclast formation, but also exert antiapoptotic effects on osteocytes in vitro[158] and in animal models of osteoporosis[159,160] through a molecular mechanism involving Cx43 hemichannels and ERK1/2.[161] Provided that bisphosphonates differentially influence apoptosis in osteocytes versus osteoclasts, an unresolved question is to what extent the beneficial effects of bisphosphonates for individuals with MM is related to osteocyte survival rather than reduced osteoclast activity. The discovery of bisphosphonates, which maintain osteocyte viability without influencing osteoclast activity,[162–164] provides an opportunity to evaluate such a scenario. Other agents that promote osteocyte viability, such as the

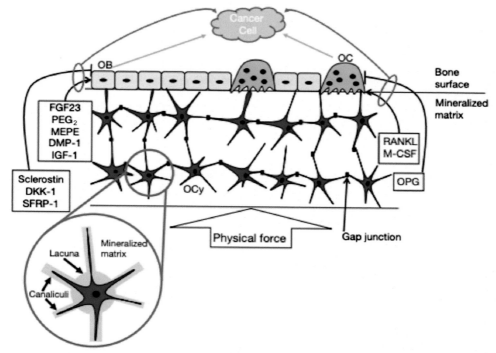

FIG. 1.1.4 Proposed model for osteocyte interaction with cancer cells. *Red arrows* indicate direct effects on cancer cells, whereas *green arrows* indicate indirect effects mediated via osteoblasts or osteoclasts. From M. Zhang, J. Dai, E.T. Keller, Multiple Roles of Osteocytes in Bone-Associated Cancers, Reference Module in Biomedical Sciences 2019, 10.1016/B978-0-12-801238-3.11246-2

proteasome inhibitor bortezomib[165] and PTH (1-34), are under investigation as a means of minimizing the skeletal impact of MM or skeletal complications to radiotherapy.[166,167]

Communication between osteocytes and neighboring cells is also subverted during tumor metastasis to bone and growth therein, which has recently been elegantly reported in the context of MM signaling through Notch receptors and breast cancer communication through connexin 43 gap junctions. Compelling evidence demonstrates that MM colonization of bone and adaptation therein involves reciprocal interactions between MM cells and osteocytes. Notch signaling is enhanced in MM cells,[168–170] and direct coculture of MM cells with MLO-A5 osteocytes promotes osteocyte apoptosis via Notch signaling.[171] As in other studies, osteocyte apoptosis increased osteoclastogenesis through enhanced Rankl expression and reductions in Opg. Furthermore, MM cells increase *Sost* expression and decrease Wnt signaling in osteocytes, leading to reductions in osteoblastogenesis. Reciprocally, Notch activation in MM cells by osteocytes promotes MM cell proliferation. Serum sclerostin levels are increased

in patients with MM and correlate with advanced disease[172] and the *Sost* expression is increased in response to MM-osteocyte interactions,[171] suggesting that manipulation of *Sost* expression or function may provide a means to reduce MM tumor burden and impact. Indeed, *Sost* deletion or pharmacologic inhibition of sclerostin reduced MM-induced osteolysis and bone loss, although it did not decrease tumor volume,[173] providing a promising opportunity to mitigate the skeletal sequelae of MM.

Osteocyte function—or, minimally, markers of osteocyte function—is also implicated in tumor burden and growth in PCa that has metastasized to the skeleton. Prostate carcinoma stimulates bone formation, both through the secretion of pro-osteogenic factors such as Wnts and through osteomimicry, wherein cancer tumor cells can secrete matrix, mineral, and osteotropic factors, as well as respond to osteotropic factors. For example, expression of the Wnt antagonist Dkk1 is observed in focal lesions of MM cells,[174] sclerostin is secreted by breast cancer cells,[175] and increased β-catenin levels are observed in individuals with PCa.[176,177] Functionally, overexpressing *Sost*, but not *Dkk1*, in osteolytic PC3 cells reduces osteolysis in xenograft-derived tumor lesions,[178] whereas increasing Wnt signaling in mixed osteolytic/osteoblast C4-2B PCa cells, through knockdown Dkk1, shifts tumor phenotype toward osteolysis.[179] Similarly, osteocyte-derived GDF15, a member of the TGF-β superfamily, promotes PCa metastasis to the skeleton.[180]

Disseminated tumor cell adaptation to the bone microenvironment also subverts gap junctions. Connexins are considered antitumorigenic, as loss of connexin 43 expression is a marker for breast tumors,[181] and connexin expression and intercellular communication are reduced during tumorigenesis, whereas transfection of cancer cells with connexins reduces cell proliferation.[182] Connexins exert multipronged influences on tumor growth via GJIC, hemichannels, and channel-independent connexin function. Wang et al.[183] found that $Ca^{2+}_i$-dependent transcription factor activation of NFAT and MEF2 was enriched in breast carcinomas that have metastasized to bones. Such observations were recapitulated in both in vivo and ex vivo models of bone metastasis, such that inhibition of calcium signaling with dominant-negative calcineurin or calmodulin-dependent protein (CaM) kinase II function eliminated tumor colonization of bone. Cx43 expression was greater in skeletally metastatic breast carcinomas, correlated with calcium-dependent transcription factor enrichment, and established calcium flux from osteogenic to tumor cells; reducing Cx43 expression or inhibiting GJIC with carbenoxolone reduced bone colonization and skeletal tumor burden. This association between connexins, GJIC, and calcium-dependent transcription factor enrichment was not unique to breast carcinoma, as prostate carcinoma and PCa cell lines demonstrated equivalent interactions between calcium signaling, Cx43 expression, and tumor burden. Notably, mesenchymal stem cells, but not osteocytes, were capable of altering calcium signaling in tumor cells, further suggesting that unique bone microenvironments contribute to tumor cell colonization and survival. The influence of connexins on tumor cell adaptation to the skeletal microenvironment is not restricted to GJIC, as connexin hemichannels are also implicated as mediators of tumor fate. Osteocytic hemichannel opening in response to fluid shear stress[48] or the bisphosphonate alendronate[184] promotes the release of cytosolic ATP, which reduces the migratory capacity of breast cancer cells and reduces tumor growth. The antitumor effect of osteocytic Cx43 appears restricted to hemichannels: growth of mammary carcinoma cells in the skeleton is greater in mice lacking osteocytic *Gja1* or expressing dominant-negative Cx43

($Gja1^{\Delta130-136}$), which reduces both GJIC and hemichannel activity, relative to control animals. In contrast, tumor growth in osteocytic $Gja1^{R76W}$ mice, which can form hemichannels but not participate in GJIC, was phenotypically similar to wild-type animals. Furthermore, the inhibitory effect of zolendronate on tumor growth was absent in osteocytic *Gja1* cKO or $Gja1^{\Delta130-136}$ mice but not $Gja1^{R76W}$ animals, indicating a specific inhibitory role of osteocytic Cx43 hemichannel-mediated ATP release in tumor metastasis and growth in the skeleton.[185]

## 5.2. Vascular Health

A consequence of multicellular growth is the insufficiency of simple diffusion to provide sufficient oxygen for host respiration; oxygen delivery, as well as nutrient provision and waste removal, is provided by the vascular system. Bone development, homeostasis, and repair are each exquisitely coupled to the vasculature (reviewed in Ref. [186]). Furthermore, blood vessels provide the necessary conduit for calcium and phosphate, liberated from the matrix, to reach target tissues for homeostatic function. A sufficient vasculature is fundamental to osteogenesis, but vessel function is often compromised by calcification as a consequence of aging and lifestyle. Calcification may occur throughout the vascular tree, with distinct molecular origins and pathologic consequences. For example, calcification of medial arteries increases vessel stiffness and decreases compliance, thereby impairing Windkessel physiology and increasing vessel pressure and cardiac workload.[187,188] Similarly, calcification of atherosclerotic plaques reduces vessel lumen size and increases tissue ischemia and thromboembolic events,[188] whereas aortic valve calcification reduces valve leaflet mobility and impairs leaflet closing, promotes left ventricular hypertrophy, and results in aortic valve regurgitation.[189]

Ectopic calcification within the vasculature and cardiac valves was long considered to be a passive, degenerative process. However, the biochemical and structural similarity of these ectopic calcifications to orthotopic bone hinted that an active process may be involved (Fig. 1.1.5).[190–192] This was initially confirmed in the study by Boström et al.,[193] wherein the potent osteogenic morphogen BMP-2 (then known as BMP-2a) immunolocalized within human atherosclerotic lesions. Because BMP-2 stimulates orthotopic calcification by increasing matrix synthesis, this finding suggested an active, cell-mediated process for vessel calcification. Further studies revealed the presence of Wnts, osteoblast-related transcription factors Runx2[194] and Msx2,[195] and matrix vesicles within arterial calcifications,[196,197] suggesting that a

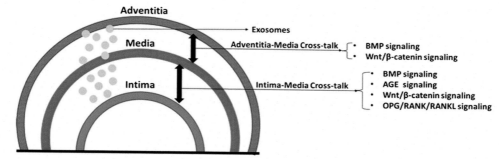

**FIG. 1.1.5** Osteogenic networks implicated in vessel wall calcification. A.S.A. Bardeesi, J. Gao, K. Zhang, S. Yu, M. Wei, P. Liu, H. Huang, **A novel role of cellular interactions in vascular calcification**, J Transl Med (2017) 15: 95. https://doi.org/10.1186/s12967-017-1190-z *AGE*, advanced glycation end product; *BMP*, bone morphogenetic protein; *DKK-1*, dickkopf-1; *DMP-1*, dentin matrix protein 1; *FGF23*, fibroblast growth factor 23; *IGF-1*, insulinlike growth factor 1; *M-CSF*, macrophage colony-stimulating factor; *MEPE*, matrix extracellular phosphoglycoprotein; *OB*, osteoblast; *OC*, osteoclast; *OCy*, osteocyte; *OPG*, osteoprotegerin; *PGE₂*, prostaglandin E₂; *RANKL*, receptor activator of nuclear factor κB ligand; *SFRP-1*, secreted frizzled-related protein 1.

subpopulation of cells within the vessel wall are capable of active mineralization processes described in chondrocytes and osteoblasts.[188,193] Altering Wnt signaling, through the overexpression of the Wnt antagonist sclerostin or the addition of recombinant sclerostin, attenuates angiotensin-II-induced atherosclerosis.[198] A similar osteogenic mechanism is involved in cardiac valve calcification, wherein the Wnt/Lrp5/β-catenin signaling pathway actively promotes calcium deposition. Caira et al.[199] revealed increased Lrp5 and osteogenic marker gene expression in calcified aortic valves compared with control valves. These results may at least partially explain the increased risk of cardiovascular disease in individuals receiving antisclerostin antibody to treat osteoporosis. However, serum sclerostin levels are higher in individuals with cardiovascular disease, emphasizing the complicated nature of sclerostin function in nonbone tissue.

For active ectopic calcification to occur, these ectopic sites must contain osteoblast(-like) cells. Both the vasculature and cardiac valves contain resident progenitor cells capable of osteogenic differentiation. Boström et al.[193] isolated a subpopulation of aortic medial cells, termed calcifying vascular cells (CVCs), that express osteoblastic markers and are capable of osteogenic differentiation in vitro. A similar effector cell, the adventitial myofibroblast (pericyte), has been implicated in medial artery calcification, as occurs in diabetic patients. Both CVCs and adventitial myofibroblasts express the smooth muscle cell markers SM22 and α-SMA, which are routinely used as the basis for confirmation of the SMC phenotype for ex vivo cultures. Similarly, the heterogeneous group of valve interstitial cells contains mesenchymal precursors with osteogenic and chondrogenic potential.[200,201]

## 6. CONCLUSIONS AND FUTURE DIRECTIONS

Although clearly the most abundant cell in bone, osteocytes have been understudied until the past decades, owing largely to technical reasons. However, recent advances in cell isolation, the characterization of specific gene transcripts in osteocytes, and the use of novel murine models of altered osteocyte gene expression have revealed the central role that osteocytes play in surprisingly diverse areas of bone, and possibly host, metabolism. Osteocytes not only are the orchestrator of osteoblast and osteoclast activity in response to mechanical, hormonal, or cytokine signals but also clearly play a central role in bone cancer, vascular disease, and age-related bone loss. However, there remain clear gaps in our understanding of osteocyte biology. For example, osteocyte function in osteoimmunology is poorly understood, the interactions of osteocytes with the peripheral and central nervous systems have only been suggested, and the role of osteocytes in development, maintenance, and activation of stem cell niches under health and disease is unresolved. Furthermore, the role that genetic variability plays in osteocyte biology and how osteocytes may be exploited to regenerate lost bone tissue and mass after injury or aging have not been explored. These are just some of the gaps in osteocyte biology whose questions we avidly await answers.

## REFERENCES

1. Steves CJ, Bird S, Williams FMK, Spector TD. The microbiome and musculoskeletal conditions of aging: a review of evidence for impact and potential therapeutics. *J Bone Miner Res.* 2016;31(2):261–269. https://doi.org/10.1002/jbmr.2765.

2. Lafortune G, Balestat G. *Trends in Severe Disability Among Elderly People.* Vol. 26. 2007. https://doi.org/10.1787/217072070078.

3. United States Bone and Joint Initiative. *The Burden of Musculoskeletal Diseases in the United States.* American Academy of Orthopaedic; 2015.

4. Schnell S, Friedman SM, Mendelson DA, Bingham KW, Kates SL. The 1-year mortality of patients treated in a hip fracture program for elders. *Geriatr Orthop Surg Rehabil.* 2010;1(1):6–14. https://doi.org/10.1177/2151458510378105.

5. Veronese N, Maggi S. Epidemiology and social costs of hip fracture. *Injury.* 2018;49(8):1458–1460. https://doi.org/10.1016/j.injury.2018.04.015.

6. Martin RB. The importance of mechanical loading in bone biology and medicine. *J Musculoskelet Neuronal Interact.* 2007;7(1):48–53.

7. FROST HM. Muscle, bone, and the Utah paradigm: a 1999 overview. *Med Sci Sport Exerc.* 2000;32(5):911.

8. Martin R, Burr D, Sharkey N. *Skeletal Tissue Mechanics.* 1998.

9. Robling AG, Castillo AB, Turner CH. Biomechanical and molecular regulation of bone remodeling. *Annu Rev Biomed Eng.* 2006;8:455–498. https://doi.org/10.1146/annurev.bioeng.8.061505.095721.

10. Frost HM. Bone "mass" and the "mechanostat": a proposal. *Anat Rec.* 1987;219(1):1–9. https://doi.org/10.1002/ar.1092190104.

11. Jee WS. Principles in bone physiology. *J Musculoskelet Neuronal Interact.* 2000;1(1):11–13.

12. Duncan RL, Turner CH. Mechanotransduction and the functional response of bone to mechanical strain. *Calcif Tissue Int.* 1995;57(5):344–358.

13. Thompson WR, Rubin CT, Rubin J. Mechanical regulation of signaling pathways in bone. *Gene.* 2012;503(2):179–193. https://doi.org/10.1016/j.gene.2012.04.076.

14. Rosa N, Simoes R, Magalhães FD, Marques AT. From mechanical stimulus to bone formation: a review. *Med Eng Phys.* 2015;37(8):719–728. https://doi.org/10.1016/j.medengphy.2015.05.015.

15. Spyropoulou A, Karamesinis K, Basdra EK. Mechanotransduction pathways in bone pathobiology. *Biochim Biophys Acta (BBA) – Mol Basis Dis.* 2015;1852(9):1700–1708. https://doi.org/10.1016/j.bbadis.2015.05.010.

16. Plotkin LI, Bellido T. Osteocytic signalling pathways as therapeutic targets for bone fragility. *Nat Rev Endocrinol.* 2016;12(10):593–605. https://doi.org/10.1038/nrendo.2016.71.

17. Paul GR, Malhotra A, Müller R. Mechanical stimuli in the local in vivo environment in bone: computational approaches linking organ-scale loads to cellular signals. *Curr Osteoporos Rep.* 2018;16(4):395–403. https://doi.org/10.1007/s11914-018-0448-6.

18. Owan I, Burr DB, Turner CH, et al. Mechanotransduction in bone: osteoblasts are more responsive to fluid forces than mechanical strain. *Am J Physiol.* 1997;273(3 Pt 1):C810–C815.

19. You J, Yellowley CE, Donahue HJ, Zhang Y, Chen Q, Jacobs CR. Substrate deformation levels associated with routine physical activity are less stimulatory to bone cells relative to loading-induced oscillatory fluid flow. *J Biomech Eng.* 2000;122(4):387–393.

20. You L, Cowin S, Schaffler M, Weinbaum S. A model for strain amplification in the actin cytoskeleton of osteocytes due to fluid drag on pericellular matrix. *J Biomech.* 2001;34(11):1375–1386.

21. You L-D, Weinbaum S, Cowin SC, Schaffler MB. Ultrastructure of the osteocyte process and its pericellular matrix. *Anat Rec A Discov Mol Cell Evol Biol.* 2004;278(2):505–513. https://doi.org/10.1002/ar.a.20050.

22. Han Y, Cowin SC, Schaffler MB, Weinbaum S. Mechanotransduction and strain amplification in osteocyte cell processes. *Proc Natl Acad Sci U S A.* 2004;101(47):16689–16694. https://doi.org/10.1073/pnas.0407429101.

23. Wang Y, McNamara LM, Schaffler MB, Weinbaum S. A model for the role of integrins in flow induced mechanotransduction in osteocytes. *Proc Natl Acad Sci U S A.* 2007;104(40):15941–15946. https://doi.org/10.1073/pnas.0707246104.

24. Thi MM, Suadicani SO, Schaffler MB, Weinbaum S, Spray DC. Mechanosensory responses of osteocytes to physiological forces occur along processes and not cell body and require αVβ3 integrin. *Proc Natl Acad Sci U S A.* 2013;110(52):21012–21017. https://doi.org/10.1073/pnas.1321210110.

25. Hamamura K, Swarnkar G, Tanjung N, et al. RhoA-mediated signaling in mechanotransduction of osteoblasts. *Connect Tissue Res.* 2012;53(5):398–406. https://doi.org/10.3109/03008207.2012.671398.

26. Uda Y, Azab E, Sun N, Shi C, Pajevic PD. Osteocyte mechanobiology. *Curr Osteoporos Rep.* 2017;15(4):318–325. https://doi.org/10.1007/s11914-017-0373-0.

27. Cabahug-Zuckerman P, Stout RF, Majeska RJ, et al. Potential role for a specialized β3 integrin-based structure on osteocyte processes in bone mechanosensation. *J Orthop Res.* 2018;36(2):642–652. https://doi.org/10.1002/jor.23792.

28. Spasic M, Jacobs CR. Lengthening primary cilia enhances cellular mechanosensitivity. *Eur Cells Mater.* 2017;33:158–168. https://doi.org/10.22203/eCM.v033a12.

29. Xiao Z, Zhang S, Mahlios J, et al. Cilia-like structures and polycystin-1 in osteoblasts/osteocytes and associated abnormalities in skeletogenesis and Runx2 expression. *J Biol Chem.* 2006;281(41):30884–30895. https://doi.org/10.1074/jbc.M604772200.

30. Xiao Z, Dallas M, Qiu N, et al. Conditional deletion of Pkd1 in osteocytes disrupts skeletal mechanosensing in mice. *FASEB J.* 2011;25(7):2418–2432. https://doi.org/10.1096/fj.10-180299.

31. Qiu N, Xiao Z, Cao L, et al. Disruption of Kif3a in osteoblasts results in defective bone formation and osteopenia. *J Cell Sci.* 2012;125(Pt 8):1945−1957. https://doi.org/10.1242/jcs.095893.
32. Li J, Duncan RL, Burr DB, Turner CH. L-type calcium channels mediate mechanically induced bone formation in vivo. *J Bone Miner Res.* 2002;17(10):1795−1800. https://doi.org/10.1359/jbmr.2002.17.10.1795.
33. Shao Y, Alicknavitch M, Farach-Carson MC. Expression of voltage sensitive calcium channel (VSCC) L-type Cav1.2 (alpha1C) and T-type Cav3.2 (alpha1H) subunits during mouse bone development. *Dev Dynam.* 2005;234(1):54−62. https://doi.org/10.1002/dvdy.20517.
34. Plotkin LI, Manolagas SC, Bellido T. Glucocorticoids induce osteocyte apoptosis by blocking focal adhesion kinase-mediated survival. Evidence for inside-out signaling leading to anoikis. *J Biol Chem.* 2007;282(33):24120−24130. https://doi.org/10.1074/jbc.M611435200.
35. Suzuki T, Notomi T, Miyajima D, et al. Osteoblastic differentiation enhances expression of TRPV4 that is required for calcium oscillation induced by mechanical force. *Bone.* 2013;54(1):172−178. https://doi.org/10.1016/j.bone.2013.01.001.
36. Yavropoulou MP, Yovos JG. The molecular basis of bone mechanotransduction. *J Musculoskelet Neuronal Interact.* 2016;16(3):221−236.
37. Brown GN, Leong PL, Guo XE. T-Type voltage-sensitive calcium channels mediate mechanically-induced intracellular calcium oscillations in osteocytes by regulating endoplasmic reticulum calcium dynamics. *Bone.* 2016;88:56−63. https://doi.org/10.1016/j.bone.2016.04.018.
38. Miyauchi A, Notoya K, Mikuni-Takagaki Y, et al. Parathyroid hormone-activated volume-sensitive calcium influx pathways in mechanically loaded osteocytes. *J Biol Chem.* 2000;275(5):3335−3342.
39. Lyons JS, Joca HC, Law RA, et al. Microtubules tune mechanotransduction through NOX2 and TRPV4 to decrease sclerostin abundance in osteocytes. *Sci Signal.* 2017;10(506). https://doi.org/10.1126/scisignal.aan5748.
40. Saunders MM, You J, Trosko JE, et al. Gap junctions and fluid flow response in MC3T3-E1 cells. *Am J Physiol Cell Physiol.* 2001;281(6):C1917−C1925.
41. Yellowley CE, Li Z, Zhou Z, Jacobs CR, Donahue HJ. Functional gap junctions between osteocytic and osteoblastic cells. *J Bone Miner Res.* 2000;15(2):209−217. https://doi.org/10.1359/jbmr.2000.15.2.209.
42. Grimston SK, Brodt MD, Silva MJ, Civitelli R. Attenuated response to in vivo mechanical loading in mice with conditional osteoblast ablation of the connexin43 gene (Gja1). *J Bone Miner Res.* 2008;23(6):879−886. https://doi.org/10.1359/jbmr.080222.
43. Bivi N, Pacheco-Costa R, Brun LR, et al. Absence of Cx43 selectively from osteocytes enhances responsiveness to mechanical force in mice. *J Orthop Res.* 2013;31(7):1075−1081. https://doi.org/10.1002/jor.22341.
44. Lloyd SA, Lewis GS, Zhang Y, Paul EM, Donahue HJ. Connexin 43 deficiency attenuates loss of trabecular bone and prevents suppression of cortical bone formation during unloading. *J Bone Miner Res.* 2012;27(11):2359−2372. https://doi.org/10.1002/jbmr.1687.
45. Lloyd SA, Loiselle AE, Zhang Y, Donahue HJ. Connexin 43 deficiency desensitizes bone to the effects of mechanical unloading through modulation of both arms of bone remodeling. *Bone.* 2013;57(1):76−83. https://doi.org/10.1016/j.bone.2013.07.022.
46. Li J, Li C, Liang D, et al. LRP6 acts as a scaffold protein in cardiac gap junction assembly. *Nat Commun.* 2016;7(1):11775. https://doi.org/10.1038/ncomms11775.
47. Moorer MC, Hebert C, Tomlinson RE, Iyer SR, Chason M, Stains JP. Defective signaling, osteoblastogenesis and bone remodeling in a mouse model of connexin 43 C-terminal truncation. *J Cell Sci.* 2017;130(3):531−540. https://doi.org/10.1242/jcs.197285.
48. Genetos DC, Kephart CJ, Zhang Y, Yellowley CE, Donahue HJ. Oscillating fluid flow activation of gap junction hemichannels induces ATP release from MLO-Y4 osteocytes. *J Cell Physiol.* 2007;212(1):207−214. https://doi.org/10.1002/jcp.21021.
49. Cherian PP, Siller-Jackson AJ, Gu S, et al. Mechanical strain opens connexin 43 hemichannels in osteocytes: a novel mechanism for the release of prostaglandin. *Mol Biol Cell.* 2005;16(7):3100−3106. https://doi.org/10.1091/mbc.E04-10-0912.
50. Batra N, Riquelme MA, Burra S, Kar R, Gu S, Jiang JX. Direct regulation of osteocytic connexin 43 hemichannels through AKT kinase activated by mechanical stimulation. *J Biol Chem.* 2014;289(15):10582−10591. https://doi.org/10.1074/jbc.M114.550608.
51. Thi MM, Islam S, Suadicani SO, Spray DC. Connexin43 and pannexin1 channels in osteoblasts: who is the "hemichannel"? *J Membr Biol.* 2012;245(7):401−409. https://doi.org/10.1007/s00232-012-9462-2.
52. Kamioka H, Sugawara Y, Murshid SA, Ishihara Y, Honjo T, Takano-Yamamoto T. Fluid shear stress induces less calcium response in a single primary osteocyte than in a single osteoblast: implication of different focal adhesion formation. *J Bone Miner Res.* 2006;21(7):1012−1021. https://doi.org/10.1359/jbmr.060408.
53. Thompson WR, Majid AS, Czymmek KJ, et al. Association of the α(2)δ(1) subunit with Ca(v)3.2 enhances membrane expression and regulates mechanically induced ATP release in MLO-Y4 osteocytes. *J Bone Miner Res.* 2011;26(9):2125−2139. https://doi.org/10.1002/jbmr.437.
54. Warden SJ, Fuchs RK, Castillo AB, Nelson IR, Turner CH. Exercise when young provides lifelong benefits to bone structure and strength. *J Bone Miner Res.* 2007;22(2):251−259. https://doi.org/10.1359/jbmr.061107.
55. Gunter K, Baxter-Jones AD, Mirwald RL, et al. Impact exercise increases BMC during growth: an 8-year longitudinal study. *J Bone Miner Res.* 2008;23(7):986−993. https://doi.org/10.1359/jbmr.071201.
56. Yeh JK, Liu CC, Aloia JF. Effects of exercise and immobilization on bone formation and resorption in young rats. *Am J Physiol.* 1993;264(2 Pt 1):E182−E189. https://doi.org/10.1152/ajpendo.1993.264.2.E182.

57. Raab DM, Smith EL, Crenshaw TD, Thomas DP. Bone mechanical properties after exercise training in young and old rats. *J Appl Physiol*. 1990;68(1):130–134. https://doi.org/10.1152/jappl.1990.68.1.130.

58. Turner CH, Robling AG. Designing exercise regimens to increase bone strength. *Exerc Sport Sci Rev*. 2003;31(1): 45–50.

59. Hsieh YF, Turner CH. Effects of loading frequency on mechanically induced bone formation. *J Bone Miner Res*. 2001; 16(5):918–924. https://doi.org/10.1359/jbmr.2001.16.5 .918.

60. Robling AG, Burr DB, Turner CH. Partitioning a daily mechanical stimulus into discrete loading bouts improves the osteogenic response to loading. *J Bone Miner Res*. 2000; 15(8):1596–1602. https://doi.org/10.1359/jbmr.2000.15.8. 1596.

61. Kannus P, Haapasalo H, Sankelo M, et al. Effect of starting age of physical activity on bone mass in the dominant arm of tennis and squash players. *Ann Intern Med*. 1995; 123(1):27–31.

62. Wallace BA, Cumming RG. Systematic review of randomized trials of the effect of exercise on bone mass in pre- and postmenopausal women. *Calcif Tissue Int*. 2000; 67(1):10–18.

63. LeBlanc A, Schneider V, Shackelford L, et al. Bone mineral and lean tissue loss after long duration space flight. *J Musculoskelet Neuronal Interact*. 2000;1(2):157–160.

64. Lang T, LeBlanc A, Evans H, Lu Y, Genant H, Yu A. Cortical and trabecular bone mineral loss from the spine and hip in long-duration spaceflight. *J Bone Miner Res*. 2004;19(6):1006–1012. https://doi.org/10.1359/ JBMR.040307.

65. Garland DE, Stewart CA, Adkins RH, et al. Osteoporosis after spinal cord injury. *J Orthop Res*. 1992;10(3): 371–378. https://doi.org/10.1002/jor.1100100309.

66. Bikle DD, Halloran BP. The response of bone to unloading. *J Bone Miner Metab*. 1999;17(4):233–244.

67. Dehority W, Halloran BP, Bikle DD, et al. Bone and hormonal changes induced by skeletal unloading in the mature male rat. *Am J Physiol*. 1999;276(1 Pt 1): E62–E69.

68. Marotti G. The structure of bone tissues and the cellular control of their deposition. *Ital J Anat Embryol*. 1996; 101(4):25–79.

69. Martin R. Toward a unifying theory of bone remodeling. *Bone*. 2000;26(1):1–6.

70. Tatsumi S, Ishii K, Amizuka N, et al. Targeted ablation of osteocytes induces osteoporosis with defective mechanotransduction. *Cell Metabol*. 2007;5(6): 464–475. https://doi.org/10.1016/j.cmet.2007.05.001.

71. Balemans W, Ebeling M, Patel N, et al. Increased bone density in sclerosteosis is due to the deficiency of a novel secreted protein (SOST). *Hum Mol Genet*. 2001;10(5): 537–543.

72. Van Hul W, Balemans W, Van Hul E, et al. Van Buchem disease (hyperostosis corticalis generalisata) maps to chromosome 17q12-q21. *Am J Hum Genet*. 1998;62(2): 391–399. https://doi.org/10.1086/301721.

73. Balemans W, Patel N, Ebeling M, et al. Identification of a 52 kb deletion downstream of the SOST gene in patients with van Buchem disease. *J Med Genet*. 2002;39(2): 91–97.

74. Staehling-Hampton K, Proll S, Paeper BW, et al. A 52-kb deletion in the SOST-MEOX1 intergenic region on 17q12-q21 is associated with van Buchem disease in the Dutch population. *Am J Med Genet*. 2002;110(2): 144–152. https://doi.org/10.1002/ajmg.10401.

75. Winkler DG, Sutherland MK, Geoghegan JC, et al. Osteocyte control of bone formation via sclerostin, a novel BMP antagonist. *EMBO J*. 2003;22(23):6267–6276. https://doi.org/10.1093/emboj/cdg599.

76. Collette NM, Genetos DC, Economides AN, et al. Targeted deletion of Sost distal enhancer increases bone formation and bone mass. *Proc Natl Acad Sci U S A*. 2012; 109(35):14092–14097. https://doi.org/10.1073/ pnas.1207188109.

77. Robling AG, Niziolek PJ, Baldridge LA, et al. Mechanical stimulation of bone in vivo reduces osteocyte expression of Sost/sclerostin. *J Biol Chem*. 2008;283(9):5866–5875. https://doi.org/10.1074/jbc.M705092200.

78. Spatz JM, Wein MN, Gooi JH, et al. The Wnt-inhibitor sclerostin is up-regulated by mechanical unloading in osteocytes in-vitro. *J Biol Chem*. May 2015;290(27):16744–16758. https://doi.org/10.1074/jbc.M114.628313.

79. Tu X, Rhee Y, Condon K, et al. Sost downregulation and local Wnt signaling are required for the osteogenic response to mechanical loading. *Bone*. October 2011. https://doi.org/10.1016/j.bone.2011.10.025.

80. Lin C, Jiang X, Dai Z, et al. Sclerostin mediates bone response to mechanical unloading through antagonizing Wnt/beta-catenin signaling. *J Bone Miner Res*. 2009; 24(10):1651–1661. https://doi.org/10.1359/ jbmr.090411.

81. Spatz JM, Ellman R, Cloutier AM, et al. Sclerostin antibody inhibits skeletal deterioration due to reduced mechanical loading. *J Bone Miner Res*. 2013;28(4): 865–874. https://doi.org/10.1002/jbmr.1807.

82. Robling AG, Kang KS, Bullock WA, et al. Sost, independent of the non-coding enhancer ECR5, is required for bone mechanoadaptation. *Bone*. 2016;92:180–188. https://doi.org/10.1016/j.bone.2016.09.001.

83. Leupin O, Kramer I, Collette NM, et al. Control of the SOST bone enhancer by PTH using MEF2 transcription factors. *J Bone Miner Res*. 2007;22(12):1957–1967. https://doi.org/10.1359/jbmr.070804.

84. Yoshida H, Hayashi S, Kunisada T, et al. The murine mutation osteopetrosis is in the coding region of the macrophage colony stimulating factor gene. *Nature*. 1990; 345(6274):442–444. https://doi.org/10.1038/345442a0.

85. Simonet WS, Lacey DL, Dunstan CR, et al. Osteoprotegerin: a novel secreted protein involved in the regulation of bone density. *Cell*. 1997;89(2):309–319.

86. Kramer I, Halleux C, Keller H, et al. Osteocyte Wnt/beta-catenin signaling is required for normal bone homeostasis. *Mol Cell Biol*. 2010;30(12):3071–3085. https://doi.org/10.1128/MCB.01428-09.

87. Schaffler MB, Kennedy OD. Osteocyte signaling in bone. *Curr Osteoporos Rep.* 2012;10(2):118–125. https://doi.org/10.1007/s11914-012-0105-4.

88. Kennedy OD, Herman BC, Laudier DM, Majeska RJ, Sun HB, Schaffler MB. Activation of resorption in fatigue-loaded bone involves both apoptosis and active pro-osteoclastogenic signaling by distinct osteocyte populations. *Bone.* 2012;50(5):1115–1122. https://doi.org/10.1016/j.bone.2012.01.025.

89. Cardoso L, Herman BC, Verborgt O, Laudier D, Majeska RJ, Schaffler MB. Osteocyte apoptosis controls activation of intracortical resorption in response to bone fatigue. *J Bone Miner Res.* 2009;24(4):597–605. https://doi.org/10.1359/jbmr.081210.

90. Emerton KB, Hu B, Woo AA, et al. Osteocyte apoptosis and control of bone resorption following ovariectomy in mice. *Bone.* 2010;46(3):577–583. https://doi.org/10.1016/j.bone.2009.11.006.

91. Marie PJ, Glorieux FH. Relation between hypomineralized periosteocytic lesions and bone mineralization in vitamin D-resistant rickets. *Calcif Tissue Int.* 1983;35(4–5):443–448.

92. Iagodovskiĭ VS, Triftanidi LA, Gorokhova GP. [Effect of space flight on rat skeletal bones (an optical light and electron microscopic study)]. *Kosm Biol Aviakosm Med.* 1977;11(1):14–20.

93. Haller AC, Zimny ML. Effects of hibernation on interradicular alveolar bone. *J Dent Res.* 1977;56(12):1552–1557. https://doi.org/10.1177/00220345770560122601.

94. Qing H, Ardeshirpour L, Pajevic PD, et al. Demonstration of osteocytic perilacunar/canalicular remodeling in mice during lactation. *J Bone Miner Res.* 2012;27(5):1018–1029. https://doi.org/10.1002/jbmr.1567.

95. Jähn K, Kelkar S, Zhao H, et al. Osteocytes acidify their microenvironment in response to PTHrP in vitro and in lactating mice in vivo. *J Bone Miner Res.* 2017;32(8):1761–1772. https://doi.org/10.1002/jbmr.3167.

96. Dole NS, Mazur CM, Acevedo C, et al. Osteocyte-intrinsic TGF-β signaling regulates bone quality through perilacunar/canalicular remodeling. *Cell Rep.* 2017;21(9):2585–2596. https://doi.org/10.1016/j.celrep.2017.10.115.

97. Gardinier JD, Mohamed F, Kohn DH. PTH signaling during exercise contributes to bone adaptation. *J Bone Miner Res.* 2015;30(6):1053–1063. https://doi.org/10.1002/jbmr.2432.

98. Nicolella DP, Moravits DE, Gale AM, Bonewald LF, Lankford J. Osteocyte lacunae tissue strain in cortical bone. *J Biomech.* 2006;39(9):1735–1743. https://doi.org/10.1016/j.jbiomech.2005.04.032.

99. Witcher PC, Miner SE, Horan DJ, et al. Sclerostin neutralization unleashes the osteoanabolic effects of Dkk1 inhibition. *JCI Insight.* 2018;3(11). https://doi.org/10.1172/jci.insight.98673.

100. Maurel DB, Duan P, Farr J, Cheng A-L, Johnson ML, Bonewald LF. Beta-catenin Haplo insufficient male mice do not lose bone in response to Hindlimb unloading. *PLoS One.* 2016;11(7):e0158381. https://doi.org/10.1371/journal.pone.0158381.

101. Tu X, Delgado-Calle J, Condon KW, et al. Osteocytes mediate the anabolic actions of canonical Wnt/β-catenin signaling in bone. *Proc Natl Acad Sci U S A.* January 2015;112(5):E478–E486. https://doi.org/10.1073/pnas.1409857112.

102. Lara-Castillo N, Kim-Weroha NA, Kamel MA, et al. In vivo mechanical loading rapidly activates β-catenin signaling in osteocytes through a prostaglandin mediated mechanism. *Bone.* 2015;76:58–66. https://doi.org/10.1016/j.bone.2015.03.019.

103. Sawakami K, Robling AG, Ai M, et al. The Wnt co-receptor LRP5 is essential for skeletal mechanotransduction but not for the anabolic bone response to parathyroid hormone treatment. *J Biol Chem.* 2006;281(33):23698–23711. https://doi.org/10.1074/jbc.M601000200.

104. Iwaniec UT, Wronski TJ, Liu J, et al. PTH stimulates bone formation in mice deficient in Lrp5. *J Bone Miner Res.* 2007;22(3):394–402. https://doi.org/10.1359/jbmr.061118.

105. Romero G, Sneddon WB, Yang Y, et al. Parathyroid hormone receptor directly interacts with dishevelled to regulate beta-Catenin signaling and osteoclastogenesis. *J Biol Chem.* 2010;285(19):14756–14763. https://doi.org/10.1074/jbc.M110.102970.

106. Xiao L, Fei Y, Hurley MM. FGF2 crosstalk with Wnt signaling in mediating the anabolic action of PTH on bone formation. *Bone Rep.* 2018;9:136–144. https://doi.org/10.1016/j.bonr.2018.09.003.

107. Bivi N, Condon KW, Allen MR, et al. Cell autonomous requirement of connexin 43 for osteocyte survival: consequences for endocortical resorption and periosteal bone formation. *J Bone Miner Res.* October 2011;27(2):374–389. https://doi.org/10.1002/jbmr.548.

108. Bivi N, Nelson MT, Faillace ME, Li J, Miller LM, Plotkin LI. Deletion of Cx43 from osteocytes results in defective bone material properties but does not decrease extrinsic strength in cortical bone. *Calcif Tissue Int.* 2012;91(3):215–224. https://doi.org/10.1007/s00223-012-9628-z.

109. Xu H, Gu S, Riquelme MA, et al. Connexin 43 channels are essential for normal bone structure and osteocyte viability. *J Bone Miner Res.* 2015;30(3):550–562. https://doi.org/10.1002/jbmr.2374.

110. Grimston SK, Watkins MP, Brodt MD, Silva MJ, Civitelli R. Enhanced periosteal and endocortical responses to axial tibial compression loading in conditional connexin43 deficient mice. *PLoS ONE.* 2012;7(9):e44222. https://doi.org/10.1371/journal.pone.0044222.

111. Canalis E, Parker K, Feng JQ, Zanotti S. Osteoblast lineage-specific effects of notch activation in the skeleton. *Endocrinology.* 2013;154(2):623–634. https://doi.org/10.1210/en.2012-1732.

112. Zanotti S, Canalis E. Notch signaling and the skeleton. *Endocr Rev.* 2016;37(3):223–253. https://doi.org/10.1210/er.2016-1002.

113. Rosen CJ, Beamer WG, Donahue LR. Defining the genetics of osteoporosis: using the mouse to understand man. *Osteoporos Int.* 2001;12(10):803–810. https://doi.org/10.1007/s001980170030.

114. Sankaran JS, Li B, Donahue L-R, Judex S. Modulation of unloading-induced bone loss in mice with altered ERK

signaling. *Mamm Genome*. 2016;27(1–2):47–61. https://doi.org/10.1007/s00335-015-9611-x.

115. Ralston SH, Uitterlinden AG. Genetics of osteoporosis. *Endocr Rev*. 2010;31(5):629–662. https://doi.org/10.1210/er.2009-0044.

116. Rubin CT, Bain SD, McLeod KJ. Suppression of the osteogenic response in the aging skeleton. *Calcif Tissue Int*. 1992;50(4):306–313.

117. Järvinen TLN, Pajamäki I, Sievänen H, et al. Femoral neck response to exercise and subsequent deconditioning in young and adult rats. *J Bone Miner Res*. 2003;18(7):1292–1299. https://doi.org/10.1359/jbmr.2003.18.7.1292.

118. Brodt MD, Silva MJ. Aged mice have enhanced endocortical response and normal periosteal response compared with young-adult mice following 1 week of axial tibial compression. *J Bone Miner Res*. 2010;25(9):2006–2015. https://doi.org/10.1002/jbmr.96.

119. Vashishth D, Verborgt O, Divine G, Schaffler MB, Fyhrie DP. Decline in osteocyte lacunar density in human cortical bone is associated with accumulation of microcracks with age. *Bone*. 2000;26(4):375–380. https://doi.org/10.1016/S8756-3282(00)00236-2.

120. Mullender MG, van der Meer DD, Huiskes R, Lips P. Osteocyte density changes in aging and osteoporosis. *Bone*. 1996;18(2):109–113.

121. Mori S, Harruff R, Ambrosius W, Burr DB. Trabecular bone volume and microdamage accumulation in the femoral heads of women with and without femoral neck fractures. *Bone*. 1997;21(6):521–526.

122. Qiu S, Rao DS, Palnitkar S, Parfitt AM. Age and distance from the surface but not menopause reduce osteocyte density in human cancellous bone. *Bone*. 2002;31(2):313–318. https://doi.org/10.1016/S8756-3282(02)00819-0.

123. Qiu S, Rao DS, Palnitkar S, Parfitt AM. Reduced iliac cancellous osteocyte density in patients with osteoporotic vertebral fracture. *J Bone Miner Res*. 2003;18(9):1657–1663. https://doi.org/10.1359/jbmr.2003.18.9.1657.

124. Busse B, Djonic D, Milovanovic P, et al. Decrease in the osteocyte lacunar density accompanied by hypermineralized lacunar occlusion reveals failure and delay of remodeling in aged human bone. *Aging Cell*. 2010;9(6):1065–1075. https://doi.org/10.1111/j.1474-9726.2010.00633.x.

125. Tami AE, Nasser P, Verborgt O, Schaffler MB, Knothe Tate ML. The role of interstitial fluid flow in the remodeling response to fatigue loading. *J Bone Miner Res*. 2002;17(11):2030–2037. https://doi.org/10.1359/jbmr.2002.17.11.2030.

126. Noble BS, Peet N, Stevens HY, et al. Mechanical loading: biphasic osteocyte survival and targeting of osteoclasts for bone destruction in rat cortical bone. *Am J Physiol Cell Physiol*. 2003;284(4):C934–C943. https://doi.org/10.1152/ajpcell.00234.2002.

127. Bonivtch AR, Bonewald LF, Nicolella DP. Tissue strain amplification at the osteocyte lacuna: a microstructural finite element analysis. *J Biomech*. 2007;40(10):2199–2206. https://doi.org/10.1016/j.jbiomech.2006.10.040.

128. Gatti V, Azoulay EM, Fritton SP. Microstructural changes associated with osteoporosis negatively affect loading-induced fluid flow around osteocytes in cortical bone. *J Biomech*. 2018;66:127–136. https://doi.org/10.1016/j.jbiomech.2017.11.011.

129. Verbruggen SW, Mc Garrigle MJ, Haugh MG, Voisin MC, McNamara LM. Altered mechanical environment of bone cells in an animal model of short- and long-term osteoporosis. *Biophys J*. 2015;108(7):1587–1598. https://doi.org/10.1016/j.bpj.2015.02.031.

130. Verbruggen SW, Vaughan TJ, McNamara LM. Mechanisms of osteocyte stimulation in osteoporosis. *J Mech Behav Biomed Mater*. 2016;62:158–168. https://doi.org/10.1016/j.jmbbm.2016.05.004.

131. Knothe Tate M, Adamson J, Tami A, Bauer T. The osteocyte. *Int J Biochem Cell Biol*. 2004;36(1):1–8.

132. Hesse B, Varga P, Langer M, et al. Canalicular network morphology is the major determinant of the spatial distribution of mass density in human bone tissue: evidence by means of synchrotron radiation phase-contrast nano-CT. *J Bone Miner Res*. 2015;30(2):346–356. https://doi.org/10.1002/jbmr.2324.

133. Hemmatian H, Jalali R, Semeins CM, et al. Mechanical loading differentially affects osteocytes in Fibulae from lactating mice compared to osteocytes in virgin mice: possible role for lacuna size. *Calcif Tissue Int*. 2018;103(6):675–685. https://doi.org/10.1007/s00223-018-0463-8.

134. Gardinier JD, Al-Omaishi S, Morris MD, Kohn DH. PTH signaling mediates perilacunar remodeling during exercise. *Matrix Biol*. 2016;52–54:162–175. https://doi.org/10.1016/j.matbio.2016.02.010.

135. Carter Y, Thomas CDL, Clement JG, Cooper DML. Femoral osteocyte lacunar density, volume and morphology in women across the lifespan. *J Struct Biol*. 2013;183(3):519–526. https://doi.org/10.1016/j.jsb.2013.07.004.

136. Heveran CM, Rauff A, King KB, Carpenter RD, Ferguson VL. A new open-source tool for measuring 3D osteocyte lacunar geometries from confocal laser scanning microscopy reveals age-related changes to lacunar size and shape in cortical mouse bone. *Bone*. 2018;110:115–127. https://doi.org/10.1016/j.bone.2018.01.018.

137. Vatsa A, Breuls RG, Semeins CM, Salmon PL, Smit TH, Klein-Nulend J. Osteocyte morphology in fibula and calvaria — is there a role for mechanosensing? *Bone*. 2008;43(3):452–458. https://doi.org/10.1016/j.bone.2008.01.030.

138. Marotti G. Osteocyte orientation in human lamellar bone and its relevance to the morphometry of periosteocytic lacunae. *Metab Bone Dis Relat Res*. 1979;1(4):325–333. https://doi.org/10.1016/0221-8747(79)90027-4.

139. Kerschnitzki M, Wagermaier W, Roschger P, et al. The organization of the osteocyte network mirrors the extracellular matrix orientation in bone. *J Struct Biol*. 2011;

173(2):303—311. https://doi.org/10.1016/j.jsb.2010.11.014.

140. Wu V, van Oers RFM, Schulten EAJM, Helder MN, Bacabac RG, Klein-Nulend J. Osteocyte morphology and orientation in relation to strain in the jaw bone. *Int J Oral Sci.* 2018;10(1):2. https://doi.org/10.1038/s41368-017-0007-5.

141. van Oers RFM, Wang H, Bacabac RG. Osteocyte shape and mechanical loading. *Curr Osteoporos Rep.* 2015;13(2):61—66. https://doi.org/10.1007/s11914-015-0256-1.

142. Bacabac RG, Mizuno D, Schmidt CF, et al. Round versus flat: bone cell morphology, elasticity, and mechanosensing. *J Biomech.* 2008;41(7):1590—1598. https://doi.org/10.1016/j.jbiomech.2008.01.031.

143. Donahue SW, Jacobs CR, Donahue HJ. Flow-induced calcium oscillations in rat osteoblasts are age, loading frequency, and shear stress dependent. *Am J Physiol Cell Physiol.* 2001;281(5):C1635—C1641. https://doi.org/10.1152/ajpcell.2001.281.5.C1635.

144. Klein-Nulend J, Sterck JGH, Semeins CM, et al. Donor age and mechanosensitivity of human bone cells. *Osteoporos Int.* 2002;13(2):137—146.

145. Bakker AD, Klein-Nulend J, Tanck E, et al. Different responsiveness to mechanical stress of bone cells from osteoporotic versus osteoarthritic donors. *Osteoporos Int.* 2006;17(6):827—833. https://doi.org/10.1007/s00198-006-0072-7.

146. Srinivasan S, Gross TS, Bain SD. Bone mechanotransduction may require augmentation in order to strengthen the senescent skeleton. *Ageing Res Rev.* 2012;11(3):353—360. https://doi.org/10.1016/j.arr.2011.12.007.

147. Singh L, Tyagi S, Myers D, Duque G. Good, bad, or ugly: the biological roles of bone marrow fat. *Curr Osteoporos Rep.* 2018;16(2):130—137. https://doi.org/10.1007/s11914-018-0427-y.

148. Farr JN, Xu M, Weivoda MM, et al. Targeting cellular senescence prevents age-related bone loss in mice. *Nat Med.* 2017;23(9):1072—1079. https://doi.org/10.1038/nm.4385.

149. Farr JN, Fraser DG, Wang H, et al. Identification of senescent cells in the bone microenvironment. *J Bone Miner Res.* 2016;31(11):1920—1929. https://doi.org/10.1002/jbmr.2892.

150. Baker DJ, Wijshake T, Tchkonia T, et al. Clearance of p16Ink4a-positive senescent cells delays ageing-associated disorders. *Nature.* 2011;479(7372):232—236. https://doi.org/10.1038/nature10600.

151. Zhu Y, Tchkonia T, Pirtskhalava T, et al. The Achilles' heel of senescent cells: from transcriptome to senolytic drugs. *Aging Cell.* 2015;14(4):644—658. https://doi.org/10.1111/acel.12344.

152. Xu M, Pirtskhalava T, Farr JN, et al. Senolytics improve physical function and increase lifespan in old age. *Nat Med.* 2018;24(8):1246—1256. https://doi.org/10.1038/s41591-018-0092-9.

153. Rajkumar SV, Kumar S. Multiple myeloma: diagnosis and treatment. *Mayo Clin Proc.* 2016;91(1):101—119. https://doi.org/10.1016/j.mayocp.2015.11.007.

154. Nørgaard M, Jensen AØ, Jacobsen JB, Cetin K, Fryzek JP, Sørensen HT. Skeletal related events, bone metastasis and survival of prostate cancer: a population based cohort study in Denmark (1999 to 2007). *J Urol.* 2010;184(1):162—167. https://doi.org/10.1016/j.juro.2010.03.034.

155. Croucher PI, McDonald MM, Martin TJ. Bone metastasis: the importance of the neighbourhood. *Nat Rev Cancer.* 2016;16(6):373—386. https://doi.org/10.1038/nrc.2016.44.

156. Giuliani N, Ferretti M, Bolzoni M, et al. Increased osteocyte death in multiple myeloma patients: role in myeloma-induced osteoclast formation. *Leukemia.* 2012;26(6):1391—1401. https://doi.org/10.1038/leu.2011.381.

157. Trotter TN, Fok M, Gibson JT, Peker D, Javed A, Yang Y. Osteocyte apoptosis attracts myeloma cells to bone and supports progression through regulation of the bone marrow microenvironment. *Blood.* 2016;128(22):484.

158. Plotkin LI, Weinstein RS, Parfitt AM, Roberson PK, Manolagas SC, Bellido T. Prevention of osteocyte and osteoblast apoptosis by bisphosphonates and calcitonin. *J Clin Invest.* 1999;104(10):1363—1374. https://doi.org/10.1172/JCI6800.

159. Watkins MP, Norris JY, Grimston SK, et al. Bisphosphonates improve trabecular bone mass and normalize cortical thickness in ovariectomized, osteoblast connexin43 deficient mice. *Bone.* 2012;51(4):787—794. https://doi.org/10.1016/j.bone.2012.06.018.

160. Brennan O, Kennedy OD, Lee TC, Rackard SM, O'Brien FJ. Effects of estrogen deficiency and bisphosphonate therapy on osteocyte viability and microdamage accumulation in an ovine model of osteoporosis. *J Orthop Res.* 2011;29(3):419—424. https://doi.org/10.1002/jor.21229.

161. Plotkin LI, Manolagas SC, Bellido T. Transduction of cell survival signals by connexin-43 hemichannels. *J Biol Chem.* 2002;277(10):8648—8657. https://doi.org/10.1074/jbc.M108625200.

162. Plotkin LI, Aguirre JI, Kousteni S, Manolagas SC, Bellido T. Bisphosphonates and estrogens inhibit osteocyte apoptosis via distinct molecular mechanisms downstream of extracellular signal-regulated kinase activation. *J Biol Chem.* 2005;280(8):7317—7325. https://doi.org/10.1074/jbc.M412817200.

163. Plotkin LI, Manolagas SC, Bellido T. Dissociation of the pro-apoptotic effects of bisphosphonates on osteoclasts from their anti-apoptotic effects on osteoblasts/osteocytes with novel analogs. *Bone.* 2006;39(3):443—452. https://doi.org/10.1016/j.bone.2006.02.060.

164. Plotkin LI, Bivi N, Bellido T. A bisphosphonate that does not affect osteoclasts prevents osteoblast and osteocyte apoptosis and the loss of bone strength induced by glucocorticoids in mice. *Bone.* 2011;49(1):122—127. https://doi.org/10.1016/j.bone.2010.08.011.

165. Toscani D, Palumbo C, Dalla Palma B, et al. The proteasome inhibitor bortezomib maintains osteocyte viability in multiple myeloma patients by reducing both

apoptosis and autophagy: a new function for proteasome inhibitors. *J Bone Miner Res.* 2016;31(4):815–827. https://doi.org/10.1002/jbmr.2741.

166. Chandra A, Lin T, Tribble MB, et al. PTH1-34 alleviates radiotherapy-induced local bone loss by improving osteoblast and osteocyte survival. *Bone.* 2014;67:33–40. https://doi.org/10.1016/j.bone.2014.06.030.

167. Chandra A, Lin T, Zhu J, et al. PTH1-34 blocks radiation-induced osteoblast apoptosis by enhancing DNA repair through canonical Wnt pathway. *J Biol Chem.* 2015;290(1):157–167. https://doi.org/10.1074/jbc.M114.608158.

168. De Vos J, Couderc G, Tarte K, et al. Identifying intercellular signaling genes expressed in malignant plasma cells by using complementary DNA arrays. *Blood.* 2001;98(3):771–780.

169. Jundt F, Pröbsting KS, Anagnostopoulos I, et al. Jagged1-induced Notch signaling drives proliferation of multiple myeloma cells. *Blood.* 2004;103(9):3511–3515. https://doi.org/10.1182/blood-2003-07-2254.

170. Houde C, Li Y, Song L, et al. Overexpression of the NOTCH ligand JAG2 in malignant plasma cells from multiple myeloma patients and cell lines. *Blood.* 2004;104(12):3697–3704. https://doi.org/10.1182/blood-2003-12-4114.

171. Delgado-Calle J, Anderson J, Cregor MD, et al. Bidirectional notch signaling and osteocyte-derived factors in the bone marrow microenvironment promote tumor cell proliferation and bone destruction in multiple myeloma. *Cancer Res.* 2016;76(5):1089–1100. https://doi.org/10.1158/0008-5472.CAN-15-1703.

172. Terpos E, Christoulas D, Katodritou E, et al. Elevated circulating sclerostin correlates with advanced disease features and abnormal bone remodeling in symptomatic myeloma: reduction post-bortezomib monotherapy. *Int J Cancer.* 2012;131(6):1466–1471. https://doi.org/10.1002/ijc.27342.

173. Delgado-Calle J, Anderson J, Cregor MD, et al. Genetic deletion of Sost or pharmacological inhibition of sclerostin prevent multiple myeloma-induced bone disease without affecting tumor growth. *Leukemia.* June 2017;31(12):2686–2694. https://doi.org/10.1038/leu.2017.152.

174. Tian E, Zhan F, Walker R, et al. The role of the Wnt-signaling antagonist DKK1 in the development of osteolytic lesions in multiple myeloma. *N Engl J Med.* 2003;349(26):2483–2494. https://doi.org/10.1056/NEJMoa030847.

175. Hesse E, Schröder S, Brandt D, Pamperin J, Saito H, Taipaleenmaki H. Sclerostin inhibition alleviates breast cancer-induced bone metastases and muscle weakness. *JCI Insight.* 2019;5. https://doi.org/10.1172/jci.insight.125543.

176. Verras M, Sun Z. Roles and regulation of Wnt signaling and beta-catenin in prostate cancer. *Cancer Lett.* 2006;237(1):22–32. https://doi.org/10.1016/j.canlet.2005.06.004.

177. Chesire DR, Ewing CM, Gage WR, Isaacs WB. In vitro evidence for complex modes of nuclear beta-catenin signaling during prostate growth and tumorigenesis. *Oncogene.* 2002;21(17):2679–2694. https://doi.org/10.1038/sj.onc.1205352.

178. Hudson BD, Hum NR, Thomas CB, et al. SOST inhibits prostate cancer invasion. *PLoS One.* 2015;10(11):e0142058. https://doi.org/10.1371/journal.pone.0142058.

179. Hall CL, Bafico A, Dai J, Aaronson SA, Keller ET. Prostate cancer cells promote osteoblastic bone metastases through Wnts. *Cancer Res.* 2005;65(17):7554–7560. https://doi.org/10.1158/0008-5472.CAN-05-1317.

180. Wang W, Yang X, Dai J, Lu Y, Zhang J, Keller ET. Prostate cancer promotes a vicious cycle of bone metastasis progression through inducing osteocytes to secrete GDF15 that stimulates prostate cancer growth and invasion. *Oncogene.* 2019;67:7. https://doi.org/10.1038/s41388-019-0736-3.

181. Laird DW, Fistouris P, Batist G, et al. Deficiency of connexin43 gap junctions is an independent marker for breast tumors. *Cancer Res.* 1999;59(16):4104–4110.

182. Kar R, Batra N, Riquelme MA, Jiang JX. Biological role of connexin intercellular channels and hemichannels. *Arch Biochem Biophys.* 2012;524(1):2–15. https://doi.org/10.1016/j.abb.2012.03.008.

183. Wang H, Tian L, Liu J, et al. The osteogenic niche is a calcium reservoir of bone micrometastases and confers unexpected therapeutic vulnerability. *Cancer Cell.* 2018;34(5):823. https://doi.org/10.1016/j.ccell.2018.10.002.

184. Zhou JZ, Riquelme MA, Gao X, Ellies LG, Sun LZ, Jiang JX. Differential impact of adenosine nucleotides released by osteocytes on breast cancer growth and bone metastasis. *Oncogene.* 2015;34(14):1831–1842. https://doi.org/10.1038/onc.2014.113.

185. Zhou JZ, Riquelme MA, Gu S, et al. Osteocytic connexin hemichannels suppress breast cancer growth and bone metastasis. *Oncogene.* 2016;35(43):5597–5607. https://doi.org/10.1038/onc.2016.101.

186. Yellowley CE, Genetos DC. Hypoxia signaling in the skeleton: implications for bone health. *Curr Osteoporos Rep.* 2019;17:26–35. https://doi.org/10.1007/s11914-019-00500-6.

187. Westerhof N, Lankhaar J-W, Westerhof BE. The arterial Windkessel. *Med Biol Eng Comput.* 2009;47(2):131–141. https://doi.org/10.1007/s11517-008-0359-2.

188. Thompson B, Towler DA. Arterial calcification and bone physiology: role of the bone–vascular axis. *Nat Rev Endocrinol.* April 2012;8(9):529–543. https://doi.org/10.1038/nrendo.2012.36.

189. Grande KJ, Cochran RP, Reinhall PG, Kunzelman KS. Mechanisms of aortic valve incompetence in aging: a finite element model. *J Heart Valve Dis.* 1999;8(2):149–156.

190. Morony S, Tintut Y, Zhang Z, et al. Osteoprotegerin inhibits vascular calcification without affecting atherosclerosis in ldlr(−/−) mice. *Circulation.* 2008;117(3):411–420.

191. Schmid K, McSharry W, Pameijer C, Binette J. Chemical and physicochemical studies on the mineral deposits of the human atherosclerotic aorta. *Atherosclerosis.* 1980;37(2):199–210.

192. Anderson HC. Calcific diseases. A concept. *Arch Pathol Lab Med.* 1983;107(7):341–348.

193. Boström K, Watson K, Horn S, Wortham C, Herman I, Demer L. Bone morphogenetic protein expression in human atherosclerotic lesions. *J Clin Invest.* 1993;91(4): 1800–1809.

194. Johnson KA, Polewski M, Terkeltaub RA. Transglutaminase 2 is central to induction of the arterial calcification program by smooth muscle cells. *Circ Res.* 2008;102(5):529–537. https://doi.org/10.1161/CIRCRESAHA .107.154260.

195. Shao J-S, Cheng S-L, Pingsterhaus JM, Charlton-Kachigian N, Loewy AP, Towler DA. Msx2 promotes cardiovascular calcification by activating paracrine Wnt signals. *J Clin Invest.* 2005;115(5):1210–1220. https:// doi.org/10.1172/JCI24140.

196. Tanimura A, McGregor DH, Anderson HC. Calcification in atherosclerosis. I. Human studies. *J Exp Pathol.* 1986; 2(4):261–273.

197. Rogers KM, Stehbens WE. The morphology of matrix vesicles produced in experimental arterial aneurysms of rabbits. *Pathology.* 1986;18(1):64–71.

198. Krishna SM, Seto S-W, Jose RJ, et al. Wnt signaling pathway inhibitor sclerostin inhibits angiotensin II-induced aortic aneurysm and atherosclerosis. *Arterioscler Thromb Vasc Biol.* 2017;37(3):553–566. https://doi.org/ 10.1161/ATVBAHA.116.308723.

199. Caira FC, Stock SR, Gleason TG, et al. Human degenerative valve disease is associated with up-regulation of low-density lipoprotein receptor-related protein 5 receptor-mediated bone formation. *J Am Coll Cardiol.* 2006;47(8):1707–1712. https://doi.org/10.1016/ j.jacc.2006.02.040.

200. Chen J-H, Yip CYY, Sone ED, Simmons CA. Identification and characterization of aortic valve mesenchymal progenitor cells with robust osteogenic calcification potential. *Am J Pathol.* 2009;174(3):1109–1119. https://doi.org/10.2353/ajpath.2009.080750.

201. Yip CYY, Chen J-H, Zhao R, Simmons CA. Calcification by valve interstitial cells is regulated by the stiffness of the extracellular matrix. *Arterioscler Thromb Vasc Biol.* 2009;29(6):936–942. https://doi.org/10.1161/ ATVBAHA.108.182394.

# CHAPTER 1.2

# Cardiovascular Mechanics and Disease

PHILIPPE SUCOSKY, PHD • JASON A. SHAR, BS • JOHANA BARRIENTOS

## 1. INTRODUCTION

The cardiovascular system consists of the heart, the network of blood vessels, and the blood, which all act in concert to regulate blood flow according to organ demand. Its major functions are to provide organs, tissues, and cells with necessary nutrients and oxygen and to remove metabolic waste and carbon dioxide. These functions are achieved through two circulatory systems operating in series, the pulmonary and systemic circulations, which are both driven by the pressure gradient generated by the heart between the arterial and venous sides of the circulation. The right side of the heart (right atrium [RA] and right ventricle [RV]) drives blood flow in the pulmonary circulation, whereas the left side (left atrium [LA] and left ventricle [LV]) provides blood flow to the systemic circulation. Forward blood flow is achieved via four heart valves: two atrioventricular valves located between each ventricle and atrium and two semilunar valves located between each ventricle and its connected artery. The pulsatile and forward blood flow results from the sophisticated and coordinated opening and closing of the four heart valves and the periodic contraction and relaxation of the heart muscle.

Under normal conditions, an adult heart beats approximately 70 times per minute and achieves a cardiac output of 5 L/min. The need to maintain this flow rate and the differences in vascular resistance in the pulmonary and systemic circulations explain the contrasted pressure environments on both sides of the heart (15−120 mm Hg in the left side of the heart, 5−25 mm Hg in the right side of the heart).[1] These conditions combined with the wide range of scales and anatomies present throughout the vasculature generate a complex fluid dynamics environment marked by turbulence, pulsatility, and large temporal and spatial gradients. Cardiac structures not only experience these hemodynamic stresses but also interact with them to maintain function and regulate their biology. In fact, the expressions of particular cardiovascular cellular phenotypes depend not only on the intrinsic genetically programmed biology but also on the local mechanical stresses induced by blood flow[2] (Fig. 1.2.1). The discovery and study of this mechanical regulation of cardiovascular biology, also known as mechanobiology, has shed new perspectives on cardiovascular disease cause and management. Since the early 80s and the demonstration of morphologic changes in endothelial cells (ECs) subjected to altered flow,[3] mechanobiological studies have evolved toward more and more sophisticated and multidisciplinary investigations integrating the characterization of the native hemodynamic environment and its impact on the cellular, tissue, and full-organ biological response.

This chapter presents a broad review of cardiovascular mechanobiology in both physiology and pathology, with an emphasis on heart valves and blood vessels. The chapter is divided into five main sections. The first section provides a mechanical description of cardiovascular hemodynamic forces and stresses. The second section describes the different devices and experimental strategies that have been developed toward the elucidation of cardiovascular mechanobiology. The third section reviews the main mechanobiological results obtained on heart valve tissue and their impact on valvular calcification and degeneration. The fourth section addresses blood vessel mechanobiology in the context of normal function, atherosclerosis, and aneurysm formation. Lastly, the fifth section introduces future research directions and describes future needs to bridge the gap between basic mechanobiological science and its possible application toward cardiovascular disease diagnosis and management.

## 2. CARDIOVASCULAR HEMODYNAMICS
### 2.1. Overview

The dynamics of blood flow is complex, as it is affected by the heterogeneous composition of blood (red blood cells, white blood cells, platelets, plasma) and the drastic changes in its rheologic behavior as it travels through the

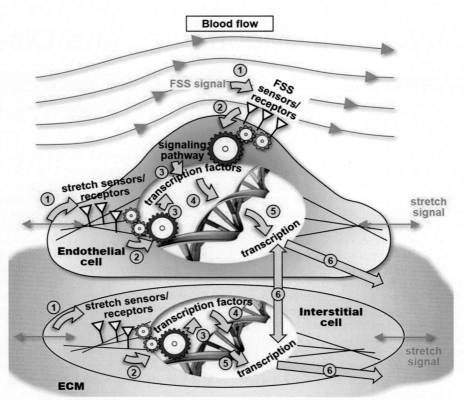

FIG. 1.2.1 Cardiovascular mechanobiology: fluid shear stress (FSS) and stretch are sensed by ECs and interstitial cells through specific receptors and stress fibers (1); the mechanical signals propagate throughout the cell (2) and are transduced to the cell nucleus (3), which results in the upregulation/downregulation of transcription factors (3), the regulation of gene expression and DNA synthesis (4), and the modulation of protein expression (5); and the resulting biological signals can affect both the local extracellular matrix (ECM) and the behavior of surrounding cells through two-way communication pathways (6). These responses are essential for the maintenance of homeostasis and the initiation/development of disease states.

different scales of the vasculature. The fluid dynamics of blood is typically characterized in terms of two dimensionless numbers. The Reynolds number is defined as

$$Re = UL/\nu,$$

where $U$ is the characteristic velocity, $L$ is the characteristic length scale (typically the representative diameter of the structure), and $\nu$ is the kinematic viscosity of blood. The Reynolds number characterizes the relative contributions of inertial versus viscous effects on the flow. The second dimensionless parameter is the Womersley number, which is defined as

$$\alpha = \sqrt{\omega L^2/\nu},$$

where $\omega$ is the characteristic pulsation frequency of the flow. This number characterizes the ratio of transient inertial forces to viscous forces. The large variations exhibited by these dimensionless parameters between different sites of the vasculature ($1 < Re < 6000$ and $4.0 < \alpha < 22.2$ across the arterial side of the circulation) suggest the existence of numerous flow regimes, from nearly steady and viscosity-dominated flows in the smallest blood vessels to highly pulsatile and in transition to turbulence in the largest arteries.[4,5]

## 2.2. Fluid Shear Stress

Blood flow generates a friction force on the surface of cardiovascular structures, which arises from the relative velocity between the fluid and the surface (Fig. 1.2.2A). The frictional force experienced by a unit area of surface is referred to as fluid shear stress (FSS) and is calculated as

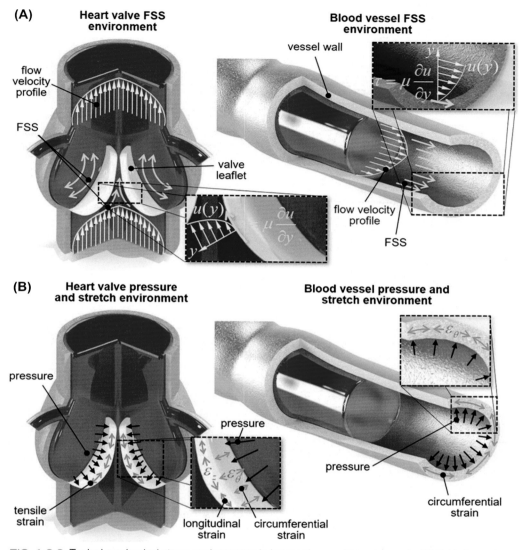

**FIG. 1.2.2** Typical mechanical stress environments in heart valves and blood vessels: **(A)** fluid shear stress (FSS) and **(B)** tensile stress.

$$\tau = \mu \frac{\partial u}{\partial y},$$

where $u$ is the velocity component tangent to the surface, $y$ is the coordinate in the normal direction to the surface, and $\mu$ is the fluid dynamic viscosity. Unlike Newtonian fluids, which are characterized by a constant viscosity, blood is non-Newtonian, as its viscosity decreases as the rate of deformation increases.[6] This complex viscous behavior combined with the wide range of

scales and flow regimes throughout the vasculature subject ECs to FSS magnitudes spanning several orders of magnitude, from 1 dyn/cm$^2$ in veins to 100 dyn/cm$^2$ on heart valve leaflets.[7-10]

## 2.3. Reynolds Stresses

The existence of transitional flow in different parts of the vasculature (e.g., left ventricular outflow tract (LVOT), aortic root, ascending aorta) gives rise to the Reynolds stresses, which characterize the impact of turbulent motion on the mean flow. The Reynolds shear stress $\tau'$ can be defined as

$$\tau' = -\rho\overline{u'v'},$$

where $u'$ and $v'$ are the velocity fluctuations in the $x$ and $y$ directions, respectively; $\rho$ is the fluid density; and the bar denotes time- or ensemble-average. Although turbulent stresses are not actual mechanical stresses experienced by ECs, they have been shown to play an important role in hemolysis.[11]

## 2.4. Cyclic Stretch

Cardiovascular tissue is exposed to cyclic pressure, which results from the periodic contraction and relaxation of the ventricles, by transferring this normal load into multiple directions.[12] In blood vessels, the tissue layers elongate in the circumferential and longitudinal directions when acted on by the intraluminal pressure (Fig. 1.2.2B). Valve leaflets elongate along the circumferential and radial directions during coaptation when the diastolic transvalvular pressure gradient is imposed across the valve. The magnitude of tissue elongation in one direction is quantified by the engineering strain (or Cauchy strain) characterizing the increase in tissue length per unit length and is defined as

$$\varepsilon = \frac{l-L}{L},$$

where $l$ and $L$ represent the deformed and initial lengths, respectively, of a line element aligned along a particular direction. Stretch levels and temporal variations are strongly dependent on the tissue, its anatomic position, and its mechanical properties. Stretch levels of up to 30% and 8% have been reported on aortic valve (AV) leaflets[13] and the anterior mitral valve (MV) leaflet,[14] respectively. In the vasculature, circumferential stretch levels of 8%–10% have been reported for the aorta[15] and 5% for peripheral arteries[16] under physiologic conditions.

## 3. BIOREACTORS FOR CARDIOVASCULAR MECHANOBIOLOGICAL STUDIES

The elucidation of the cause-and-effect relationships between cardiovascular biology and hemodynamic stresses requires the replication of desired mechanical stress environments in a well-controlled sterile environment capable of maintaining both the structural integrity and the viability of the tissue/cells. Different bioreactors have been designed to achieve this objective. This section describes three devices that have been used to subject cells and tissue to mechanical stimulation in vitro.

## 3.1. Parallel Plate Devices

Parallel plate devices have been used in early mechanobiological studies aimed at investigating the effects of FSS magnitude or directionality on vascular and valvular ECs in vitro. Such systems typically consist of two stationary parallel plates (Fig. 1.2.3A). Cells are seeded along the bottom plate, and a pump drives a flow of culture medium between the two plates by generating a pressure differential between the inlet and outlet sections of the device, which exposes the cells to FSS. The device is generally operated to produce a fully developed laminar flow. The volume flow rate $Q$ of the culture medium achieved by the pump and the dimensional characteristics of the chamber cross section (height $h$ and width $b$) control the magnitude of the FSS, $\tau$, produced on the surface of the plates as

$$\tau = \frac{6\mu Q}{bh^2}$$

Parallel plate systems are typically limited to the production of steady uniform FSS due to the inertia of their driving components and the large volume of working fluid. Several modifications have been made in the recent years to generate pulsatile/oscillatory shear stress[17,18] and shear stress in microenvironments.[19,20]

## 3.2. Cone-and-Plate Bioreactors

Previous investigations of the biological response of vascular and valvular ECs to FSS have implemented cone-and-plate bioreactors. These devices, which are derived from cone-and-plate viscometers, consist of an inverted cone rotating above a stationary plate supporting the cells (Fig. 1.2.3B). The rotation of the cone generates a flow of culture medium above the plate, resulting in the generation of FSS on the cell monolayer. Under certain conditions (low Reynolds number, small cone-plate angle $\alpha$, small gap $h$ between the cone apex and the plate), the device generates a circumferential FSS on the plate, whose local magnitude $\tau$ at any radial location $r$ is proportional to the cone angular velocity $\omega$, and the fluid dynamic viscosity $\mu$,

$$\tau = \mu\omega\frac{r}{h+r\alpha},$$

making it a convenient tool for the production of a time-varying FSS environment. Such devices have been used to study the response of vascular ECs to physiologic and atherosclerotic FSS[21–23] and of valvular ECs to physiologic and disturbed flow environments.[24] The need to elucidate the role of cell-extracellular matrix (ECM) communication in FSS signaling has motivated the design of more sophisticated cone-and-plate devices

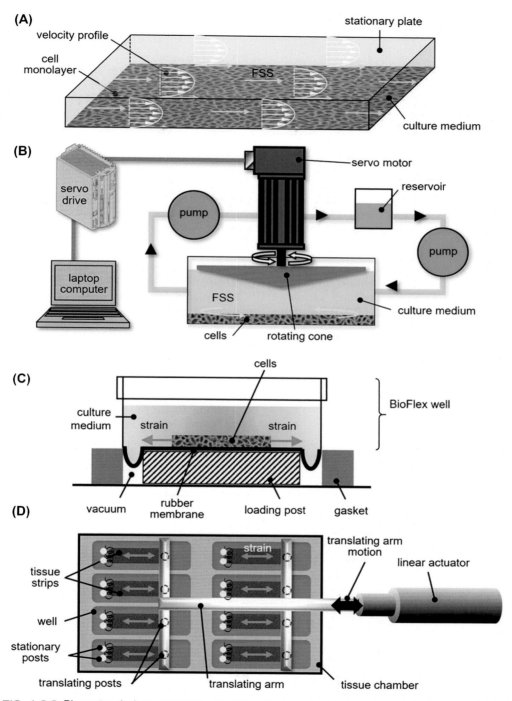

**(A)**

velocity profile

stationary plate

cell monolayer

FSS

culture medium

**(B)**

servo drive

servo motor

reservoir

pump

pump

laptop computer

FSS

culture medium

cells    rotating cone

**(C)**

cells

culture medium

strain    strain

BioFlex well

vacuum    rubber membrane    loading post    gasket

**(D)**

strain

translating arm motion

linear actuator

tissue strips

well

stationary posts

translating posts    translating arm    tissue chamber

FIG. 1.2.3 Bioreactor devices commonly used to subject cardiovascular cells and tissue to mechanical stresses: **(A)** parallel plate flow chamber, **(B)** cone-and-plate bioreactor, **(C)** Flexcell device, and **(D)** uniaxial stretch bioreactor. *FSS*, fluid shear stress.

capable of subjecting whole pieces of arterial or venous tissue to FSS[25] and valvular leaflet tissue to side-specific FSS.[26]

### 3.3. Stretch Bioreactors

The sensitivity of cardiovascular structures to stretch or tensile stress has been investigated on cells and tissue using two types of devices. The Flexcell system (Flexcell International, Hillsborough, NC) is a commercial device used in the application of a biaxial strain field to cells[27] (Fig. 1.2.3C). Cells are seeded on a compliant membrane resting on a stationary post. Vacuum pressure and positive air pressure are used to deform the membrane, subjecting the cells to a cyclic strain of up to 33%. Depending on the loading post geometry, cells can be subjected to equibiaxial, gradient biaxial, or uniaxial strain.

The tensile stretch bioreactor is another device designed to subject intact cardiovascular tissue to different levels and frequencies of tensile stress.[28,29] The device consists of two sealed chambers, each accommodating eight tissue wells (Fig. 1.2.3D). The two ends of each rectangular tissue strip are each attached to one stationary post and one translating post. The displacement of the translating post is achieved by a linear actuator, which results in the cyclic stretching of the tissue sample. The programming of the servo drive provides control over the frequency and magnitude of the stretch signal.

## 4. HEART VALVE MECHANOBIOLOGY

### 4.1. Valvular Anatomy, Function, and Structure

Unidirectional blood flow from the heart ventricles to the arterial system, the venous system, and the heart atria is achieved by the timed opening and closure of the four heart valves. The mitral and tricuspid valves are atrioventricular valves and regulate the passage of blood from the LA to the LV, and from the RA to the RV, respectively. The AV and pulmonary valve are semilunar valves located in the LVOT and right ventricular outflow tract, respectively, and enforce unidirectional blood flow between each ventricle and its connected artery (i.e., aorta on the left side, pulmonary artery on the right side). Heart valves typically undergo 3 billion opening/closing cycles during their lifespan, and their histologic structure is tailored to withstand the bending, tensile stress, and FSS imposed by the blood flow.[30,31] The harsh pressure conditions present in the left heart and the high susceptibility to disease of cardiovascular structures in the systemic circulation provide some

motivation for focusing on the two valves on the systemic side, namely, the AV and the MV.

### 4.1.1. Aortic valve

The AV consists of three leaflets housed at the base of the aorta in the aortic root. The root is formed by three hemispheric cavities (i.e., sinuses of Valsalva), two of which feature a coronary ostium connected to a coronary artery.[32,33] AV leaflets have a trilayered structure, consisting of the ventricularis facing the LV, the fibrosa facing the aorta, and the spongiosa sandwiched between the other two layers.[34] The ventricularis and the fibrosa contain radially and circumferentially aligned collagen fibers, respectively, intertwined with elastin fibers, whereas the spongiosa contains mainly glycosaminoglycans (GAGs). The cell population within the AV includes valvular interstitial cells (ICs), which consist of myofibroblasts, fibroblasts, and smooth muscle cells (SMCs) embedded in the ECM, and valvular ECs lining the surface of the leaflets.

The dynamics and deformation of AV leaflets are regulated by the variations in transvalvular pressure between the LV and the aorta during the cardiac cycle. At the beginning of systole when the LV is filled with blood and the AV is closed, the static pressure in the aorta is ~80 mm Hg. During systole (ventricular contraction), blood pressure increases in the LV until it exceeds the aortic pressure, which opens the AV. The valve remains open during the ejection phase as a result of the small (i.e., a few millimeters of mercury) transvalvular pressure established between the LV and the aorta during that phase, and blood flow is ejected from the LV toward the systemic circulation. As the LV volume decreases at the end of systole, the LV pressure decreases and the AV closes. During diastole (ventricular relaxation), the LV pressure remains lower than the aortic pressure and the AV remains closed, preventing any retrograde blood flow from the aorta.

### 4.1.2. Mitral valve

The MV is the atrioventricular valve of the left side of the heart. It consists of an anterior and a posterior leaflet, which attach to a D-shaped annulus and are tethered to the two papillary muscles via thin fibrous chordae tendineae.[35,36] The anterior (aortic) leaflet is rounded and has a continuity to the anterior annulus, whereas the posterior (mural) leaflet is long and narrow, consists of three or more scallops, and contours two-thirds of the mitral annulus. MVs are composed of four layers: the atrialis, spongiosa, fibrosa, and ventricularis.[37] Each layer consists of various cellular components, which contribute to the overall function.[38] The atrialis contains ECs with aligned elastic and collagen fibers. The spongiosa is primarily made up of an ECM of proteoglycans, GAGs, and elastic fibers. The fibrosa is the major load-bearing layer and is composed of aligned and compact collagen fibers surrounded by GAGs and proteoglycans. Finally, the ventricularis is made up of ECs covering a collagen and elastin matrix. The heterogeneous cell population in MV leaflets includes ICs consisting of fibroblasts, myocytes and SMCs, and ECs covering the leaflet surface. Proper MV function involves the sophisticated and coordinated deformations of the papillary muscles, the annulus, the chordae tendineae, and the leaflets. The opening and closing of the MV are governed by the transmitral pressure gradient established between the LA and the LV. During isovolumetric relaxation, LV pressure falls below LA pressure (~5 mm Hg), causing the MV to open and initiating ventricular filling. During ventricular filling, LV pressure progressively rises, causing the flow to decelerate. This stage is immediately followed by atrial contraction, which forces any residual blood in the LA to flow into the LV. The end of the filling phase and the associated LV pressure rise initiate MV closure while the tethering from the chordae prevents the leaflets from prolapsing into the LA during coaptation.

## 4.2. Valvular Hemodynamics
### 4.2.1. Aortic valve
The AV functions in a harsh flow environment characterized by pulsatility, three-dimensionality, and turbulence.[39,40] During the opening phase, the AV experiences blood flow velocities ramping from 0 to 1.35 m/s in less than 15 ms,[41] with a peak-systolic Reynolds number close to 6000 characterizing blood flow transition from the laminar to the turbulent regime. The deceleration of blood flow in late systole results in the formation of vortical structures in the aortic sinus, which push the tip of the leaflets toward the closed position, enabling fast and efficient leaflet coaptation.[42,43] The transfer of momentum between the valve and blood flow subjects the leaflets to a time-varying and side-specific stress environment.[44] During systole, the stress environment is dominated by the high FSS resulting from the high-momentum jet emanating from the LVOT. The distinct flow environments generated on both sides of AV leaflets[45] give rise to a side-, site-, and leaflet-specific FSS environment.[39,40,46] Computational simulations[9,47] and experimental measurements[10,48] of AV flow have quantified FSS magnitude and pulsatility characteristics in terms of temporal shear magnitude (TSM, i.e., the time-averaged FSS magnitude over one cycle) and oscillatory

shear index (OSI, i.e., degree of FSS bidirectionality). The ventricularis FSS exhibits high magnitude (TSM: 7.5–22.3 dyn/cm²) and pulsatility (OSI: 0.06–0.17), while the fibrosa FSS is marked by low magnitude (TSM: 0.9–3.9 dyn/cm²) and bidirectionality (OSI: 0.26–0.44). In addition, the transition of the flow near peak systole generates substantial turbulent stresses downstream of the leaflets. Laser Doppler anemometric measurements in a pulsatile flow loop revealed Reynolds stress levels approaching 1650 dyn/cm² at peak systole.[49] Stretch is another important component of the AV hemodynamic environment. Although tensile stresses remain relatively low when the valve is open, they increase dramatically during leaflet coaptation. During diastole, the large pressure gradient generated across the leaflets subjects the ventricularis to tension and the fibrosa to compression, causing the leaflets to stretch both radially and circumferentially.[44] In vitro measurements on porcine AV leaflets revealed the existence of elevated strain levels along the leaflet base and coaptation line, with substantial temporal variations during the cardiac cycle (radial strain: 5%–7% in late systole and early diastole, 15%–23% at peak diastole and in mid-diastole; circumferential strain: 3%–5% in late systole and early diastole, 12%–23% at peak diastole and in mid-diastole).[50] The strain distribution on the leaflets dynamically adapts to the imposed transvalvular pressure, increasing by 5% for every 40 mm Hg of added pressure, in order to maintain the valve in a coapted seal configuration even under hypertensive conditions.[50]

### 4.2.2. Mitral valve
MV hemodynamics exhibits time-dependent flow characteristics and regimes during the cardiac cycle. The flow through the MV experiences two sets of acceleration/deceleration phases.[51–53] The first flow acceleration occurs in early diastole when LA pressure exceeds LV pressure. During this opening phase, blood flow velocity through the MV annulus accelerates from 0 to 0.5–0.8 m/s.[54] In late diastole, the contraction of the LA at the end of the filling phase is accompanied by a second flow acceleration through the annulus to a peak velocity of 0.3–0.5 m/s.[54] These velocity characteristics translate in a range of Reynolds numbers from 0 to 4000–6000,[55] suggesting the existence of transitional flow conditions. During early ventricular filling, the separation of the high-momentum blood flow from the anterior leaflet surface generates a shear layer, which results in the formation of a large vortex in the LV.[56–58] Although MV closure is mostly determined by the adverse transmitral pressure imposed

during mid-diastole,[53] this vortex dynamics has been shown to assist MV closure, promote ventricular filling, and enhance ventricular ejection during systole.[53,57,59,60] The variations in transmitral pressure and the specific elastin-collagen network arrangement in each leaflet layer subjects the anterior leaflet to a highly anisotropic strain field characterized by a peak circumferential strain of 2.5%–3.3% and a peak radial strain of 16%–22%.[14,61,62] The posterior leaflet experiences substantially lower strain levels.[63]

## 4.3. Valvular Disease
### 4.3.1. Calcific aortic valve disease
Calcific aortic valve disease (CAVD) is the most common form of AV disease. It consists of the formation of calcific lesions on the leaflet fibrosa, which contribute to the obstruction of the LVOT (stenosis) and progressive heart failure.[64] CAVD pathogenesis progresses through three consecutive stages: inflammation, fibrosis, and calcification.[65–69] The inflammatory stage is marked by increased oxidative stress,[70,71] the upregulation of cell adhesion molecules, and the accumulation of lipids[72,73] that result in the infiltration of inflammatory cells such as T lymphocytes and macrophages in the fibrosa.[69,74,75] Those processes are accompanied by increased expressions of inflammatory mediators such as transforming growth factor (TGF)-β1[74,76] and bone morphogenetic proteins (BMPs).[74,77] During the fibrotic stage, ICs are switched from their normal fibroblastlike phenotype to an activated myofibroblast- or osteoblastlike phenotype, and the loss in balance between matrix metalloproteinases (MMPs) and tissue inhibitors of MMPs results in abnormal ECM remodeling and disorganized ECM fibers. Lastly, during the calcification stage, osteoblastlike ICs secrete chondrogenic and osteogenic growth factors that lead to calcific lesion formation on the leaflet fibrosa.[68,78]

CAVD results in improper leaflet dynamics and increased valvular resistance to flow, which in turn generates substantial alterations in valvular hemodynamics. As shown in vitro, stenotic conditions tend to generate a skewed and asymmetric valvular jet, increase the peak-systolic jet velocity (up to 4 m/s under mild stenosis and up to 7 m/s under severe stenosis), and promote turbulence.[49,79–82] Fluid-structure interaction simulations of blood flow through a moderately stenotic AV reported an increase in ventricularis FSS from 45 to 70 dyn/cm$^2$ and a decrease in fibrosa FSS from 12 to 5 dyn/cm$^2$ relative to the normal valve.[83] Laser Doppler anemometric measurements of turbulent stresses in mildly and severely stenotic AVs revealed dramatic increases in streamwise Reynolds stress levels

(3800 dyn/cm$^2$ and 14,500 dyn/cm$^2$, respectively) as compared with a normal valve (1650 dyn/cm$^2$).[49]

### 4.3.2. Myxomatous valve disease
Myxomatous mitral valve disease (MMVD) is the leading cause of leaflet prolapse and mitral regurgitation.[84] The bulging of the MV leaflets into the LA during systole results in blood leakage into the LA, which decreases overall LV performance and may lead to life-threatening conditions. MMVD is a degenerative disorder characterized by the overall mechanical weakening of the valve components, which results in tissue thickening and enlargement, annular dilation, and chordae elongation.[85–87] Myxomatous MV leaflets and chordae exhibit microstructural alterations marked by GAG and proteoglycan accumulation in the leaflet spongiosa, type III collagen synthesis, collagen and elastin network fragmentation, and MMP upregulation.[88–93] These ECM alterations are caused by the transdifferentiation of the valve ICs into an activated myofibroblast phenotype.[90]

Myxomatous MV leaflets are different both morphologically and mechanically from normal leaflets. Clinical observations based on three-dimensional transesophageal echocardiography have revealed an overall increase in leaflet surface area, annular diameter, and area and circumference in myxomatous MVs.[94] The leaflet prolapse resulting from those morphologic changes is associated with a characteristic eccentric regurgitant jet.[95] Mechanical tests performed on explanted MVs also revealed that the adverse remodeling affecting the leaflet ECM results in an overall increase in extensibility marked by a 2.4-fold increase in circumferential strain and a 2.5-fold increase in radial strain.[96]

## 4.4. Valvular Response to Hemodynamic Stresses
The structure and cellular composition of heart valves are tailored to sustain and interact with the surrounding mechanical stress environment.[97–101] For example, valvular ICs populating the valves of the left side of the heart are stiffer than ICs found in the valves of the right side of the heart, which provides the AV and MV greater load-bearing capabilities to sustain the harsher pressure environment of the systemic circulation.[102] Similarly, ECs on valvular surfaces express different phenotypes to adapt to the local FSS environment and to protect the underlying tissue from the constant shearing force caused by blood flow.[103,104] This section reviews the main mechanisms of AV and MV adaptation to their two dominant fluid and structural stresses, namely, FSS and tensile stress.

### 4.4.1. Fluid shear stress mechanotransduction

Valvular ECs alter their morphology and alignment in response to FSS. Scanning electron microscopic measurements on canine AV leaflets and in vitro experiments on porcine valvular ECs under steady flow have revealed that AV ECs tend to align circumferentially with the free edge of the leaflet.[103,105] EC alignment in AV leaflets was shown to be driven by the spatial arrangement of several focal adhesion components (e.g., β1 integrins, vinculin, focal adhesion kinase), which is modulated by the Rho-kinase signaling pathway.[103]

Early studies in parallel plate flow chamber on the effects of steady FSS on porcine AV biology evidenced the key role of FSS in the maintenance of the leaflet biosynthetic activity, in terms of protein, GAG, and DNA synthesis.[106] As further evidence of the sensitivity of AV ECs to FSS, in vitro studies on denuded and intact porcine AV leaflets also revealed substantial differences in collagen and GAG content.[31] The presence of the endothelium promoted collagen synthesis and maintained the GAG content, whereas denudation resulted in opposite trends. The ability of FSS to modulate the expression of key proteins in the leaflets revealed by these studies also suggests the existence of a communication network between ECs and ICs, by which the FSS signal experienced on the leaflet could be transduced into changes in ECM biology. FSS also affects ECM turnover in AV leaflets, which is an important mechanism in the maintenance of valvular homeostasis and disease. Exposure of porcine leaflets to different levels of steady FSS in a parallel plate flow chamber resulted in an overall upregulation in ECM synthesis, the inhibition of cathepsin L activity and expression, and the increase in MMP-2/9 activity.[107] The contrasted response in the expression and activity of those proteinases to FSS reveals the complexity of the mechanobiological pathways contributing to the modulation of ECM remodeling in AV leaflets.

The early pathogenesis of CAVD, which involves inflammatory processes typically localized to the leaflet endothelium, motivated investigations of the potential role of FSS abnormalities in valvular inflammation and adverse remodeling. An in vitro study using a modified cone-and-plate bioreactor subjected each surface of porcine AV leaflets to the FSS environment experienced by the opposite surface.[104] Exposure of the fibrosa to the high unidirectional FSS environment of the ventricularis resulted in a dramatic increase in intercellular adhesion molecule (ICAM)-1, vascular cell adhesion molecule (VCAM)-1, BMP-4, and TGF-β1 expression relative to the fresh controls and specimens subjected

to their native low oscillatory FSS. In contrast, expression of these biomarkers was maintained at normal levels when the ventricularis was subjected to normal (i.e., high pulsatile) and abnormal (i.e., low oscillatory) FSS. The availability of more sophisticated bioreactors capable of replicating the native side-specific FSS environment of AV leaflets enabled the investigation of the contribution of combined abnormalities in FSS magnitude and frequency to early valvular pathogenesis.[108] The results revealed that cytokine expression was stimulated under elevated FSS magnitude at normal frequency, whereas ECM degradation was stimulated under both elevated FSS magnitude at normal frequency and physiologic FSS magnitude at abnormal frequency. The response was time dependent and peaked after 48 hours of conditioning. A subsequent investigation suggested that endothelial activation is weakly regulated by BMP-4 in response to FSS abnormalities, TGF-β1 silencing attenuates FSS-induced ECM degradation via MMP-9 downregulation, and BMP-4 and TGF-β1 do not synergistically interact in response to FSS abnormalities.[109]

The insights provided by the fundamental studies into the mechanobiology of FSS in AV tissue are particularly relevant in the context of valvular disease.[110,111] CAVD, which generates stenotic conditions across the AV, subjects the leaflets to supraphysiologic FSS magnitudes. As shown in vitro in porcine leaflets, stenotic FSS levels increase the expression of inflammatory biomarkers (BMP-4, TGF-β1) and ECM remodeling biomarkers (MMP-2/9, cathepsin L/S) in an FSS-magnitude-dependent manner, suggesting the key role played by FSS abnormalities in disease initiation and progression.[112] The bicuspid aortic valve (BAV), which is a congenital disorder consisting of the formation of a valve with only two leaflets, is another source of FSS abnormalities for AV leaflets and a major risk factor for CAVD. Interestingly, computational and experimental studies have identified regions of FSS overloads and increased FSS bidirectionality on BAV leaflets (174% increase in TSM, 0.10 increase in OSI relative to normal AV leaflets).[82,113–118] As evidenced in vitro using cone-and-plate bioreactors, those hemodynamic alterations promote endothelial activation, BMP-4 and TGF-1 paracrine signaling, catabolic enzyme (MMP-2/9, cathepsin L/S) secretion and activity, and bone matrix synthesis via osteocalcin upregulation.[119,120] Interestingly, the extent of the pathologic response correlates with the degree of hemodynamic abnormality experienced by the leaflets.

In contrast to the well-documented effects of FSS on AV biology, the mechanotransduction of FSS in MV leaflets remains largely unexplored. To the best of the authors' knowledge, the only suggestion of FSS mechanobiology in MV leaflets was done in the context of the effects of mitral regurgitation hemodynamics on ECM remodeling.[95] The exposure of porcine MV leaflets to flow conditions mimicking MV prolapse promoted collagen III and GAG synthesis, which resulted in overall weakening and decrease in tissue stiffness.

### 4.4.2. Tensile stress mechanotransduction

Collagen, which is a key component of the valve ECM, is the primary tensile-stress-bearing fiber. The primary collagen subtypes present in valve leaflets are collagen I, III, and V, and their approximate ratios are 85:10:5 and 74:24:2 in the AV and MV, respectively.[34,93] One of the early observations of valvular response to stretch consisted of the demonstration of the reorganization of collagen fiber orientation under increasing pressure using small-angle light scattering.[121] Side-specific observations under low-pressure conditions (0–1 mm Hg) revealed a higher degree of orientation on the fibrosa and a more random fiber alignment on the ventricularis. Fiber orientation discrepancies between both sides were attenuated under increasing load, suggesting the complex, highly heterogeneous structural response of the valve ECM to pressure-induced stretch. The implementation of a similar technique on porcine MV leaflet and chordal tissue revealed a preferential collagen alignment parallel to the annulus in the central part of both leaflets and a more orthogonal arrangement near the strut chordal insertion.[122] The dynamic alterations in collagen architecture in AV and MV leaflet regions dominated by high strain concentrations[13,50] suggest the sensitivity of valvular tissue to stretch and its ability to adapt to the loading conditions in order to maintain valvular function.

As opposed to FSS, which is only sensed by the leaflet endothelium, stretch affects both IC and EC populations. Studies have suggested that the biological responses triggered by cyclic stretch on both cell types may contribute to the load-bearing capabilities of the tissue and may also be involved in the initiation and progression of pathologic states. Early stretch studies on isolated ICs using a Flexcell device demonstrated the upregulation of collagen synthesis in a stretch-magnitude- and duration-dependent manner.[123] A subsequent study on porcine leaflet tissue subjected to different levels and frequency of cyclic stretch in vitro confirmed these findings and evidenced the upregulation of the contractile phenotype of the valve ICs via increased α-smooth muscle actin (α-SMA) expression in the ventricularis.[29] These findings suggest the key

role played by AV ICs in the maintenance of tissue homeostasis through the regulation of ECM biosynthesis. Lending more support to this mechanobiological regulation and adaptation, an in vitro micropipette aspiration study on ovine ICs reported that the higher stiffness of MV and AV ICs relative to that in heart valves from the right side of the heart correlated with a higher content of α-SMA and heat shock protein 47[102] Together, these studies suggest the sensitivity of valve ICs to local ECM structural stresses and their ability to alter matrix stiffness and collagen synthesis. Although most studies have focused on the biosynthetic response of ICs to cyclic stretch, valvular ECs also exhibit a particular susceptibility to sense this mechanical signal to adapt to it dynamically. *En face* imaging of live AV leaflets subjected to cyclic stretch for 72 hours revealed increased cellular viability as compared to static specimens, and a highly side-specific endothelial response marked by increased apoptosis on the ventricularis relative to the fibrosa.[124]

The synergies among the valvular stretch environment, the leaflet biosynthetic activity, and ECM composition have motivated investigations on the potential role played by stretch abnormalities in pathologic remodeling. Porcine AV leaflets subjected to 15% circumferential cyclic stretch mimicking the effects of hypertensive pressure loading exhibited an upregulation of MMP-1, -2, and -9 and cathepsin S and K activities and a downregulation in cathepsin L activity relative to fresh leaflets.[125] Although 20% stretch resulted in similar trends, the response was more moderate and accompanied by increased cell proliferation and apoptosis. Placed in the context of CAVD, these results suggest that hypertensive levels of stretch may contribute to ECM degeneration via alterations in proteolytic enzyme expression and activity. A follow-up in vitro study on porcine AV leaflets provided more mechanistic insights into the mode of stretch mechanotransduction by demonstrating increased cellular proliferation, collagen and GAG synthesis, and upregulation of serotonin (5HT) receptors in response to normotensive stretch (10%) relative to static conditions.[126] Besides their effects on ECM remodeling, stretch abnormalities have been shown to promote osteogenesis in vitro. A study investigating the effects of stretch magnitude on CAVD pathogenesis reported evidence of tissue mineralization and calcification on the fibrosa of porcine AV leaflets subjected to increasing levels of stretch in osteogenic medium.[127] The attenuation of the mechanoresponse following medium supplementation with the BMP antagonist noggin suggested the key role played by BMP signaling in stretch-induced AV calcification. Lastly, the phenotypic changes undergone by AV ICs have been documented in porcine AV leaflets subjected to combinations of pathologic stretch (15%) and hypertensive pressure (140/100 mm Hg) using a modified stretch bioreactor.[128] The study reported a downregulation of α-SMA, vimentin, and calponin, suggesting the inhibition of the IC contractile and myofibroblast phenotypes and the switch of ICs to a more synthetic phenotype. These results correlate with the typical thickening of the spongiosa and fibrosa in the early stage of CAVD.

There are few reports on the effects of stretch on MV biology and disease. Porcine MV anterior leaflets subjected to 10% radial strain in a uniaxial stretch device exhibited decreased expressions in cytoskeletal proteins and proteins involved in energy metabolism such as glycolysis and oxidoreductase activity.[129] Similar experiments conducted on denuded leaflets resulted in a downregulation of ECM and cell-matrix adhesion proteins and an upregulation of translation-related and chaperone proteins, suggesting the protective role played by the MV endothelium under physiologic stretch. Lastly, the effects of stretch have been investigated in the context of MMVD *in vitro*[130] Porcine MV leaflets subjected to steady stretch exhibited increased transcription of endothelin-B receptors relative to specimens maintained under static conditions, without affecting messenger RNA expression of endothelin-B receptors and endothelin-1. The stretch regulation of endothelin-1 signaling, which is a key player in ECM remodeling alterations during MMVD pathogenesis, provides another evidence for a mechanopotential cause in MMVD.

## 5. BLOOD VESSEL MECHANOBIOLOGY
### 5.1. Blood Vessel Anatomy, Function, and Structure

Blood vessels are viscoelastic multilayered conduits that transport blood throughout the body.[131] The transport of blood throughout the vasculature involves a complex network of blood vessels ranging in size, structure, and function.[5,132] Arteries are the largest and thickest blood vessels (mean diameter: 4.0 mm; mean thickness: 1.0 mm) and carry blood from the heart to the microvasculature. Arteries connect to smaller arterioles (mean diameter: 30 μm; mean thickness: 6.0 μm) that connect in turn to capillaries, the smallest blood vessels in the vasculature (mean diameter: 8.0 μm; mean thickness: 0.5 μm), which are responsible for gas and metabolite exchange between the blood and the surrounding tissues. Downstream from the capillaries are the venules

(mean diameter: 20 μm; mean thickness: 1.0 μm), and further downstream are the veins (mean diameter: 5.0 mm; mean thickness: 0.5 mm), which bring blood back to the atria.

Arteries and veins have a trilayered structure consisting of (from inner surface to outer surface) the tunica intima, the tunica media, and the tunica adventitia. Although the composition of each layer is tailored to the function of each blood vessel type, the main cellular and ECM components include ECs, SMCs, elastin, and collagen fibers. Smooth muscle and elastin, the main components of arterial tissue, provide arteries with a low compliance, which enables the maintenance of the driving pressure during diastole. Although the same two components are found in veins, the large-diameter, thin-walled structure of veins result in higher compliance. Lastly, blood vessels in the microcirculation exhibit more simple structures that typically consist of a thin EC monolayer to maximize biotransport between the blood and surrounding tissues.

### 5.2. Vascular Hemodynamics

Vascular hemodynamics varies both spatially due to variations in geometric features (branching, tapering, bifurcation, bends) and material properties within the vasculature and temporally due to blood flow pulsatility. As a result, the vasculature exhibits large variations in blood flow patterns: anterograde and laminar flows in straight sections of arteries and disorganized and irregular flow patterns in curvatures, bifurcations, and other complex geometry and vortical structures in the vicinity of venous veins.[133,134] This diversity in blood flow patterns is associated with a spectrum of flow regimes, as suggested by the wide range of Reynolds and Womersley numbers that have been identified throughout the vasculature (4500 and 13.2 in the ascending aorta, 1250 and 8 in the abdominal aorta, 0.09 and 0.04 in arterioles, and 0.001 and 0.005 in capillaries, respectively).[135,136] The propagation of the pressure pulse wave along the vascular tree and its interactions with the anatomic features of the vasculature (e.g., curvature, bifurcations, branching) generate large variations in blood pressure throughout the systemic circulation: 90−93 mm Hg in the aorta, 70−90 mm Hg in arteries, 40−70 mm Hg in arterioles, 25−40 mm Hg in capillaries, 15−25 mm Hg in venules, 8−10 mm Hg in the vena cava, and 0−5 mm Hg at the entrance of the RA.[1] As a result, ECs on the intimal layer of blood vessels experience an array of FSS levels, which vary from 1−5 dyn/cm$^2$ in venules[137,138] and 2−5 dyn/cm$^2$ in the common femoral artery[139] to 2−13 dyn/cm$^2$ in the ascending aorta,[140] 7–10 dyn/cm$^2$ in the abdominal aorta,[141] and 41–67 dyn/cm$^2$ in retinal arterioles.[137,142] The local intraluminal pressure also subjects blood vessels to structural strains in both the circumferential and longitudinal directions, which have been reported as 8.0% and 7.6% in the ascending aorta,[143] 13.2% and −1.3% in the abdominal aorta,[144] 3.4% and 3.8% in the common carotid arteries,[145] 1.5% and 2.4% in the brachial arteries,[145] and 1.2% and 2.9% in the femoral arteries, respectively.[145]

### 5.3. Vascular Disease
#### 5.3.1. Atherosclerosis
Atherosclerosis is a common chronic inflammatory disease affecting the intima of large- and medium-sized arteries.[146–149] It is characterized by the formation of an atheromatous plaque resulting from the deposition and accumulation of lipid molecules within the arterial wall and underlying smooth muscle,[147,150] preferentially in vascular bifurcations and curvatures. As the plaque becomes larger, stenotic flow conditions develop, increasing the risk for plaque rupture, thrombosis, and complete obstruction of the arterial lumen. The development of an atherosclerotic plaque on the arterial wall is the result of an inflammatory response to endothelial injury, characterized by the downregulation in the endothelial production of nitric oxide, the upregulation of cell adhesion molecules (ICAM-1, VCAM-1), and the release of proinflammatory cytokines.[21,151] This inflammatory response promotes the adhesion of inflammatory cells (monocytes, T cells) and low-density lipoprotein to the injury site and their migration to the subendothelial space. These processes are facilitated by MMP pathways,[152] the degradation of the basement membrane, and the migration of proliferating SMCs and result ultimately in intimal thickening and plaque formation.

From a fluid mechanics perspective, the resulting reduction in luminal diameter generates flow conditions mimicking those of flows past a nozzle or orifice, which are marked by a substantial pressure drop, the formation of a vena contracta, flow transition to turbulence, flow separation from the vessel wall, and flow reattachment further downstream.[153,154] This complex hemodynamics subjects the vessel wall to a spatially varying FSS environment marked by high magnitude and unidirectionality upstream of the plaque and low magnitude and oscillation immediately downstream of the plaque.[155] While the spatial distribution and local magnitude of the FSS imposed on the wall of the stenosis greatly vary with the degree of occlusion and blood flow rate,[156] flow simulations in a stenotic

carotid bifurcation have reported up to 465% increase in FSS magnitude at the stenosis relative to the baseline levels captured at downstream and upstream locations.[157] In addition to its impact on fluid stresses, atherosclerosis affects structural strains within the vessel. As reported by computational fluid-structure interaction simulations in a stenotic carotid artery, the increase in jet velocity at the throat of the plaque generates a local negative flow pressure (−25 to −40 mm Hg) that translates into a negative transmural pressure and subjects the arterial wall to elevated circumferential compressive strains (−39% to −45%) in a stenosis-severity- and eccentricity-dependent manner.[158] Together, these studies suggest the existence of strong flow alterations under atherosclerotic conditions. However, the heterogeneity of atherosclerotic plaque compositions, material properties, and morphologies combined with the diversity of vessel wall anatomies prevent the identification of a baseline stress environment in atherosclerosis.

#### 5.3.2. Aneurysm
In contrast to atherosclerosis, which results in an effective reduction in arterial lumen diameter, an aneurysm consists of the abnormal enlargement of the arterial wall as a result of the degradation of its structural components.[155,159,160] Common sites of aneurysm formation include the cerebral vasculature, the abdominal aorta, the ascending aorta, and the coronary arteries.[134,159,161] Although the mechanisms of aneurysm formation are site specific,[9] initial loss of elastin and progressive vessel wall degradation are the hallmarks of the disease.[134,160,162] As with other cardiovascular disorders, inflammation and tissue remodeling are the major key players in aneurysm pathogenesis. An initial hemodynamic insult mediates endothelial activation via increased expressions of ICAM-1, VCAM-1, and E-selectin through the inflammatory cytokine tumor necrosis factor α.[163,164] These inflammatory processes trigger MMP pathways, which promote ECM degradation and SMC apoptosis, weaken the arterial wall, and result in dilation and aneurysm formation.[155,162,165,166] Ultimately, aneurysm rupture is a life-threatening event and its cause remains under debate.[134,155]

The local increase in luminal diameter associated with an aneurysm results in substantial hemodynamic alterations such as flow separation, stagnation, increased vorticity, decreased velocity, and increased transmural pressures.[4,155,167] While these characteristics are strongly patient and site specific, common features of aneurysms include the existence of high FSS at the base of the enlargement and low FSS at the tip, which is the typical site of aneurysm rupture.[167,168] Regardless of the anatomic diversity, the point of aneurysm rupture often correlates with regions of low FSS within the dilation.[169,170] Flow simulations in a cerebral artery have predicted large FSS concentrations ($82-206$ $dyn/cm^2$) at the apex of the bifurcation, whereas the tip of the aneurysm, which was consistently reported as the site of rupture, experienced low FSS ($5-28$ $dyn/cm^2$) and complex secondary flow patterns.[167] The loss in structural components in the arterial wall also results in stretch alterations at the site of aneurysm formation. Average circumferential and longitudinal strains of $2.7\%-5.8\%$ and $2.1\%-3.6\%$, respectively, have been reported in abdominal aortic aneurysms.[171] However, there is very limited data available on the mechanical strains in aneurysms owing to their hemodynamic and anatomic diversity.

## 5.4. Vascular Response to Hemodynamic Stresses

Hemodynamic forces play a crucial role in vascular biology.[133,172] Fluid and structural stresses generated under pulsatile flow conditions are sensed by vascular ECs and SMCs and trigger complex mechanobiological cascades contributing to the maintenance of vascular homeostasis or pathogenesis. This section reviews the adaptive responses of the vascular wall to mechanical stresses in both physiology and disease.

### 5.4.1. Fluid shear stress mechanotransduction

Vascular ECs interface between the tunica intima and blood flow. They are able to sense hemodynamic alterations and respond to them by altering their morphology, phenotype, and biology.[2,133] Early in vivo studies in rabbits and in vitro studies in parallel plate flow chambers evidenced the ability of vascular ECs to orient their major axis with the direction of flow and to reorganize their F-actin structure, in an FSS-magnitude- and oscillation-dependent manner.[3,173,174] When exposed to a unidirectional laminar blood flow, vascular ECs align parallel to the flow direction and form long and well-organized actin stress fibers in the central cell region.[2,3,17,173,175-177] Conversely, disturbed flow patterns (e.g., oscillatory flow, turbulence, flow separation) initiate adaptive cellular responses marked by the progressive loss of overall cellular alignment, the transition from elongated to round/polygonal cellular morphology, and the formation of F-actin in dense bands at the periphery of the cell.[2,133,175,178] The dynamic reorganization of the stress fiber network in response to FSS could play a key role in the maintenance of vascular integrity under adverse hemodynamics in vivo.[175,179]

Cellular turnover is also modulated by FSS. In vitro studies on bovine aortic ECs subjected to FSS in parallel plate flow chambers reported a decreasing rate of EC proliferation as the steady FSS level increased from 15 to 90 $dyn/cm^2$, an amplification of this response under pulsatile FSS, and an overall reduction in EC apoptosis regardless of the FSS regimen (i.e., steady/pulsatile/oscillatory)[180,181]. In addition, the vascular endothelium is able to modulate EC proliferation, migration, and loss depending on local FSS spatial gradients.[182] As demonstrated in vitro on human umbilical vein ECs, high FSS gradients (34 $dyn/cm^2 \cdot mm$) promoted cellular division and migration toward regions of lower FSS gradients, while regions of uniform FSS were associated with increased cell loss, suggesting the possible regulation of endothelial remodeling on spatial FSS characteristics.

The response of vascular tissue to FSS is not limited to endothelial processes. The entire vessel wall undergoes large-scale geometric remodeling (e.g., wall thickness and luminal diameter alterations) to maintain vascular integrity in response to FSS alterations,[133,183] suggesting the existence of a complex communication network between the FSS sensing ECs and SMCs within the matrix. An in vivo study in rabbits reported that a 70% reduction in blood flow rate in ligated external carotid arteries caused a 21% decrease in luminal diameter with an intact endothelium, but no change in endothelium-depleted arteries.[184] Similarly, generation of supraphysiologic blood flow rates (i.e., 1.4- to 2.7-fold increase vs. basal level) in rat mesenteric arteries resulted in a 12%–38% increase in luminal diameter, and an increase in medial wall thickness resulting from SMC hyperplasia and cellular hypertrophy.[185]

The sensitivity and adaptation of vascular tissue to the local FSS environment play an important role in pathogenesis. The preferential formation of atherosclerotic lesions and aneurysms in vascular bifurcations and curvatures, which are dominated by complex and disturbed flow patterns, suggests a role for hemodynamics in vascular disease.[2,133,155,178] Low oscillatory FSS conditions, which are typically found in disturbed flow regions,[133,134] promote atherogenesis in vascular ECs.[133,172] In contrast, high laminar FSS results in the downregulation of atherogenic genes and the release of nitric oxide (NO), which play a protective role against plaque formation.[133] Some mechanobiological pathways involved in the transduction of disturbed FSS conditions into an atherogenic response have been identified in vitro. The switch of the FSS environment from a pulsatile to an oscillatory regimen on the porcine left common carotid arterial segments conditioned in a perfusion system resulted in an upregulation in MMP-2 and -9 and a downregulation of endothelial nitric oxide synthase (eNOS).[172] In contrast, unidirectional FSS resulted in an atheroprotective response marked by the maintenance of eNOS gene expression and endothelial function. Similar adverse remodeling responses have been identified in vascular regions prone to aneurysm formation. Histologic and computational fluid dynamic analyses performed on surgically reconstructed canine carotid bifurcations revealed hyperplasialike vascular remodeling at the bifurcation apex and ECM degeneration (i.e., elastic lamina disruption, SMC and fibronectin depletion) in adjacent wall regions subjected to high FSS magnitude and spatial gradients.[186] These regions were also characterized by a decreased eNOS expression relative to the healthy surrounding vascular segments and a marked increase in biomarkers associated with intracranial aneurysms, such as MMP-2 and -9, inducible nitric oxide synthase, and nitrotyrosine.[187] These results underscore the significance of FSS in aneurysm formation and atherosclerosis and suggest the existence of synergies between the spatial/temporal hemodynamic characteristics of blood flow and vascular pathogenesis.

### 5.4.2. Tensile stress mechanotransduction

In the vasculature, both ECs and SMCs are sensitive to stretch and adapt their morphology and structure via cytoskeletal and focal adhesion complex alterations.[188,189] A typical effect of stretch on vascular ECs is the emergence of a bundle of actin filaments, which contributes to the structural integrity of the cells and participates in the transduction of mechanical signals in nonmuscle cells.[17,190,191] When subjected to cyclic stretch loading, vascular ECs and SMCs elongate and realign perpendicular to the axis of deformation to maintain their structure and minimize alterations in intracellular strain.[17,192–197] Cyclic stretch also promotes EC proliferation by upregulating the secretion of vascular endothelial growth factor (VEGF) and its receptor, VEGF-R2.[198] In vitro experiments using a Flexcell bioreactor have revealed a similar response in vascular SMCs subjected to 20% cyclic stretch, as illustrated by the net increase in cell count and proliferating cell nuclear antigen expression within 24–48 hours.[199] Another in vitro study on porcine carotid arteries subjected to tensile strain in a perfusion organ culture system revealed an increase in SMC proliferation in response to axial stretch in denuded arterial segments, whereas this response was inhibited in arteries with an intact endothelium.[200] Lastly, as demonstrated in vitro on bovine aortic ECs using a Flexcell bioreactor, normal stretch levels (6%–10%) maintain a physiologic balance between cellular proliferation and apoptosis, whereas pathologic levels (20%) break this intricate balance by promoting cell death.[201]

Besides its impact on cells, stretch loading also modulates ECM composition and turnover. Hypertensive conditions, which subject the vasculature to increased circumferential tensile loading, have been associated with alterations in vascular wall thickness via neointima formation and hypertrophy and in vascular wall composition via SMC migration and phenotypic changes.[202,203] MMPs are key players in ECM turnover, and their expression and activity have been shown to be modulated under stretch. In vitro experiments on bovine aortic ECs reported increased MMP-2 expression and activity in response to physiologic (0%–10%) cyclic strain in a magnitude- and time-dependent manner.[204] This response was driven by both *P*38- and ERK (extracellular signal-regulated kinase)-dependent pathways, suggesting that strain-induced changes in endothelial MAPK (mitogen-activated protein kinase) signaling may regulate EC phenotype in vivo via MMP production. Another in vitro study on human vascular ECs examined the effects of normal and pathologic dynamic strain levels on ECM degradation.[131] Subphysiologic stretch (5%) resulted in the upregulation of MMP-9. In contrast, physiologic loading (10%) was able to downregulate MMP-9 expression, even in the presence of the inflammatory mediator tumor necrosis factor α, suggesting the protective effects of physiologic stretch on ECM remodeling.

Lastly, stretch abnormalities are also able to elicit inflammatory responses similar to those observed in vascular disease. Human ECs subjected to subphysiologic uniaxial cyclic stretch in a proprietary cell stretching device exhibited increased expression in proinflammatory mediator nuclear factor (NF)-κB as compared with cells conditioned under a physiologic stretch level.[205] The response was amplified under a multidirectional stretch mimicking atherosclerotic states, suggesting the potential role of pathologic wall stretch alterations in early atherogenesis. Given the susceptibility of NF-κB to activate proinflammatory cell adhesion molecules (e.g., VCAM-1, ICAM-1), which are involved in early aneurysmal pathogenesis, these results may not be limited to atherogenesis and may suggest the possible involvement of flow-induced stretch abnormalities in both atherosclerotic plaque and aneurysm formation.[21,151,163,164]

## 6. FROM BENCH TO BEDSIDE: RESEARCH NEEDS AND FUTURE DIRECTIONS

Advances made over the past 40 years in the understanding of how mechanical forces interact with the biology of the cardiovascular system have shed new light on the cardiovascular function and disease. However, the translation of the basic science of mechanobiology into clinical solutions (e.g., diagnostic techniques, cell-based therapies) is still hampered by the complex analysis of mechanobiological data, the difficult isolation of the role played by mechanosensitive molecules, and the challenging development of molecules capable of blocking the mechanosensitive response with limited side effects.[110] Addressing these issues will enable the development of novel treatments that will take advantage of the ability of cells to switch on or off specific pathways contributing to disease via mechanical signals.

In addition to these requirements, new studies are needed to expand the knowledge of cardiovascular mechanobiology to other tissues. While the vascular and valvular responses to mechanical signals have been documented extensively, the mechanosensitivity of the heart muscle remains largely unexplored. The ability to understand and isolate the biological response of the myocardium to LV structural and hemodynamic stress alterations has the potential to advance our understanding of myocardial disease. Discrete subaortic stenosis (DSS), for example, which involves the formation of a fibromuscular ring of tissue immediately below the AV and is particularly prevalent in the pediatric population, can only be treated via surgical resection with mixed results because of its unknown cause.[206] The formation of the DSS membrane seems to involve the same typical pathways as those observed in other vascular and valvular diseases, including cell proliferation, fibrosis, and ECM remodeling.[207] Interestingly, a common anatomic abnormality associated with DSS is the existence of a steepened aortoseptal angle (AoSA), which describes the angle between the long axis of the aorta and the septal wall. The realization that such anatomic abnormality may result in mechanical stress alterations in the myocardial wall motivates the investigation of the cause-and-effect relationships between steep AoSA LV hemodynamics and myocardium biology. Although a preliminary computational study has already demonstrated the existence of deranged flow features,[206] the elucidation of mechanobiological synergies between the endocardium and LV hemodynamics is needed to gain new insights into DSS formation.

Lastly, the complexity of the cardiovascular hemodynamic environment is still hampering the elucidation of the cause-and-effect relationships between cardiovascular biology and mechanics. Although progress has been made in the design of more sophisticated bioreactors that closely mimic the native stress environment, devices are still needed to selectively explore the combined effects of different mechanical signals on full organ function and tissue and cell biology.

## 7. CONCLUSION

Mechanobiology has provided a new way to think about the function of cells, tissues, and organs and is now considered a potential tool to elucidate disease mechanisms. Mechanobiological studies have demonstrated that cardiovascular biology is modulated by external signals that are transduced to the tissue via a complex cell-ECM communication network. Recent progress in bioreactor designs and in experimental and computational methodologies for the characterization of cardiovascular mechanics provide new opportunities toward the elucidation of cardiovascular mechanobiology. More research efforts are needed in order to bridge the gap between the fundamental science of cardiovascular mechanobiology and the development of clinical solutions, noninvasive pharmacologic therapies, and cell-targeted modalities.

## REFERENCES

1. Guyton AC, Hall JE. *A Textbook of Medical Physiology*. Philadelphia: W.B. Saunders Company; 2000.
2. Davies PF. Flow-mediated endothelial mechanotransduction. *Physiol Rev*. 1995;75(3):519–560. https://doi.org/10.1152/physrev.1995.75.3.519.
3. Nerem RM, Levesque MJ, Cornhill JF. Vascular endothelial morphology as an indicator of the pattern of blood flow. *J Biomech Eng*. 1981;103(3):172–176. https://doi.org/10.1115/1.3138275.
4. Ku DN. Blood flow in arteries. *Annu Rev Fluid Mech*. 1997;29:399–434. https://doi.org/10.1146/annurev.fluid.29.1.399.
5. Milnor WR. *Hemodynamics*. 2nd ed. Baltimore, MD, MD: Williams & Wilkins; 1989.
6. Whitmore R. *Rheology of Circulation*. New York: Pergamon Press; 1968.
7. Koutsiaris AG, Tachmitzi SV, Batis N, et al. Volume flow and wall shear stress quantification in the human conjunctival capillaries and post-capillary venules in vivo. *Biorheology*. 2007;44:375–386. https://doi.org/10.1016/j.athoracsur.2006.03.117.
8. Koutsiaris AG, Tachmitzi SV, Batis N. Wall shear stress quantification in the human conjunctival pre-capillary arterioles in vivo. *Microvasc Res*. 2013;85(1):34–39. https://doi.org/10.1016/j.mvr.2012.11.003.
9. Cao K, Bukač M, Sucosky P. Three-dimensional macroscale assessment of regional and temporal wall shear

stress characteristics on aortic valve leaflets. *Comput Methods Biomech Biomed Eng.* 2016;19(6):603—613. https://doi.org/10.1080/10255842.2015.1052419.

10. Yap CH, Saikrishnan N, Tamilselvan G, Yoganathan AP. Experimental measurement of dynamic fluid shear stress on the ventricular surface of the aortic valve leaflet. *Biomech Model Mechanobiol.* 2011;11(1—2):171—182. https://doi.org/10.1007/s10237-011-0301-7.

11. Yen J-H, Chen S-F, Chern M-K, Lu P-C. The effect of turbulent viscous shear stress on red blood cell hemolysis. *J Artif Organs.* 2014;17(2):178—185. https://doi.org/10.1007/s10047-014-0755-3.

12. Lehoux S, Tedgui A. Cellular mechanics and gene expression in blood vessels. *J Biomech.* 2003;36(5):631—643. https://doi.org/10.1016/S0021-9290(02)00441-4.

13. Weiler M, Yap CH, Balachandran K, Padala M, Yoganathan AP. Regional analysis of dynamic deformation characteristics of native aortic valve leaflets. *J Biomech.* 2011;44(8):1459—1465. https://doi.org/10.1016/j.jbiomech.2011.03.017.

14. Sacks MS, Enomoto Y, Graybill JR, et al. In-vivo dynamic deformation of the mitral valve anterior leaflet. *Ann Thorac Surg.* 2006;82(4):1369—1377. https://doi.org/10.1016/j.athoracsur.2006.03.117.

15. O'Rourke M, O'Rourke M. Mechanical principles in arterial disease. *Hypertension.* 1995;26(1):2—9. https://doi.org/10.1161/01.HYP.26.1.2.

16. Boutouyrie P, Laurent S, Benetos A, Girerd XJ, Hoeks AP, Safar ME. Opposing effects of ageing on distal and proximal large arteries in hypertensives. *J Hypertens Suppl.* 1992;10(6):S87—S91. https://doi.org/10.1097/00004872-199208001-00023.

17. Chien S. Mechanotransduction and endothelial cell homeostasis: the wisdom of the cell. *Am J Physiol Heart Circ Physiol.* 2007;292(3):H1209—H1224. https://doi.org/10.1152/ajpheart.01047.2006.

18. Hsu P-P, Li S, Li Y-S, et al. Effects of flow patterns on endothelial cell migration into a zone of mechanical denudation. *Biochem Biophys Res Commun.* 2001;285(3):751—759. https://doi.org/10.1006/bbrc.2001.5221.

19. Khan OF, Sefton MV. Endothelial cell behaviour within a microfluidic mimic of the flow channels of a modular tissue engineered construct. *Biomed Microdevices.* 2011;13(1):69—87. https://doi.org/10.1007/s10544-010-9472-8.

20. Urbaczek AC, Leão PAGC, de Souza FZR, et al. Endothelial cell culture under perfusion on a polyester-toner microfluidic device. *Sci Rep.* 2017;7(1):10466. https://doi.org/10.1038/s41598-017-11043-0.

21. Sykes M, Vega JD, Boo YC, et al. Bone morphogenic protein 4 produced in endothelial cells by oscillatory shear stress stimulates an inflammatory response. *J Biol Chem.* 2003;278(33):31128—31135. https://doi.org/10.1074/jbc.m300703200.

22. Tressel SL, Huang R-P, Tomsen N, Jo H. Laminar shear inhibits tubule formation and migration of endothelial cells by an angiopoietin-2 dependent mechanism.

*Arterioscler Thromb Vasc Biol.* 2007;27(10):2150—2156. https://doi.org/10.1161/ATVBAHA.107.150920.

23. Dai G, Natarajan S, Zhang Y, et al. Distinct endothelial phenotypes evoked by arterial waveforms derived from atherosclerosis-susceptible and -resistant regions of human vasculature. *Proc Natl Acad Sci.* 2004;101:14871—14876. https://doi.org/10.1073/pnas.0406073101.

24. Holliday CJ, Ankeny RF, Jo H, Nerem RM. Discovery of shear- and side-specific mRNAs and miRNAs in human aortic valvular endothelial cells. *Am J Physiol Heart Circ Physiol.* 2011;301(3):H856. https://doi.org/10.1152/AJPHEART.00117.2011.

25. Sucosky P, Padala M, Elhammali A, Balachandran K, Jo H, Yoganathan AP. Design of an ex vivo culture system to investigate the effects of shear stress on cardiovascular tissue. *J Biomech Eng.* 2008;130(3):35001—35008. https://doi.org/10.1115/1.2907753.

26. Sun L, Rajamannan NM, Sucosky P. Design and validation of a novel bioreactor to subject aortic valve leaflets to side-specific shear stress. *Ann Biomed Eng.* 2011;39(8):2174—2185. https://doi.org/10.1007/s10439-011-0305-6.

27. Banes AJ. *Flexible Bottom Culture Plate for Applying Mechanical Load to Cell Cultures.* 2000.

28. Engelmayr GC, Hildebrand DK, Sutherland FWH, Mayer JE, Sacks MS. A novel bioreactor for the dynamic flexural stimulation of tissue engineered heart valve biomaterials. *Biomaterials.* 2003;24(14):2523—2532. https://doi.org/10.1016/S0142-9612(03)00051-6.

29. Balachandran K, Konduri S, Sucosky P, Jo H, Yoganathan AP. An ex vivo study of the biological properties of porcine aortic valves in response to circumferential cyclic stretch. *Ann Biomed Eng.* 2006;34(11):1655—1665. https://doi.org/10.1007/s10439-006-9167-8.

30. Schoen FJ. Evolving concepts of cardiac valve dynamics: the continuum of development, functional structure, pathobiology, and tissue engineering. *Circulation.* 2008;118(18):1864—1880. https://doi.org/10.1161/CIRCULATIONAHA.108.805911.

31. Balachandran K, Sucosky P, Yoganathan AP. Hemodynamics and mechanobiology of aortic valve inflammation and calcification. *Int J Inflamm.* 2011;2011:263870. https://doi.org/10.4061/2011/263870.

32. Sutton JP, Ho SY, Anderson RH. The forgotten interleaflet triangles: a review of the surgical anatomy of the aortic valve. *Ann Thorac Surg.* 1995;59(2):419—427. https://doi.org/10.1016/0003-4975(94)00893-C.

33. Kunihara T. Anatomy of the aortic root: implications for aortic root reconstruction. *Gen Thorac Cardiovasc Surg.* 2017;65(9):488—499. https://doi.org/10.1007/s11748-017-0792-y.

34. Thubrikar M. *The Aortic Valve.* Boca Raton: CRC Press; 1990.

35. Perloff JK, Roberts WC. The mitral apparatus. Functional anatomy of mitral regurgitation. *Circulation.* 1972;46(2):227—239. https://doi.org/10.1161/01.CIR.46.2.227.

36. Ho SY. Anatomy of the mitral valve. *Heart.* 2002;88(Suppl 4):iv5—10. https://doi.org/10.1136/heart.88.suppl_4.iv5.

37. Gross L, Kugel MA. Topographic anatomy and histology of the valves in the human heart. *Am J Pathol.* 1931;7(5), 445−474.7.

38. McCarthy KP, Ring L, Rana BS. Anatomy of the mitral valve: understanding the mitral valve complex in mitral regurgitation. *Eur J Echocardiogr.* 2010;11(10):3−9. https://doi.org/10.1093/ejechocard/jeq153.

39. Dasi LP, Simon HA, Sucosky P, Yoganathan AP. Fluid mechanics of artificial heart valves. *Clin Exp Pharmacol Physiol.* 2009;36(2):225−237. https://doi.org/10.1111/j.1440-1681.2008.05099.x.

40. Dasi LP, Sucosky P, de Zelicourt D, Sundareswaran K, Jimenez J, Yoganathan AP. Advances in cardiovascular fluid mechanics: bench to bedside. *Ann N Y Acad Sci.* 2009;1161:1−25. https://doi.org/10.1111/j.1749-6632.2008.04320.x.

41. Otto CM. Evaluation and management of chronic mitral regurgitation. *N Engl J Med.* 2001;345(10):740−746. https://doi.org/10.1056/NEJMcp003331.

42. Reul H, Talukder N. Heart valve mechanics. In: Hwang NHC, Gross DR, Patel DJ, eds. *Quantitative Cardiovascular Studies Clinical and Research Applications of Engineering Principles.* Baltimore, MD: University Park Press; 1979:527−564.

43. Robicsek F. Leonardo da Vinci and the sinuses of Valsalva. *Ann Thorac Surg.* 1991;52(2):328−335. https://doi.org/10.1016/0003-4975(91)91371-2.

44. Arjunon S, Rathan S, Jo H, Yoganathan AP. Aortic valve: mechanical environment and mechanobiology. *Ann Biomed Eng.* 2013;41(7):1331−1346. https://doi.org/10.1007/s10439-013-0785-7.

45. Kilner PJ, Yang GZ, Wilkes AJ, Mohiaddin RH, Firmin DN, Yacoub MH. Asymmetric redirection of flow through the heart. *Nature.* 2000;404(6779):759−761. https://doi.org/10.1038/35008075.

46. Yoganathan AP, He Z, Casey Jones S. Fluid mechanics of heart valves. *Annu Rev Biomed Eng.* 2004;6:331−362. https://doi.org/10.1146/annurev.bioeng.6.040803.140111.

47. Cao K, Sucosky P. Aortic valve leaflet wall shear stress characterization revisited: impact of coronary flow. *Comput Methods Biomech Biomed Eng.* 2017;20(5):468−470. https://doi.org/10.1080/10255842.2016.1244266.

48. Yap CH, Saikrishnan N, Tamilselvan G, Yoganathan AP. Experimental technique of measuring dynamic fluid shear stress on the aortic surface of the aortic valve leaflet. *J Biomech Eng.* 2011;133(6):061007. https://doi.org/10.1115/1.4004232.

49. Yearwood TL, Misbacht GA, Chandran KB. Experimental fluid dynamics of aortic stenosis in a model of the human aorta. *Clin Phys Physiol Meas.* 1989;10(1):11−24. https://doi.org/10.1088/0143-0815/10/1/002.

50. Yap CH, Kim HS, Balachandran K, Weiler M, Haj-Ali R, Yoganathan AP. Dynamic deformation characteristics of porcine aortic valve leaflet under normal and hypertensive conditions. *Am J Physiol Circ Physiol.* 2010;298(2):H395−H405. https://doi.org/10.1152/ajpheart.00040.2009.

51. Kalmanson D, Bernier A, Veyrat C, Witchitz S, Savier CH, Chiche P. Normal pattern and physiological significance of mitral valve flow velocity recorded using transseptal directional Doppler ultrasound catheterization. *Br Heart J.* 1975;37(3):249−256. https://doi.org/10.1136/hrt.37.3.249.

52. Desser KB, Benchimol A. Blood flow velocity measured at the mitral valve of man. *Am J Cardiol.* 1974;33(4):541−545. https://doi.org/10.1016/0002-9149(74)90614-6.

53. Reul H, Talukder N, Muller E. Fluid mechanics of the natural mitral valve. *J Biomech.* 1981;14(5):361−372. https://doi.org/10.1016/0021-9290(81)90046-4.

54. Weyman AE. *Principles and Practices of Echocardiography.* Philadelphia, PA: Lea & Febiger; 1994.

55. Khalafvand SS, Hung T-K, Ng EY-K, Zhong L. Kinematic, dynamic, and energy characteristics of diastolic flow in the left ventricle. *Comput Math Methods Med.* 2015;2015:1−12. https://doi.org/10.1155/2015/701945.

56. Kim WY, Walker PG, Pedersen EM, et al. Left ventricular blood flow patterns in normal subjects: a quantitative analysis by three-dimensional magnetic resonance velocity mapping. *J Am Coll Cardiol.* 1995;26(1):224−238. https://doi.org/10.1016/0735-1097(95)00141-L.

57. Bellhouse BJ. Fluid mechanics of a model mitral valve and left ventricle. *Cardiovasc Res.* 1972;6(2):199−210. https://doi.org/10.1093/cvr/6.2.199.

58. Gharib M, Rambod E, Kheradvar A, Sahn DJ, JO D. Optimal vortex formation as an index of cardiac health. *Proc Natl Acad Sci.* 2006;103(16):6305−6308. https://doi.org/10.1073/pnas.0600520103.

59. Fyrenius A, Wigström L, Ebbers T, Karlsson M, Engvall J, Bolger AF. Three dimensional flow in the human left atrium. *Heart.* 2001;86(4):448−455. https://doi.org/10.1136/heart.86.4.448.

60. Bazilevs Y, del Alamo JC, Humphrey JD. From imaging to prediction: emerging non-invasive methods in pediatric cardiology. *Prog Pediatr Cardiol.* 2010;30(1−2):81−89. https://doi.org/10.1016/j.ppedcard.2010.09.010.

61. Sacks MS, He Z, Baijens L, et al. Surface strains in the anterior leaflet of the functioning mitral valve. *Ann Biomed Eng.* 2002;30(10):1281−1290. https://doi.org/10.1114/1.1529194.

62. Toma M, Einstein DR, Bloodworth CH, et al. Fluid-structure interaction and structural analyses using a comprehensive mitral valve model with 3D chordal structure. *Int J Numer Method Biomed Eng.* 2017;33(4). https://doi.org/10.1002/cnm.2815.

63. He Z, Ritchie J, Grashow JS, Sacks MS, Yoganathan AP. In vitro dynamic strain behavior of the mitral valve posterior leaflet. *J Biomech Eng.* 2005;127(3):504−511. https://doi.org/10.1115/1.1894385.

64. Stewart BF, Siscovick D, Lind BK, Gardin JM, Gottdiener JS, Smith VE. Clinical factors associated with calcific aortic valve disease. Cardiovascular Health Study. *J Am Coll Cardiol.* 1997;29(3):630−634. https://doi.org/10.1016/S0735-1097(96)00563-3.

65. Rajamannan NM. Calcific aortic stenosis: lessons learned from experimental and clinical studies. *Arterioscler Thromb Vasc Biol.* 2009;29(2):162−168. https://doi.org/10.1161/ATVBAHA.107.156752.

66. Carabello BA, Paulus WJ. Aortic stenosis. *Lancet.* 2009;373(9667):956−966. https://doi.org/10.1016/S0140-6736(09)60211-7.

67. O'Brien KD. Pathogenesis of calcific aortic valve disease: a disease process comes of age (and a good deal more). *Arterioscler Thromb Vasc Biol.* 2006;26(8):1721−1728. https://doi.org/10.1161/01.ATV.0000227513.13697.ac.

68. Rajamannan NM, Evans FJ, Aikawa E, et al. Calcific aortic valve disease: not simply a degenerative process: a review and agenda for research from the National Heart and Lung and Blood Institute Aortic Stenosis Working Group. Executive summary: calcific aortic valve disease-2011 update. *Circulation.* 2011;124(16):1783−1791. https://doi.org/10.1161/CIRCULATIONAHA.110.006767.

69. Otto CM, Kuusisto J, Reichenbach DD. Characterization of the early lesion of "degenerative" valvular aortic stenosis. Histological and immunohistochemical studies. *Circulation.* 1994;90(2):844−853. https://doi.org/10.1161/01.cir.90.2.844.

70. Miller JD, Chu Y, Brooks RM, Richenbacher WE, Pena-Silva R, Heistad DD. Dysregulation of antioxidant mechanisms contributes to increased oxidative stress in calcific aortic valvular stenosis in humans. *J Am Coll Cardiol.* 2008;52(10):843−850. https://doi.org/10.1016/j.jacc.2008.05.043.

71. Rajamannan NM, Subramaniam M, Stock SR, et al. Atorvastatin inhibits calcification and enhances nitric oxide synthase production in the hypercholesterolaemic aortic valve. *Heart.* 2005;91(6):806−810. https://doi.org/10.1136/hrt.2003.029785.

72. Freeman RV, Otto CM. Spectrum of calcific aortic valve disease: pathogenesis, disease progression, and treatment strategies. *Circulation.* 2005;111(24):3316−3326. https://doi.org/10.1161/CIRCULATIONAHA.104.486738.

73. Olsson M, Thyberg J, Nilsson J. Presence of oxidized low density lipoprotein in nonrheumatic stenotic aortic valves. *Arterioscler Thromb Vasc Biol.* 1999;19(5):1218−1222. https://doi.org/10.1161/01.ATV.19.5.1218.

74. Mohler 3rd ER, Gannon F, Reynolds C, Zimmerman R, Keane MG, Kaplan FS. Bone formation and inflammation in cardiac valves. *Circulation.* 2001;103(11):1522−1528. https://doi.org/10.1161/01.CIR.103.11.1522.

75. Olsson M, Dalsgaard CJ, Haegerstrand A, Rosenqvist M, Ryden L, Nilsson J. Accumulation of T lymphocytes and expression of interleukin-2 receptors in nonrheumatic stenotic aortic valves. *J Am Coll Cardiol.* 1994;23(5):1162−1170. https://doi.org/10.1016/0735-1097(94)90606-8.

76. Jian B, Narula N, Li QY, Mohler 3rd ER, Levy RJ. Progression of aortic valve stenosis: TGF-beta1 is present in calcified aortic valve cusps and promotes aortic valve interstitial cell calcification via apoptosis. *Ann Thorac Surg.* 2003;75(2):456−457. https://doi.org/10.1016/S0003-4975(02)04312-6.

77. Kaden JJ, Bickelhaupt S, Grobholz R, et al. Expression of bone sialoprotein and bone morphogenetic protein-2 in calcific aortic stenosis. *J Heart Valve Dis.* 2004;13(4):560−566.

78. Victoria Gomez-Stallons M, Wirrig-Schwendeman EE, Hassel KR, Conway SJ, Yutzey KE. BMP signaling is required for aortic valve calcification HHS public access. *Arterioscler Thromb Vasc Biol.* 2016. https://doi.org/10.1161/ATVBAHA.116.307526.

79. Yoganathan AP. Fluid mechanics of aortic stenosis. *Eur Heart J.* 1988;9(Suppl E):13−17. https://doi.org/10.1093/eurheartj/9.suppl_E.13.

80. Clark C. The fluid mechanics of aortic stenosis − II. Unsteady flow experiments. *J Biomech.* 1976;9(9):567−573. https://doi.org/10.1016/0021-9290(76)90097-X.

81. Seaman C, McNally A, Biddle S, Jankowski L, Sucosky P. Generation of simulated calcific lesions in valve leaflets for flow studies. *J Heart Valve Dis.* 2015;24(1):1−11.

82. Seaman C, Akingba AG, Sucosky P. Steady flow hemodynamic and energy loss measurements in normal and simulated calcified tricuspid and bicuspid aortic valves. *J Biomech Eng.* 2014;136(4):1−11. https://doi.org/10.1115/1.4026575.

83. Sadeghpour F, Fatouraee N, Navidbakhsh M. Haemodynamic of blood flow through stenotic aortic valve. *J Med Eng Technol.* 2017;41(2):108−114. https://doi.org/10.1080/03091902.2016.1226439.

84. de Marchena E, Badiye A, Robalino G, et al. Respective prevalence of the different carpentier classes of mitral regurgitation: a stepping stone for future therapeutic research and development. *J Card Surg.* 2011;26(4):385−392. https://doi.org/10.1111/j.1540-8191.2011.01274.x.

85. Cosgrove DM, Stewart WJ. Mitral valvuloplasty. *Curr Probl Cardiol.* 1989;14(7):359−415. https://doi.org/10.1016/S0146-2806(89)80002-7.

86. Barber JE, Ratliff NB, Cosgrove DM, Griffin BP, Vesely I. Myxomatous mitral valve chordae. I: mechanical properties. *J Heart Valve Dis.* 2001;10(3):320−324.

87. Delling FN, Vasan RS. Epidemiology and pathophysiology of mitral valve prolapse: new insights into disease progression, genetics, and molecular basis. *Circulation.* 2014;129(21):2158−2170. https://doi.org/10.1161/CIRCULATIONAHA.113.006702.

88. Grande-Allen KJ, Griffin BP, Ratliff NB, Cosgrove DM, Vesely I. Glycosaminoglycan profiles of myxomatous mitral leaflets and chordae parallel the severity of mechanical alterations. *J Am Coll Cardiol.* 2003;42(2):271−277. https://doi.org/10.1016/S0735-1097(03)00626-0.

89. Whittaker P, Boughner DR, Perkins DG, Canham PB. Quantitative structural analysis of collagen in chordae tendineae and its relation to floppy mitral valves and proteoglycan infiltration. *Br Heart J.* 1987;57(3):264−269. https://doi.org/10.1136/hrt.57.3.264.

90. Rabkin E, Aikawa M, Stone J, Fukumoto Y, Libby P, Schoen F. Activated interstitial myofibroblasts express catabolic enzymes and mediate matrix remodeling in

myxomatous heart valves. *Circulation.* 2000;104: 2525–2532. https://doi.org/10.1161/hc4601.099489.

91. Tamura K, Fukuda Y, Ishizaki M, Masuda Y, Yamanaka N, Ferrans VJ. Abnormalities in elastic fibers and other connective-tissue components of floppy mitral valve. *Am Heart J.* 1995;129(6):1149–1158. https://doi.org/10.1016/0002-8703(95)90397-6.

92. Akhtar S, Meek KM, James V. Ultrastructure abnormalities in proteoglycans, collagen fibrils, and elastic fibers in normal and myxomatous mitral valve chordae tendineae. *Cardiovasc Pathol.* 1999;8(4):191–201. https://doi.org/10.1016/S1054-8807(99)00004-6.

93. Cole WG, Chan D, Hickey AJ, Wilcken DE. Collagen composition of normal and myxomatous human mitral heart valves. *Biochem J.* 1984;219(2):451–460. https://doi.org/10.1042/bj2190451.

94. Clavel M-A, Mantovani F, Malouf J, et al. Dynamic phenotypes of degenerative myxomatous mitral valve disease. *Circ Cardiovasc Imaging.* 2015;8(5). https://doi.org/10.1161/CIRCIMAGING.114.002989.

95. Connell PS, Azimuddin AF, Kim SE, et al. Regurgitation hemodynamics alone cause mitral valve remodeling characteristic of clinical disease states in vitro. *Ann Biomed Eng.* 2016;44(4):954–967. https://doi.org/10.1007/s10439-015-1398-0.

96. Barber JE, Kasper FK, Ratliff NB, Cosgrove DM, Griffin BP, Vesely I. Mechanical properties of myxomatous mitral valves. *J Thorac Cardiovasc Surg.* 2001;122(5):955–962. https://doi.org/10.1067/mtc.2001.117621.

97. Weber KT, Sun Y, Katwa LC, Cleutjens JP, Zhou G. Connective tissue and repair in the heart. Potential regulatory mechanisms. *Ann N Y Acad Sci.* 1995;752:286–299. https://doi.org/10.1111/j.1749-6632.1995.tb17438.x.

98. Schneider PJ, Deck JD. Tissue and cell renewal in the natural aortic valve of rats: an autoradiographic study. *Cardiovasc Res.* 1981;15(4):181–189. https://doi.org/10.1093/cvr/15.4.181.

99. Willems IE, Havenith MG, Smits JF, Daemen MJ. Structural alterations in heart valves during left ventricular pressure overload in the rat. *Lab Invest.* 1994;71(1):127–133.

100. Chaput M, Handschumacher MD, Tournoux F, et al. Mitral leaflet adaptation to ventricular remodeling occurrence and adequacy in patients with functional mitral regurgitation. *Circulation.* 2008. https://doi.org/10.1161/CIRCULATIONAHA.107.749440.

101. Dal-Bianco JP, Aikawa E, Bischoff J, et al. Active adaptation of the tethered mitral valve: insights into a compensatory mechanism for functional mitral regurgitation. *Circulation.* 2009. https://doi.org/10.1161/CIRCULATIONAHA.108.846782.

102. Merryman WD, Youn I, Lukoff HD, et al. Correlation between heart valve interstitial cell stiffness and transvalvular pressure: implications for collagen biosynthesis. *Am J Physiol Circ Physiol.* 2006;290(1):H224–H231. https://doi.org/10.1152/ajpheart.00521.2005.

103. Butcher JT, Penrod AM, Garcia AJ, Nerem RM. Unique morphology and focal adhesion development of valvular endothelial cells in static and fluid flow environments.

*Arterioscler Thromb Vasc Biol.* 2004;24:1429–1434. https://doi.org/10.1161/01.ATV.0000130462.50769.5a.

104. Sucosky P, Balachandran K, Elhammali A, Jo H, Yoganathan AP. Altered shear stress stimulates up-regulation of endothelial VCAM-1 and ICAM-1 in a BMP-4- and TGF-β1-dependent pathway. *Arterioscler Thromb Vasc Biol.* 2009;29(2):254–260. https://doi.org/10.1161/ATVBAHA.108.176347.

105. Deck JD. Endothelial cell orientation on aortic valve leaflets. *Cardiovasc Res.* 1986;20(10):760–767. https://doi.org/10.1093/cvr/20.10.760.

106. Weston MW, Yoganathan AP. Biosynthetic activity in heart valve leaflets in response to in vitro flow environments. *Ann Biomed Eng.* 2001;29:752–763. https://doi.org/10.1114/1.1397794.

107. Platt MO, Xing Y, Jo H, Yoganathan AP. Cyclic pressure and shear stress regulate matrix metalloproteinases and cathepsin activity in porcine aortic valves. *J Heart Valve Dis.* 2006;15(5):622–629.

108. Sun L, Rajamannan NM, Sucosky P. Defining the role of fluid shear stress in the expression of early signaling markers for calcific aortic valve disease. *PLoS One.* 2013;8(12):e84433. https://doi.org/10.1371/journal.pone.0084433.

109. Sun L, Sucosky P. Bone morphogenetic protein-4 and transforming growth factor-beta1 mechanisms in acute valvular response to supra-physiologic hemodynamic stresses. *World J Cardiol.* 2015;7(6):331–343. https://doi.org/10.4330/wjc.v7.i6.331.

110. Atkins SK, McNally A, Sucosky P. Mechanobiology in cardiovascular disease management: potential strategies and current needs. *Front Bioeng Biotechnol.* 2016;4:79. https://doi.org/10.3389/fbioe.2016.00079.

111. Merryman WD. Mechano-potential etiologies of aortic valve disease. *J Biomech.* 2010;43(1):87–92. https://doi.org/10.1016/j.jbiomech.2009.09.013.

112. Hoehn D, Sun L, Sucosky P. Role of pathologic shear stress alterations in aortic valve endothelial activation. *Cardiovasc Eng Technol.* 2010;1(2):165–178. https://doi.org/10.1007/s13239-010-0015-5.

113. Cao K, Sucosky P. Computational comparison of regional stress and deformation characteristics in tricuspid and bicuspid aortic valve leaflets. *Int J Numer Method Biomed Eng.* 2017;33(3):e02798. https://doi.org/10.1002/cnm.2798.

114. Chandra S, Rajamannan NM, Sucosky P. Computational assessment of bicuspid aortic valve wall-shear stress: implications for calcific aortic valve disease. *Biomech Model Mechanobiol.* 2012;11(7):1085–1096. https://doi.org/10.1007/s10237-012-0375-x.

115. McNally A, Madan A, Sucosky P. Morphotype-dependent flow characteristics in bicuspid aortic valve ascending aortas: a benchtop particle image velocimetry study. *Front Physiol.* 2017;8:44. https://doi.org/10.3389/fphys.2017.00044.

116. Saikrishnan N, Yap C-H, Milligan NC, Vasilyev NV, Yoganathan AP. In vitro characterization of bicuspid aortic valve hemodynamics using particle image

velocimetry. *Ann Biomed Eng.* 2012;40(8):1760—1775. https://doi.org/10.1007/s10439-012-0527-2.

117. Yap CH, Saikrishnan N, Tamilselvan G, Vasilyev N, Yoganathan AP, Vasiliyev NV. The congenital bicuspid aortic valve can experience high frequency unsteady shear stresses on its leaflet surface. *Am J Physiol Heart Circ Physiol.* 2012;303(6):H721—H731. https://doi.org/10.1152/ajpheart.00829.2011.

118. Seaman C, Sucosky P. Anatomic versus effective orifice area in a bicuspid aortic valve. *Echocardiography.* 2014; 31(8):1028. https://doi.org/10.1111/echo.12720.

119. Sun L, Chandra S, Sucosky P. Ex vivo evidence for the contribution of hemodynamic shear stress abnormalities to the early pathogenesis of calcific bicuspid aortic valve disease. *PLoS One.* 2012;7(10):e48843. https://doi.org/10.1371/journal.pone.0048843.

120. Atkins SK, Sucosky P. The etiology of bicuspid aortic valve disease: focus on hemodynamics. *World J Cardiol.* 2014;12(12):1227—1233. https://doi.org/10.4330/wjc.v6.i12.1227.

121. Sacks MS, Smith DB, Hiester ED. The aortic valve microstructure: effects of transvalvular pressure. *J Biomed Mater Res.* 1998;41(1):131—141. https://doi.org/10.1002/(SICI)1097-4636(199807)41:1<131::AID-JBM16>3.0.CO;2-Q.

122. Cochran RP, Kunzelman KS, Chuong CJ, Sacks MS, Eberhart RC. Nondestructive analysis of mitral valve collagen fiber orientation. *ASAIO Trans.* 1991;37(3): M447—M448.

123. Ku CH, Johnson PH, Batten P, et al. Collagen synthesis by mesenchymal stem cells and aortic valve interstitial cells in response to mechanical stretch. *Cardiovasc Res.* 2006;71(3):548—556. https://doi.org/10.1016/j.cardiores.2006.03.022.

124. Metzler SA, Waller SC, Warnock JN. Quantitative characterization of aortic valve endothelial cell viability and morphology in situ under cyclic stretch. *Cardiovasc Eng Technol.* 2019;10(1):173—180. https://doi.org/10.1007/s13239-018-00375-1.

125. Balachandran K, Sucosky P, Jo H, Yoganathan AP. Elevated cyclic stretch alters matrix remodeling in aortic valve cusps — implications for degenerative aortic valve disease? *Am J Physiol Heart Circ Physiol.* 2009;296(3):H756—H764. https://doi.org/10.1152/ajpheart.00900.2008.

126. Balachandran K, Bakay MA, Connolly JM, Zhang X, Yoganathan AP, Levy RJ. Aortic valve cyclic stretch causes increased remodeling activity and enhanced serotonin receptor responsiveness. *Ann Thorac Surg.* 2011;92(1): 147—153. https://doi.org/10.1016/j.athoracsur.2011.03.084.

127. Balachandran K, Sucosky P, Jo H, Yoganathan AP. Elevated cyclic stretch induces aortic valve calcification in a bone morphogenic protein-dependent manner. *Am J Pathol.* 2010;177(1):49—57. https://doi.org/10.2353/ajpath.2010.090631.

128. Thayer P, Balachandran K, Rathan S, et al. The effects of combined cyclic stretch and pressure on the aortic valve

interstitial cell phenotype. *Ann Biomed Eng.* 2011;39(6): 1654—1667. https://doi.org/10.1007/s10439-011-0273-x.

129. Ali MS, Wang X, Lacerda CM. The effect of physiological stretch and the valvular endothelium on mitral valve proteomes. *Exp Biol Med.* 2019. https://doi.org/10.1177/1535370219829006, 153537021982900.

130. Smerup M, Hasenkam JM, Yang J, et al. Increased expression of endothelin B receptor in static stretch exposed porcine mitral valve leaflets. *Res Vet Sci.* 2006;82(2): 232—238. https://doi.org/10.1016/j.rvsc.2006.07.009.

131. Porta C, Fresu L, Bertozzi G, et al. Effect of cyclic stretch on vascular endothelial cells and abdominal aortic aneurysm (AAA): role in the inflammatory response. *Int J Mol Sci.* 2019;20(2):287. https://doi.org/10.3390/ijms20020287.

132. Burton AC. Relation of structure to function of the tissues of the wall of blood vessels. *Physiol Rev.* 1954; 34(4):619—642. https://doi.org/10.1152/physrev.1954.34.4.619.

133. Chiu J-J, Chien S. Effects of disturbed flow on vascular endothelium: pathophysiological basis and clinical perspectives. *Physiol Rev.* 2011;91(1):327—387. https://doi.org/10.1152/physrev.00047.2009.

134. Meng H, Tanweer O, Wilson TA, Riina HA, Metaxa E. A comparative review of the hemodynamics and pathogenesis of cerebral and abdominal aortic aneurysms: lessons to learn from each other. *J Cerebrovasc Endovasc Neurosurg.* 2015;16(4):335. https://doi.org/10.7461/jcen.2014.16.4.335.

135. Fung YC. *Biomechanics : Motion, Flow, Stress, and Growth.* New-York: Springer-Verlag; 1990. https://doi.org/10.1007/978-1-4419-6856-2.

136. Caro CG, Pedley TJ, Schroter RC, Seed WA. *The Mechanics of the Circulation.* Oxford: Oxford University Press; 1978.

137. Nagaoka T, Yoshida A. Noninvasive evaluation of wall shear stress on retinal microcirculation in humans. *Investig Ophthalmol Vis Sci.* 2006;47(3):1113—1119. https://doi.org/10.1167/iovs.05-0218.

138. Ichioka S, Shibata M, Kosaki K, Sato Y, Harii K, Kamiya A. Effects of shear stress on wound-healing angiogenesis in the rabbit ear chamber. *J Surg Res.* 1997;72(1):29—35. https://doi.org/10.1006/jsre.1997.5170.

139. Kornet L, Hoeks APG, Lambregts J, Reneman RS. Mean wall shear stress in the femoral arterial bifurcation is low and independent of age at rest. *J Vasc Res.* 2000; 37(2):112—122. https://doi.org/10.1159/000025722.

140. Farag ES, van Ooij P, Planken RN, et al. Aortic valve stenosis and aortic diameters determine the extent of increased wall shear stress in bicuspid aortic valve disease. *J Magn Reson Imaging.* 2018;48(2):522—530. https://doi.org/10.1002/jmri.25956.

141. Tang BT, Cheng CP, Draney MT, et al. Abdominal aortic hemodynamics in young healthy adults at rest and during lower limb exercise: quantification using image-based computer modeling. *Am J Physiol Circ Physiol.* 2006;291(2):H668—H676. https://doi.org/10.1152/ajpheart.01301.2005.

142. Reneman RS, Hoeks APG. Wall shear stress as measured in vivo: consequences for the design of the arterial system. *Med Biol Eng Comput.* 2008;46(5):499−507. https://doi.org/10.1007/s11517-008-0330-2.

143. Bell V, Mitchell WA, Sigurdsson S, et al. Longitudinal and circumferential strain of the proximal aorta. *J Am Heart Assoc.* 2014;3(6):1−11. https://doi.org/10.1161/JAHA. 114.001536.

144. Karatolios K, Wittek A, Nwe TH, et al. Method for aortic wall strain measurement with three-dimensional ultrasound speckle tracking and fitted finite element analysis. *Ann Thorac Surg.* 2013;96(5):1664−1671. https://doi.org/10.1016/j.athoracsur.2013.06.037.

145. Stumpf MJ, Schaefer CA, Krycki J, et al. Impairment of vascular strain in patients with obstructive sleep apnea. *PLoS One.* 2018;13(2):1−13. https://doi.org/10.1371/journal.pone.0193397.

146. Kumar S, Kim CW, Simmons RD, Jo H. Role of flow-sensitive microRNAs in endothelial dysfunction and atherosclerosis mechanosensitive athero-miRs. *Arterioscler Thromb Vasc Biol.* 2014;34(10):2206−2216. https://doi.org/10.1161/ATVBAHA.114.303425.

147. Rafieian-Kopaei M, Setorki M, Doudi M, Baradaran A, Nasri H. Atherosclerosis : process , indicators , risk factors and new hopes. *Int J Prev Med.* 2018;5(8):927−946.

148. Ross R. Atherosclerosis is an inflammatory disease. *Am Heart J.* 1999;138(5 Pt 2):S419−S420.

149. Montecucco F, Mach F. Atherosclerosis is an inflammatory disease. *Semin Immunopathol.* 2009;31(1):1−3. https://doi.org/10.1007/s00281-009-0146-7.

150. Ross R. Atherosclerosis — an inflammatory disease. *N Engl J Med.* 1999;340(2):12.

151. Chiu JJ, Chen CN, Lee PL, et al. Analysis of the effect of disturbed flow on monocytic adhesion to endothelial cells. *J Biomech.* 2003;36(12):1883−1895. https://doi.org/10.1016/S0021-9290(03)00210-0.

152. Bassiouny HS, Song RH, Hong XF, Singh A, Kocharyan H, Glagov S. Flow regulation of 72-kD collagenase IV (MMP-2) after experimental arterial injury. *Circulation.* 1998;98(2):157−163. https://doi.org/10.1161/01.CIR. 98.2.157.

153. Bluestein D, Niu L, Schoephoerster RT, Dewanjee MK. Fluid mechanics of arterial stenosis: relationship to the development of mural thrombus. *Ann Biomed Eng.* 1997; 25(2):344−356. https://doi.org/10.1007/BF02648048.

154. Glagov S, Zarins C, Giddens DP, Ku DN. Hemodynamics and atherosclerosis. Insights and perspectives gained from studies of human arteries. *Arch Pathol Lab Med.* 1988;112(10):1018−1031.

155. Chandran KKB, Rittgers SES, Yoganathan AAP. *Biofluid Mechanics: The Human Circulation.* 2 ed. Boca Raton: CRC Press/Taylor and Francis; 2007.

156. Malota Z, Glowacki J, Sadowski W, Kostur M. Numerical analysis of the impact of flow rate, heart rate, vessel geometry, and degree of stenosis on coronary hemodynamic indices. *BMC Cardiovasc Disord.* 2018;18(1). https://doi.org/10.1186/s12872-018-0865-6.

157. Schirmer CM, Malek AM. Computational fluid dynamic characterization of carotid bifurcation stenosis in patient-based geometries. *Brain Behav.* 2012;2(1):42−52. https://doi.org/10.1002/brb3.25.

158. Tang D, Yang C, Kobayashi S, Zheng J, Vito RP. Effect of stenosis asymmetry on blood flow and artery Compression: a three- dimensional fluid-structure interaction model. *Ann Biomed Eng.* 2003;31:1182−1193. https://doi.org/10.1114/1.1615577.

159. Nichols L, Lagana S, Parwani A. Coronary artery aneurysm: a review and hypothesis regarding etiology. *Arch Pathol Lab Med.* 2008;132(5):823−828. https://doi.org/10.1136/bmj.2.5209.1344.

160. Thompson RW, Parks WC, Wassef M, et al. Pathogenesis of abdominal aortic aneurysms: a multidisciplinary research program supported by the National Heart, Lung, and Blood Institute. *J Vasc Surg.* 2002;34(4): 730−738. https://doi.org/10.1067/mva.2001.116966.

161. Aggarwal S, Qamar A, Sharma V, Sharma A. Abdominal aortic aneurysm: a comprehensive review. *Exp Clin Cardiol.* 2011;16(1):11−15.

162. Raut SS, Chandra S, Shum J, et al. Biological, geometric and biomechanical factors influencing abdominal aortic aneurysm rupture risk: a comprehensive review. *Recent Pat Med Imaging.* 2013;3(1):44−59. https://doi.org/10.2174/1877613211303010006.

163. Hui-Yuen JS, Duong TT, Yeung RSM. TNF-alpha is necessary for induction of coronary artery inflammation and aneurysm formation in an animal model of kawasaki disease. *J Immunol.* 2014;176(10):6294−6301. https://doi.org/10.4049/jimmunol.176.10.6294.

164. Syed M, Lesch M. Coronary artery aneurysm: a review. *Clin Cardiol.* 2006;29(10):439−443.

165. Chalouhi N, Hoh BL, Hasan D. Review of cerebral aneurysm formation, growth, and rupture. *Stroke.* 2013;44(12):3613−3622. https://doi.org/10.1161/STROKEAHA.113.002390.

166. Abou Sherif S, Ozden Tok O, Taşköylü Ö, Goktekin O, Kilic ID. Coronary artery aneurysms: a review of the epidemiology, pathophysiology, diagnosis, and treatment. *Front Cardiovasc Med.* 2017;4:1−12. https://doi.org/10.3389/fcvm.2017.00024.

167. Shojima M, Oshima M, Takagi K, et al. Magnitude and role of wall shear stress on cerebral aneurysm: computational fluid dynamic study of 20 middle cerebral artery aneurysms. *Stroke.* 2004;35(11):2500−2505. https://doi.org/10.1161/01.STR.0000144648.89172.0f.

168. Fukazawa K, Ishida F, Umeda Y, et al. Using computational fluid dynamics analysis to characterize local hemodynamic features of middle cerebral artery aneurysm rupture points. *World Neurosurg.* 2015; 83(1):80−86. https://doi.org/10.1016/j.wneu.2013. 02.012.

169. Boyd AJ, Kuhn DCS, Lozowy RJ, Kulbisky GP. Low wall shear stress predominates at sites of abdominal aortic aneurysm rupture. *J Vasc Surg.* 2016;63(6):1613−1619. https://doi.org/10.1016/j.jvs.2015.01.040.

170. Goubergrits L, Schaller J, Kertzscher U, et al. Statistical wall shear stress maps of ruptured and unruptured middle cerebral artery aneurysms. *J R Soc Interface*. 2012; 9(69):677–688. https://doi.org/10.1098/rsif.2011.0490.

171. Bihari P, Shelke A, Nwe TH, et al. Strain measurement of abdominal aortic aneurysm with real-time 3D ultrasound speckle tracking. *Eur J Vasc Endovasc Surg*. 2013; 45(4):315–323. https://doi.org/10.1016/j.ejvs.2013.01.004.

172. Montorzi G, Roy S, Stergiopulos N, Silacci P, Gambillara V, Chambaz C. Plaque-prone hemodynamics impair endothelial function in pig carotid arteries. *Am J Physiol Circ Physiol*. 2006;290(6):H2320–H2328. https://doi.org/10.1152/ajpheart.00486.2005.

173. Levesque MJ, Nerem RM. The elongation and orientation of cultured endothelial cells in response to shear stress. *J Biomech Eng*. 1985;107(4):341–347. https://doi.org/10.1115/1.3138567.

174. Nerem RM. Shear force and its effect on cell structure and function. *ASGSB Bull*. 1991;4(2):87–94.

175. Chien S, Chein S, Chien S. Effects of disturbed flow on endothelial cells. *Ann Biomed Eng*. 2008;36(4): 554–562. https://doi.org/10.1007/s10439-007-9426-3.

176. Metaxa E, Meng H, Kaluvala SR, et al. Nitric oxide-dependent stimulation of endothelial cell proliferation by sustained high flow. *Am J Physiol Circ Physiol*. 2008; 295(2):H736–H742. https://doi.org/10.1152/ajpheart.01156.2007.

177. Dewey Jr CF, Bussolari SR, Gimbrone MA, Davies PF. The dynamic response of vascular endothelial cells to fluid shear stress. *J Biomech Eng*. 1981;103(3):177–185. https://doi.org/10.1115/1.3138276.

178. Chakraborty A, Chakraborty S, Jala VR, Haribabu B, Sharp MK, Berson RE. Effects of biaxial oscillatory shear stress on endothelial cell proliferation and morphology. *Biotechnol Bioeng*. 2012;109(3):695–707. https://doi.org/10.1002/bit.24352.

179. Wong A, Pollard T, Herman I. Actin filament stress fibers in vascular endothelial cells in vivo. *Science*. 1983; 219(4586):867–869.

180. Levesque MJ, Nerem RM, Sprague EA. Vascular endothelial cell proliferation in culture and the influence of flow. *Biomaterials*. 1990;11(9):702–707. https://doi.org/10.1016/0142-9612(90)90031-K.

181. Kadohama T, Nishimura K, Hoshino Y, Sasajima T, Sumpio B. Effects of different types of fluid shear stress on endothelial cell proliferation and survival. *J Cell Physiol*. 2007;731(August 2006):720–731. https://doi.org/10.1002/jcp.

182. Tardy Y, Resnick N, Nagel T, Gimbrone MA, Dewey CF. Shear stress gradients remodel endothelial monolayers in vitro via a cell proliferation-migration-loss cycle. *Arterioscler Thromb Vasc Biol*. 1997;17(11):3102–3106. https://doi.org/10.1161/01.ATV.17.11.3102.

183. Kamiya A, Togawa T. Adaptive regulation of wall shear stress to flow change in the canine carotid artery. *Am J Physiol*. 1980;239(1):H14–H21. https://doi.org/10.1152/ajpheart.1980.239.1.H14.

184. Langille B, O'Donnell F. Reductions in arterial diameter produced by chronic decreases in blood flow are endothelium-dependent. *Science*. 1986;231(4736): 405–407. https://doi.org/10.1126/science.3941904.

185. Tulis DA, Unthank JL, Prewitt RL. Flow-induced arterial remodeling in rat mesenteric vasculature. *Am J Physiol Circ Physiol*. 2017;274(3):H874–H882. https://doi.org/10.1152/ajpheart.1998.274.3.h874.

186. Meng H, Wang Z, Hoi Y, et al. Complex hemodynamics at the apex of an arterial bifurcation induces vascular remodeling resembling cerebral aneurysm initiation. *Stroke*. 2007;38(6):1924–1931. https://doi.org/10.1161/STROKEAHA.106.481234.

187. Wang Z, Kolega J, Hoi Y, et al. Molecular alterations associated with aneurysmal remodeling are localized in the high hemodynamic stress region of a created carotid bifurcation. *Neurosurgery*. 2009;65(1):169–178. https://doi.org/10.1227/01.NEU.0000343541.85713.01.

188. Jufri NF, Mohamedali A, Avolio A, Baker MS. Mechanical stretch : physiological and pathological implications for human vascular endothelial cells. *Vasc Cell*. 2015:1–12. https://doi.org/10.1186/s13221-015-0033-z.

189. Qiu J, Zheng Y, Hu J, et al. Biomechanical regulation of vascular smooth muscle cell functions: from in vitro to in vivo understanding. *J R Soc Interface*. 2014;11(90). https://doi.org/10.1098/rsif.2013.0852.

190. Tojkander S, Gateva G, Lappalainen P. Actin stress fibers – assembly, dynamics and biological roles. *J Cell Sci*. 2012;125(8):1855–1864. https://doi.org/10.1242/jcs.098087.

191. Takemasa T, Yamaguchi T, Yamamoto Y, Sugimoto K, Yamashita K. Oblique alignment of stress fibers in cells reduces the mechanical stress in cyclically deforming fields. *Eur J Cell Biol*. 1998;77(2):91–99. https://doi.org/10.1016/S0171-9335(98)80076-9.

192. Naruse K, Yamada T, Sokabe M. Involvement of SA channels in orienting response of cultured endothelial cells to cyclic stretch. *Am J Physiol Circ Physiol*. 2017;274(5): H1532–H1538. https://doi.org/10.1152/ajpheart.1998.274.5.h1532.

193. Yoshigi M, Clark EB, Yost HJ. Quantification of stretch-induced cytoskeletal remodeling in vascular endothelial cells by image processing. *Cytometry*. 2003;55A(2): 109–118. https://doi.org/10.1002/cyto.a.10076.

194. Ives C. Mechanical effects on endothelial cell morphology: in vitro assessment. *Vitr Cell Dev Biol*. 1986;22:500–507.

195. Zhao S, Suciu A, Ziegler T, et al. Synergistic effects of fluid shear stress and cyclic circumferential stretch on vascular endothelial cell morphology and cytoskeleton. *Arterioscler Thromb Vasc Biol*. 1995;15(10):1781–1786. https://doi.org/10.1161/01.ATV.15.10.1781.

196. Thubrikar MJ, Robicsek F. Pressure-induced arterial wall stress and atherosclerosis. *Ann Thorac Surg*. 1995;59(6): 1594–1603. https://doi.org/10.1016/0003-4975(94)01037-D.

197. Standley PR, Camaratta A, Nolan BP, Purgason CT, Stanley MA. Cyclic stretch induces vascular smooth

muscle cell alignment via NO signaling. *Am J Physiol Circ Physiol.* 2015;283(5):H1907—H1914. https://doi.org/10.1152/ajpheart.01043.2001.

198. Bocci G, Fasciani A, Danesi R, Viacava P, Genazzani AR, Del Tacca M. In-vitro evidence of autocrine secretion of vascular endothelial growth factor by endothelial cells from human placental blood vessels. *Mol Hum Reprod.* 2001;7(8):771—777. https://doi.org/10.1093/molehr/7.8.771.

199. Richard MN, Deniset JF, Kneesh AL, Blackwood D, Pierce AN, Pierce GN. Mechanical stretching stimulates smooth muscle cell growth, nuclear protein import, and nuclear pore expression through mitogen-activated protein kinase activation. *J Biol Chem.* 2007;282(32):23081—23088. https://doi.org/10.1074/jbc.M703602200.

200. Lee Y-U, Hayman D, Sprague EA, Han H-C. Effects of axial stretch on cell proliferation and intimal thickness in arteries in organ culture. *Cell Mol Bioeng.* 2010;3(3):286—295. https://doi.org/10.1007/s12195-010-0128-9.

201. Liu XM, Ensenat D, Wang H, Schafer AI, Durante W. Physiologic cyclic stretch inhibits apoptosis in vascular endothelium. *FEBS Lett.* 2003;541(1—3):52—56. https://doi.org/10.1016/S0014-5793(03)00285-0.

202. Anwar MA, Shalhoub J, Lim CS, Gohel MS, Davies AH. The effect of pressure-induced mechanical stretch on vascular wall differential gene expression. *J Vasc Res.* 2012;49(6):463—478. https://doi.org/10.1159/000339151.

203. Leung DYM, Glagov S, Mathews MB. Elastin and collagen accumulation in rabbit ascending aorta and pulmonary trunk during postnatal growth. Correlation of cellular synthetic response with medial tension. *Circ Res.* 1977;41(3):316—323. https://doi.org/10.1161/01.RES.41.3.316.

204. von Offenberg Sweeney N, Cummins PM, Birney YA, Cullen JP, Redmond EM, Cahill PA. Cyclic strain-mediated regulation of endothelial matrix metalloproteinase-2 expression and activity. *Cardiovasc Res.* 2004;63(4):625—634. https://doi.org/10.1016/j.cardiores.2004.05.008.

205. Pedrigi RM, Papadimitriou KI, Kondiboyina A, et al. Disturbed cyclical stretch of endothelial cells promotes nuclear expression of the pro-atherogenic transcription factor NF-κB. *Ann Biomed Eng.* 2017;45(4):898—909. https://doi.org/10.1007/s10439-016-1750-z.

206. Massé DD, Shar JA, Brown KN, Keswani SG, Grande-Allen KJ, Sucosky P. Discrete subaortic stenosis: perspective roadmap to a complex disease. *Front Cardiovasc Med.* 2018;5:122. https://doi.org/10.3389/fcvm.2018.00122.

207. Muna WF, Ferrans VJ, Pierce JE, Roberts WC. Ultrastructure of the fibrous subaortic "ring" in dogs with discrete subaortic stenosis. *Lab Investig.* 1978;39(5):471—482.

# Mechanobiology of the Optic Nerve Head in Primary Open-Angle Glaucoma

JR-JIUN LIOU, PHD • JASON E. SHOEMAKER, PHD •
JONATHAN P. VANDE GEEST, PHD

## 1. INTRODUCTION

Primary open-angle glaucoma (POAG) is a progressive neurodegenerative disease that affects over 60 million people globally.[1] Current treatments are laser surgeries or eye drops including β-blockers, carbonic anhydrase inhibitors, or combined medications.[2] These drugs primarily decrease the rate of aqueous humor production and mildly modulate the balance of extracellular matrix (ECM) production and degradation.[3] However, the specific mechanisms that govern the initiation and progression of POAG are currently unknown.

It is known that eventual loss of vision in glaucoma is due to the death of retinal ganglion cells (RGCs) that are responsible for transmitting visual information to the brain. While intraocular pressure (IOP) is a known risk factor for POAG, IOP alone is limited as a predictor for the initiation and progression of the disease.[4] In fact, the existence of POAG in patients with normal IOP and the presence of normal vision in patients with moderately elevated IOP suggest there are other factors playing a role in the onset and development of POAG. Structurally, the peripapillary sclera and lamina cribrosa (LC) protect RGC axons as they exit the eye through the optic nerve head (ONH). The LC is a porous collagenous disc in the ONH that undergoes significant ECM remodeling in POAG,[5–7] playing an important role in RGC death in this disease.[8]

The risk factors involved in disease progression include genetics, IOP, age, and race.[9,10] All these factors likely contribute to variations in the stiffness of posterior ocular tissues, including the peripapillary sclera and LC microstructure. However, the relationship between the stiffness of these tissues and the homeostatic ECM remodeling of the LC is currently unknown. The morphologic changes, including thinning and cupping of the glaucomatous LC and elevated levels of ECM-related growth factors in the ONH of glaucomatous tissues, suggest that homeostatic ECM remodeling of the LC is dysregulated in POAG.[11] Current research suggests that there are numerous growth factors, ECM proteins, and proteases[11] that are dysregulated in the ONH of POAG tissues, in addition to the evidence of altered oxidative stress levels.[12] For example, elevated levels of transforming growth factor (TGF) β2, which has been found in POAG samples, can lead to increased production of ECM proteins including collagen and fibronectin. TGFβ2 also increases expression of plasminogen activator inhibitor (PAI) that prevents activation of matrix metalloproteinases (MMPs) via inhibition of tissue plasminogen activator, thus leading to further matrix deposition.[11] Other factors such as bone morphogenetic protein (BMP), connective tissue growth factor, and metalloproteinases (along with their inhibitors) have been suggested to be important in POAG.[11]

The relationship among the stiffness of posterior ocular tissues, remodeling of these tissues, morphologic changes of the LC, and eventual loss of vision due to ganglion cell death remains unclear. To motivate future research focused on answering these questions, in this chapter, we describe the structure and function of posterior ocular tissues and also review what is known about how mechanical stimuli and mechanical properties affect the development of the disease.

## 2. STRUCTURE AND FUNCTION

The LC is located in the ONH and has a columnar structure that aligns to form channels where RGC axons are located.[13,14] The function of LC is to provide structural support and protect the RGC axons of the ONH as they exit the eye.[15] In primates, the LC is a dense ECM network primarily secreted by the ONH LC cells,[16–18] which are negative for glial fibrillary acidic protein (GFAP) but positive for α-smooth muscle actin (αSMA).[14,19,20] Another major cell type present in the LC is the ONH astrocytes, which are positive for GFAP (especially when activated)

Mechanobiology. https://doi.org/10.1016/B978-0-12-817931-4.00003-0

**47**

and negative for αSMA.[14,19,20] Morphologically, astrocytes have multiple long and thin cellular processes, whereas LC cells have large flat polygonal shapes.[14] Astrocytes, different from neurons, do not propagate action potentials but increase intracellular calcium concentration for astrocyte-astrocyte or astrocyte-neuron communication during signal transmission.[16] In healthy tissues, astrocytes not only regulate synaptic function and remodeling but also modulate blood flow, extracellular fluid, and neurotransmitters.[16] Given that MMPs are found to be colocalized with activated astrocytes, astrocytes play an important role in the ECM changes present in glaucoma.[21] To characterize the changes that occur in ONH astrocytes in glaucoma, Hernandez et al.[22] identified differentially expressed genes in ONH cells and found these genes were significantly associated with cell adhesion. Another study by Nikolskaya et al.[23] performed a gene ontology analysis of reactive primary optic nerve astrocytes from patients with glaucoma and found an association with pathways involving neurotoxicity regulation. Toll-like receptor signaling has also been found to be upregulated in glaucomatous astrocytes, further implicating reactive astrocytes in glaucoma pathogenesis.[24]

POAG is a consequence of the death of RGCs, a type of neuronal cell in the retina.[25] In the ONH, there are approximately 1.5 million RGC axons spanning from the Bruch's membrane opening to the scleral portions of the neural canal.[13] Glaucoma may result from disturbances of axon-astrocyte interactions at the ONH that are caused by increased IOP, leading to axonal damage and subsequent degeneration of RGCs.[16–18,26] LC cells and astrocytes sense strain through integrin receptors that tie directly to adjacent fibrillar ECM and therefore respond to local changes in mechanical strain.[15,27] The cells that support the axons of the RGCs are responsible for the normal homeostasis of the LC ECM, and as such the control mechanisms that govern these relationships are important in the development of glaucoma.

# 3. MECHANICS OF THE OPTIC NERVE HEAD IN GLAUCOMA

In this section, we cover a few biomechanical aspects in glaucoma: (1) relationships between IOP and LC remodeling, (2) chronic morphologic changes of the LC, (3) deformation and displacement of the ONH in response to acute IOP elevation, and (4) how stiffness of the peripapillary sclera and LC is associated with disease progression.

## 3.1. Intraocular Pressure and Lamina Cribrosa Remodeling

IOP is the fluid pressure inside the eye and is currently the only modifiable risk factor to treat glaucoma. With chronic elevated and fluctuating IOPs characteristic of certain subtypes of glaucoma, it is expected that glaucomatous ONHs exhibit similar phenotypes as those detailed in Section 2. On the cellular level, elevated IOP and the associated stress and strain produce changes in the ECM of the impacted ONH tissues.[28,29] Therefore such changes are also expected to be found in glaucomatous tissues. Many of these studies evaluated the structural changes of the LC in patients with POAG, but LC remodeling has been found to be dependent on the glaucoma subtype. The most commonly studied differences are between normal tension glaucoma and high-tension glaucoma subtypes. With normal tension glaucoma, patients exhibit normal ranges of IOPs; therefore they may experience a difference in LC remodeling compared with those with elevated IOP, assuming material stiffness is the same. Li et al.[28] confirmed these expectations by showing that LCs were positioned more posteriorly in high-tension glaucoma than in normal tension glaucoma and in normal tension glaucoma than in controls, which agrees with findings from previous studies of POAG eyes. Such findings potentially indicate that even with normal IOPs, some level of remodeling is occurring in the ONHs of patients with normal tension glaucoma. In a study, Lee et al.[29] further highlighted the importance of IOP in POAG LC remodeling by showing that patients with normal tension glaucoma with relatively flat LCs had lower IOPs and were associated with greater parapapillary structural alteration, thus further supporting the connection between increased LC curvature and increased IOP.

Modulating IOP via medical interventions such as trabeculectomy, in which part of the trabecular meshwork is removed, has been found to have an impact on the remodeling and deformation of the LC.[30–33] For example, Shin et al.[31] investigated POAG eyes after trabeculectomy and found that both IOP and the LC depth significantly decreased. These decreases were greater in eyes with improved microvascular structure.[31] Another recent study by Esfandiari et al.[33] further supported these findings by showing an association between anterior LC depth and baseline IOP in patients with advanced POAG who underwent trabeculectomy. Other studies have investigated changes to the curvature of the LC after trabeculectomy and found that this curvature decreased after lowering IOP, resulting in sustained flattening and shallowing of the LC.[30,32] Testing the response of the ONH to acute increases in pressure, Fazio et al.[34] found that acute increases in IOP correlated with the displacement of the LC, further strengthening the connection between IOP and LC displacement. Although these studies indicate a relationship between IOP and LC remodeling, Miki et al.,[35] were unable to find a significant difference in IOP between eyes with and without LC defects. Therefore the correlation between IOP and LC remodeling needs to be better investigated, especially with respect to the microstructure of the LC. Such discrepancies may stem from differences in the patient populations used in these studies and/or the individual variability within these populations.

Recent studies have pointed out that mean IOP fails to capture the IOP fluctuations within short periods, which may be more informative for the IOP-glaucoma relationship. For example, larger IOP fluctuations can be found in aged and glaucomatous eyes likely due to the stiffening of the sclera.[36–39] Downs et al.[40] utilized an implantable telemetric pressure transducer in a nonhuman primate model to measure IOP during the day. They reported that IOP fluctuates during the day but the profile is not repeatable from day to day within the same animals.[40] The period of IOP fluctuations is found to range between 4 and 15 seconds, which, along with mean IOP, varies with body position.[41] The transient IOP fluctuations (IOP minus IOP baseline) are positively correlated with the ocular perfusion pressure (mean blood pressure minus IOP).[42] Future studies to correlate IOP and LC remodeling should take into consideration that mean IOPs are not the same as IOP fluctuations.

## 3.2. Chronic Morphologic Changes of the Lamina Cribrosa

In response to IOP, the LC has decreased thickness with increased depth, compared with age-matched controls.[13,43] LC depth is influenced by the central retinal vessel trunk location, age, untreated IOP, and retinal nerve fiber layer thickness.[44] The mean central thickness of the LC ranges from 260 to 490 μm.[45] The ocular axial length of the eye shortens and scleral thickness increases with age in normal subjects, and the magnitude of glaucomatous LC displacement is associated with age, indicating that the changes found with age are also seen in glaucomatous tissues.[46,47] Another study found a significant increase in LC depth and scleral thickness

associated with increasing age in eyes from African donors, whereas age was not significantly associated in eyes from European donors.[41] In this study, African LCs and peripapillary sclerae were significantly thinner than tissues from European donors, suggesting there is a greater age-related remodeling in these tissues in patients of African donors compared with European donors.[41] Consistently, Fazio and colleagues[34] demonstrated that patients of African descent present with greater acute posterior bowing of the LC in response to acute elevation of IOP compared with European patients. There is also an association between increased baseline anterior lamina depth and male sex.[48] LCs of women are significantly thinner than LCs of men in the ONH,[41] which may correlate to a higher prevalence of POAG in women.[49–51] However, other researchers have reported contradicting results that the sex hormones in postmenopausal women may protect the ONH from POAG development.[52,53]

In addition to depth and thickness, studies have also found that glaucomatous LCs display increased posterior curvature when compared with normal, age-matched controls.[13,43,44,54–57] In a study, Lee et al.[29] showed that patients with normal tension glaucoma have flat LCs and lower IOPs, indicating a connection between increased LC curvature and increased IOP. Other studies have investigated changes to the curvature of the LC after trabeculectomy and found that this curvature decreased after lowering IOP, resulting in sustained flattening and shallowing of the LC.[30,32]

## 3.3. Deformation and Displacement of the Optic Nerve Head in Response to Acute Intraocular Pressure Elevation

Several studies have found that compared with normal, age-matched controls, glaucomatous LCs display posterior displacement or bowing.[13,43,44,54–57] In contrast, Wu et al. showed that glaucomatous LCs can be displaced both posteriorly and anteriorly.[46,57] Therefore there is a possibility that different mechanisms of remodeling are present within glaucomatous eyes that are not dependent on elevated IOP. The posterior LC displacement commonly found in patients with glaucoma occurred most frequently in the preperimetric glaucoma and mild-to-moderate stages of glaucoma.[55] The LC does not consistently displace posteriorly when IOP increases.[18,58,59] Lower cup-to-disc ratio and larger IOP reduction after trabeculectomy in patients with advanced POAG are correlated with larger LC displacement and improving visual field tests.[33]

Younger patients with advanced POAG show better visual field improvement after trabeculectomy and larger LC displacement than older counterparts.[33]

## 3.4. Stiffness of the Peripapillary Sclera and Lamina Cribrosa

Stiffness, defined as the ratio of stress to strain, describes how compliant a tissue is in response to mechanical load. The stiffness of the sclera surrounding the ONH increases more rapidly with age in African donors when compared with European donors, and this may be the reason why higher prevalence of glaucoma is found in aged patients and in those of African descent.[60] In our studies, we found that the relationship between tensile strain and pressure was significant for eyes of Hispanic patients, whereas it was not significant for those of African patients, suggesting the mechanism of increased prevalence may be different between these two populations.[61] When we compared the peak shear strain in the superior quadrant of the LC, the numbers were higher in those of European descent than in those of African descent and Hispanic ethnicity.[62] The regional strain heterogeneity was detected in the LCs of African and Hispanic donors while it was absent in those of European donors.[62] Many of the studies have correlated aging with increased tissue stiffness in glaucoma.[36,63] In human eyes with glaucoma, the sclera is stiffer by in vivo indirect clinical measurement.[64] Likewise, experimental IOP elevation in monkeys produces a stiffer mechanical response in the sclera.[38,39] However, whether increased stiffness is a protective response or the cause of damage is still open for debate. If increased stiffness is indeed a protective response, would sclera stiffening help improve clinical outcomes? Future research addressing these questions may develop new additive therapies to treat POAG.

## 3.5. Summary

Overall, changes in LC structure and position are present in patients with progressive glaucoma, and it is unclear whether they are the result of pressure damage or a phenotype of patients susceptible to glaucoma. To better understand the timeline for LC remodeling and its relation to disease progression, Park et al.[55] investigated differences in the LCs of different stages of treated glaucoma using optical coherence tomography and found LC displacement occurred most frequently in mild-to-moderate states of glaucoma. These results indicated that the remodeling occurs earlier in glaucoma, with changes being seen in patients before visual field defects. Another study found that the posterior movement of the glaucomatous LC occurred at the same frequency as thinning of the neuroretinal parameter, further

suggesting that remodeling of the LC and glaucoma pathogenesis occurs in a sequential, interdependent manner.[65] Additional details on the biomechanics of the ONH in glaucoma can also be found in several review articles on the topic.[66−69]

## 4. MECHANOBIOLOGY OF THE OPTIC NERVE HEAD IN GLAUCOMA

Studies have suggested ECM remodeling is initiated by specific signaling molecules including TGFs and proteases, in addition to oxidative stress and mechanical stretch, which are the key factors in the development of glaucoma. We review these factors and the relevant molecular mechanisms in this section. Cellular senescence[70] and aging[36,63,71−75] are also involved; however, these are not discussed here.

### 4.1. Transforming Growth Factor β Signaling Pathway

TGFβ plays a critical role in tissue remodeling. Plasma-derived TGFβ1, whose concentration in plasma is a magnitude higher than that in the aqueous humor, might contribute to elevated IOP.[76,77] In vitro, high IOP induces TGFβ1 expression in trabecular meshwork cells and ONH astrocytes.[78] The concentration of the most well-studied growth factor in glaucoma, TGFβ2, is elevated in the aqueous humor, trabecular meshwork, and ONH of POAG tissues,[25,79,80,] which may be mediated through the connective tissue growth factor (CTGF).[78] In fact, an increase of TGFβ3 levels in the aqueous humor has also been reported.[81] Stimuli such as TGFβ1,[82,83] TGFβ2,[79] hypoxia,[84] oxidative stress,[85] and cyclic stretch[82,83] can increase fibrotic gene expression in LC cells. Zode et al.[79] have shown that Smad2 or Smad3 small interfering RNA significantly inhibits TGFβ2-induced production of fibronectin and PAI-1 in ONH astrocytes (Fig. 1.3.1).

The prevalence of POAG is three times higher in people of African descent than in those of European descents.[86,87] Hence, investigating the differences between the ONHs of patients from African and European descent, and exploring the cellular differences between these tissues, could help better understand the mechanisms behind these observed anatomic changes. In one study, Lukas et al.[88] used microarrays to identify genes differentially expressed in primary cultures of POAG ONH astrocytes from African American and Caucasian donors, finding genes associated with promoting cell motility and migration, regulating cell adhesion, and neural degeneration-based structural tissue changes, such as *MYLK, TGFBR2, RAC2,* and *VCAN.* Another study

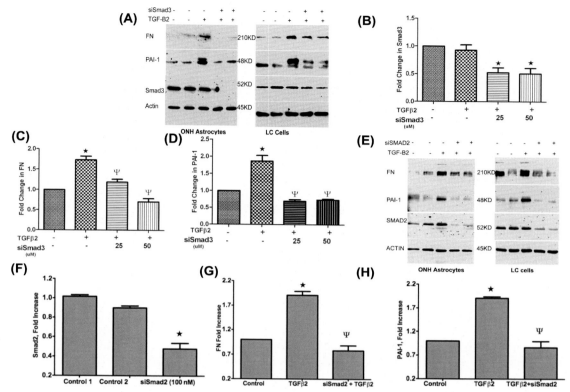

FIG. 1.3.1 Effect of Smad2 or Smad3 small interfering RNA (siRNA) on transforming growth factor (TGF) β2 stimulation of fibronectin (FN) and plasminogen activator inhibitor (PAI) 1 in optic nerve head (ONH) astrocytes and LC cells. **(A)** ONH astrocytes and lamina cribrosa (LC) cells were treated with nontargeting siRNA (lane 1), RNA-induced silencing complex (RISC)-free siRNA (lane 2), TGFβ2 only (lane 3), low concentration of Smad3 siRNA with TGFβ2 (lane 4), and high concentration of Smad3 siRNA with TGFβ2 (lane 5). Western blotting of FN, PAI-1, and Smad3 was performed for protein expression analysis. **(B)** Quantification of Western blotting confirmed decrease of Smad3 protein level in both Smad3 siRNA treated groups in ONH astrocytes. **(C, D)** Quantification of Western blotting confirmed Smad3 siRNA treatment significantly inhibits TGFβ2-induced FN or PAI-1 production in ONH astrocytes. **(E)** Similarly, ONH astrocytes and LC cells were treated with nontargeting siRNA (lane 1), RISC-free siRNA (lane 2), TGFβ2 only (lane 3), and Smad2 siRNA with TGFβ2 (lane 4—5). Western blotting was performed for protein expression analysis. **(F)** Quantification of Western blotting confirmed that Smad2 siRNA significantly decreases Smad2 protein levels. **(G, H)** Quantification of Western blotting confirmed Smad2 siRNA significantly inhibits TGFβ2-induced FN or PAI-1 production in ONH astrocytes. (Image credit Zode GS, Sethi A, Brun-Zinkernagel AM, Chang IF, Clark AF, Wordinger RJ. Transforming growth factor-beta2 increases extracellular matrix proteins in optic nerve head cells via activation of the Smad signaling pathway. *Mol Vis.* 2011;17:1745—1758.)

investigated the cyclic AMP (cAMP) levels in African American and Caucasian donor astrocytes and found that although basal levels were similar, intracellular cAMP expression levels increased in the astrocytes of African American donors compared with those of Caucasian donors after elevating hydrostatic pressure for extended periods.[89] Such results indicate that $Ca^{2+}$ signaling may differ between these populations, which

could impact the susceptibility of African American patients to glaucoma, compared to Caucasian patients. Suthanthiran et al.[90] previously reported that TGFβ1 is associated with renal insufficiency and that a higher concentration of TGFβ1 protein was detected in black patients with end-stage renal disease when compared with the expression in whites. The most commonly studied pathway involved in glaucomatous changes in ECM is the TGFβ pathway, which is highly associated with changes in IOP. As with elevated IOP, glaucomatous tissues show an increased expression of TGFβ. Based on a signaling pathway analysis algorithm, gene expression profiles are very similar between glaucomatous LC and TGFβ-treated LC, suggesting an elevated level of TGFβ in glaucomatous tissues could lead to activated profibrotic pathways, including the STAT (signal transducer and activator of transcription) pathway, potentially making the LC more susceptible to elevated IOP.[11] Regardless, it is clear that TGFβ-mediated gene expression is dysregulated in the ECM of the ONH in glaucoma.

Thrombospondins, which are known to activate the TGFβ signaling pathway, stimulate fibroblast proliferation in the cornea, induce myofibroblast transition, and alter the ECM production of the cornea.[91–93] When compared with normal LC cells, reduced gene expression of thrombospondin 1 and TGFβ has been demonstrated in POAG LC cells.[94] TGFβ pathways also interact with other molecules. For instance, TGFβ2 significantly increases pro-MMP2, rather than active MMP2, in human trabecular meshwork cell cultures, but active MMP2, MMP9, or MT1-MMP (also known as MMP14) are not influenced.[95] With PAI-1 neutralizing antibody, TGFβ2 increases active MMP2 expression, suggesting that TGFβ2-induced PAI-1 expression decreases MMP2 activity and matrix accumulation in glaucomatous eyes.[95] Steely et al.[96] compared human trabecular meshwork and LC cells and found that the protein fingerprints in these two cell types are the most similar among all ocular cells compared, suggesting a close link between the two tissues in POAG pathogenesis. In a rat glaucoma model, increased TGFβ2 deposition was colocalized with increased MMP1, tissue inhibitor of metalloproteinase (TIMP) 1, and collagen type I.[97] The involvement of MMPs and TIMPs is discussed in the following section.

BMPs also belong to the TGFβ superfamily. Studies suggest that the induction of gremlin, an extracellular BMP antagonist, is caused by the upregulation of TGFβ2.[78,98] Gremlin inhibits BMP ligand-receptor binding and the subsequent downstream signaling.[78,98] TGFβ2 primarily acts through CTGF and further increases fibronectin and αSMA production[99]; interestingly, TGFβ2 induces gremlin expression, which inhibits BMP7 activation, and knockdown of Smad7 inhibits the antagonism of BMP7 on TGFβ2 signaling,[100] suggesting that there is a negative feedback loop to modulate TGFβ2 levels.

Gremlin, encoded by GREM1, exists in both secreted and cell-membrane-associated forms with a molecular weight of 21 kDa.[101] Gremlin acts as a BMP antagonist

by binding directly to BMP2, BMP4, and BMP7[101] and therefore inhibiting ligand-receptor binding and downstream signaling. More specifically, TGFβ2 significantly induces fibronectin production and BMP4 blocks this induction in human trabecular meshwork cells; when these cells are treated with recombinant gremlin, it blocks the negative effect of BMP4 on TGFβ2-induced fibronectin production.[102] These studies suggest that elevated expression of gremlin inhibits BMP4 antagonism of TGFβ2 and leads to ECM production and elevated IOP.[78,98]

As mentioned earlier, TGFβ2 primarily acts through CTGF and also further increases fibronectin and αSMA production.[99] CTGF was discovered by Bradham et al.[103] in 1991 as a protein secreted by human umbilical vascular endothelial cells. CTGF, also known as CCN2, is a member of the CCN superfamily (CYR61, CTGF, NOV). The level of CTGF was found to be significantly higher in the aqueous humor of glaucomatous patients than in those without glaucoma.[104] In trabecular meshwork cell cultures, CTGF stimulation induces fibrillin-1 production through the activation of MAPK and JNK pathways.[105] When treated with TGFβ2, CTGF expression is increased in trabecular meshwork cell cultures.[106,107] Knockdown of CTGF via small interfering RNAs prevents TGFβ2-induced increase in fibronectin synthesis.[108] Interestingly, CTGF knockdown does not affect the TGFβ2-regulated level of tissue transglutaminase,[80] a key enzyme that catalyzes posttranslational modification of matrix protiens.[109] Tissue transglutaminases levels were found to be increased in glaucomatous samples,[110] and this makes matrix proteins more resistant to degradation. This suggests that TGFβ2-regulated matrix production has both CTGF-dependent and CTGF-independent pathways.

Studies have also shown that transglutaminase level is elevated in glaucomatous samples[110] and lysyl oxidase level is increased after TGFβ2 treatment.[111] These enzymes covalently cross-link ECM proteins and make the matrix more rigid. In our laboratory, we have found that TGFβ2 significantly induces collagen deposition in LC cells on stiff scaffolds when compared with soft scaffolds (data not shown). Understanding the roles of these proteins in glaucoma will provide information for the development of therapeutic targets in glaucoma. There are additional resources available that further detail the biology of TGFβ-mediated ECM remodeling in glaucoma.[78,94,98,112,113]

## 4.2. Proteases

MMPs are enzymes that play a primary role in ECM maintenance and degradation. Among all MMPs, MMP1, MMP2, MMP3, and MMP14 are more abundant in POAG versus normal LC tissues.[114] These four MMPs, along with MMP9, are also constitutively secreted by trabecular meshwork cells and upregulated during mechanical stretch.[115,116] MMP1 also permits migration of astrocytes throughout the LC ECM and into optic nerve bundles, contributing to matrix degradation around the axons and potentially impairing neuronal survival.[21] Studies have also observed changes in MMP levels within glaucomatous tissues, with higher expression of MMP1 and MT1-MMP (also known as MMP14) found in POAG laminar and postlaminar ONH.[21] Increased expression of MMP1 and MMP14 was also found in the ONH of a monkey model of experimental glaucoma, in which IOP is chronically elevated.[117] In this study, the changes in expression were found specifically in reactive astrocytes of the ONH, and the elevated MMP1 expression in glaucomatous ONHs was also localized to reactive astrocytes and axons.[21,117] The authors therefore concluded that the elevated expression of these MMPs was specific to tissue remodeling from elevated IOP and not a secondary loss of axons.[117] In our laboratory, we have found that MMP2 production is significantly increased in normal LC cells following TGFβ2 treatment, whereas in POAG LC cells, MMP2 production remains unaltered (data not shown).

TIMPs regulate the activity of MMPs. The level of endogenous MMP2 was significantly decreased in aqueous samples of glaucomatous patients but the ratio of MMP2 to TIMP2 was decreased, suggesting an excess of TIMP2 over MMP2 and abnormal matrix accumulation in glaucoma pathogenesis.[118,119] This was confirmed by other studies showing that TIMP2 protein levels are significantly elevated in glaucoma aqueous humor samples compared with cataract controls[120] and that concentrations of TIMP1, TIMP2, and TIMP4 are significantly increased in aqueous humor from patients with POAG.[119,121] When compared with primary angle-closure glaucoma and normal controls, the ratios of MMP2/TIMP2, MMP2/TIMP1, and (MMP2+MMP3)/(TIMP1+TIMP2) in POAG are lower than those in primary angle-closure glaucoma but higher than those in normal controls.[121] In absolute angle-closure glaucoma, MMP9 was found in proliferating glial cells surrounding the optic nerve axons in most of the glaucomatous eyes, whereas TIMP1 was only found in one-third of glaucomatous eyes.[122] Interestingly, in a rat glaucoma model, the loss of laminin and increase of TIMP1 is highly correlated with the integral of IOP elevation over time (Fig. 1.3.2).[97] In addition to TIMP1 and TIMP2, a stoichiometric analysis showed an overbalance of TIMPs over MMPs in

POAG, especially of TIMP4.[123] In the conjunctival stroma (a connective tissue that attaches to the tarsal plate and loosely arranged over the globe), using immunohistochemistry, TIMP3 was found to be significantly increased in POAG compared with control. The expression of MMPs and TIMPs is also increased in the conjunctiva of patients with POAG treated with topical antiglaucoma drops, suggesting these drops alter the ECM remodeling of the conjunctiva in POAG eyes and may affect the wound healing process and outcomes of glaucoma surgery.[124] Table 1.3.1 summarizes several studies reporting the protein expression of MMPs and TIMPs using enzyme-linked immunosorbent assays, immunohistochemistry, or Western blotting.

PAI-1 is a potent inhibitor of the activation of MMPs.[95] TGFβ2 causes an increase in the expression of PAI-1 that prevents activation of a number of MMPs via inhibition of tissue plasminogen activator and subsequently plasmin, thus leading to further increases in ECM deposition.[11,125] This is consistent with numerous studies linking a downregulation of *SERPINE1* (the gene that encodes PAI-1) with POAG.[126–128] For example, *SERPINE1* was downregulated in glaucomatous versus normal ONH astrocytes as well as in normal astrocytes exposed to elevated IOP.[126] Another study evaluating single nucleotide polymorphisms associated with POAG and metabolic syndrome-associated genes in patients with POAG found a significantly different single nucleotide polymorphism in *SERPINE1* in POAG compared to controls, translating to a missense expression in patients with POAG. Therefore patients with POAG potentially produce inactive PAI-1 protein. Our laboratory's evidence (data not shown) shows that dysregulation of TGFβ2-mediated MMP2 in POAG versus normal LC cells may, therefore, be mediated by PAI-1's interaction with the plasmin MMP system.

In a preliminary experiment, we have identified genes that are differentially expressed when comparing astrocytes seeded on decellularized LC tissues of African, European, and Hispanic ethnicities (Fig. 1.3.3). Interestingly, *GREM1* (gremlin), *PLG* (plasminogen), and *CSPG5* (chondroitin sulfate proteoglycan) are dysregulated in astrocytes seeded on LCs of African and Hispanic origin compared to those seeded on LCs of European origin. It is known that gremlin blocks BMP4, which is an inhibitor of TGFβ-mediated ECM synthesis,[102] and the level of gremlin protein is significantly increased in human glaucomatous ONH tissues.[129,130] On the other hand, plasminogen, once cleaved to become plasmin, can directly bind to the ECM, degrade

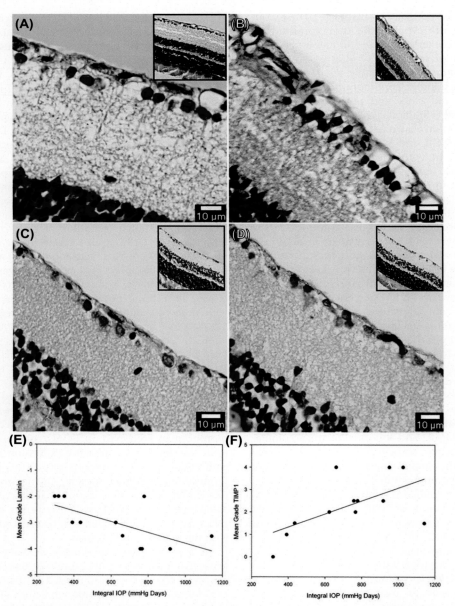

FIG. 1.3.2 Immunohistochemistry of laminin or tissue inhibitor of metalloproteinase (TIMP) 1 in control and glaucomatous eyes in a rat glaucoma model. Immunostaining demonstrated a decrease in laminin levels in glaucomatous eyes **(C)** compared with the control **(A)**. Immunostaining demonstrated an increase in TIMP1 in all glaucomatous eyes **(D)** compared with the control **(B)**. The loss of laminin **(E)** or increased TIMP1 expression **(F)** is significantly correlated with intraocular pressure (IOP) integral. (Image credit Guo L, Moss SE, Alexander RA, Ali RR, Fitzke FW, Cordeiro MF. Retinal ganglion cell apoptosis in glaucoma is related to intraocular pressure and IOP-induced effects on extracellular matrix. *Invest Ophthalmol Vis Sci*. 2005;46(1): 175–182.)

**TABLE 1.3.1**

**MMP and TIMP Protein Expression via Enzyme-Linked Immunosorbent Assay, Immunohistochemistry, or Western Blot in the Normal Versus Glaucomatous Optic Nerves.**

| | |
|---|---|
| MMP1 | • MMP1 detected and colocalized with GFAP positive astrocytes in the laminar and postlaminar region of the ONHs in POAG[21]<br>• Increased MMP1 immunoreactivity in the ONH of monkey eyes with experimental glaucoma[117]<br>• Similar intensity of MMP1 immunostaining between normal and POAG eyes[154]<br>• Increased MMP1 protein concentration in aqueous humor from patients with POAG when compared with controls[155] |
| MMP2 | • Similar MMP2 level between normal and POAG samples[21]<br>• MMP2 unchanged in the ONH of monkey eyes with experimental glaucoma[117]<br>• Increased intensity of MMP2 immunostaining in prelaminar and laminar regions of POAG eyes when compared with normal eyes[154]<br>• Increased MMP2 activity detected in tears using gelatin zymography in POAG when compared to controls[156] |
| MMP3 | • No MMP3 detected in astrocyte-located regions but detected in blood vessels throughout the region[21]<br>• Increased MMP3 immunoreactivity in the ONH at the site of transection in monkey eyes with experimental glaucoma[117]<br>• Increased intensity of MMP3 immunostaining in all regions of ONHs from POAG donors compared with normal ONH[154] |
| MMP9 | • No MMP9 detected in the LC and postlaminar region of ONHs[21]<br>• Increased MMP9 immunoreactivity in the ONH at the site of transection in monkey eyes with experimental glaucoma[117]<br>• Increased MMP9 protein concentration in aqueous humor from patients with POAG when compared with controls[155]<br>• Increased MMP9 activity detected in tears using gelatin zymography in POAG when compared with controls[156]<br>• Increased MMP9 expression in POAG confirmed via immunohistochemistry[156] |
| MMP7, MMP12, MMP14 | • No MMP7 or MMP12 detected in the LC and postlaminar region of ONHs[21]<br>• Increased MMP14 immunoreactivity and localized with reactive astrocytes in the LC and postlaminar region of ONHs in POAG[21]<br>• Increased MMP12 protein concentration in aqueous humor from patients with POAG when compared with controls[155] |
| TIMP1, TIMP2 | • Similar TIMP1 and TIMP2 levels between normal and POAG samples[21]<br>• Similar TIMP1 protein concentration in aqueous humor from patients with POAG when compared to controls[155]<br>• Decreased TIMP1 and TIMP2 immunoreactivity in monkey ONHs from eyes with experimental glaucoma[117] |

*GFAP*, glial fibrillary acidic protein; *LC*, lamina cribrosa; *MMP*, matrix metalloproteinase; *ONH*, optic nerve head; *POAG*, primary open-angle glaucoma; *TIMP*, tissue inhibitor of metalloproteinase.

multiple ECM proteins, and activate MMP2, MMP9, and MMP14.[95] Dysregulation of plasminogen in this experiment suggests that the plasminogen/plasmin proteolytic system may be differentially altered in the LC of those at increased risk of POAG. Finally, chondroitin sulfate proteoglycan, encoded by *CSPG5*, may be important in TGFβ-mediated matrix production in astrocytes.[131] Future work will need to be done to see if the genes dysregulated in glaucoma coincide with those in at-risk groups for glaucomatous damage.

### 4.3. Hypoxia and Oxidative Stress

Several studies have suggested oxidative stress is associated with the development of glaucoma. Increased immunoreactivity of hypoxia inducible factor 1 alpha was found in glaucomatous ONHs when compared to control

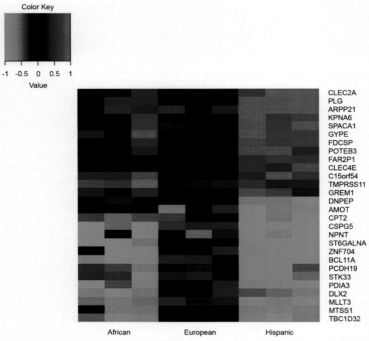

**FIG. 1.3.3** Gene microarray of human astrocytes (single donor) seeded in decellularized lamina cribrosa tissues of African, European, and Hispanic donors ($n = 3$ donors each). Those at higher risk for developing POAG (African, Hispanic) show a similar profile of dysregulated genes compared to those at lower risk (European).

eyes.[12] Some in vitro studies utilized hypoxic stress or hydrogen peroxide as an insult to induce cellular response or cellular damage. For example, hypoxic stress (1% $O_2$ for 24 hours) downregulates genes involved in apoptosis (BCL2 interacting protein 3, *BNIP3*), ECM remodeling (macrophage migration inhibitory factor, *MIF*; discoidin domain receptor family member 1, *DDR1*; insulin like growth factor 2 receptor, *IGF2R*), mitochondrion (cytochrome P450 family 1 subfamily B member 1, *CYP1B1*), and angiogenesis (vascular endothelial growth factor A, *VEGFA*) in human LC cells.[84] When treated with hydrogen peroxide, LC cells significantly increased profibrotic gene expression (fibronectin, *FN1*; fibrillin-1, *FBN1*; collagen type I, *COL1A1*; and αSMA, *ACTA2*).[132] Similarly, treatment with hydrogen peroxide for 1 hour increases TGFβ2 secretion in human astrocyte monolayers.[133]

Elevated intracellular calcium levels and decreased mitochondrial membrane potential are found in glaucomatous LC cells compared to normal LC cells.[134] In the same study, the researchers also found increased intracellular reactive oxygen species production and lower gene expression of antioxidant enzymes

(*AKR1C1* and *GCLC*) in glaucomatous LC cells, suggesting a decrease in antioxidant activity in glaucoma.[134] Different from LC cells, rat RGCs are resistant to cell death in response to reactive oxygen species generating systems when compared to other retinal cells, suggesting the presence of endogenous peroxidases protecting RGCs from axonal damage.[135] This protective response may be linked to an increased inflammatory response. Tezel et al.[136] found that reactive oxygen species increase T-cell proliferation and tumor necrosis factor α secretion in rat glial cells. Taken together, the upregulation of hypoxia-inducible genes or antioxidant response genes, alterations to antioxidant enzymes, and increased production of superoxide are indicators of the presence of hypoxia.[137]

Clinically, hypoxia is associated with reduced blood flow and reduced perfusion pressure in the ONH in patients with POAG.[138] Another indicator of glaucomatous damage is optic disc hemorrhages, which have been associated with disease progression.[139] With glaucoma, disc hemorrhages spatially corresponded with the peripheral placement of the LC and those that are colocalized with focal LC defects tended to be larger and more

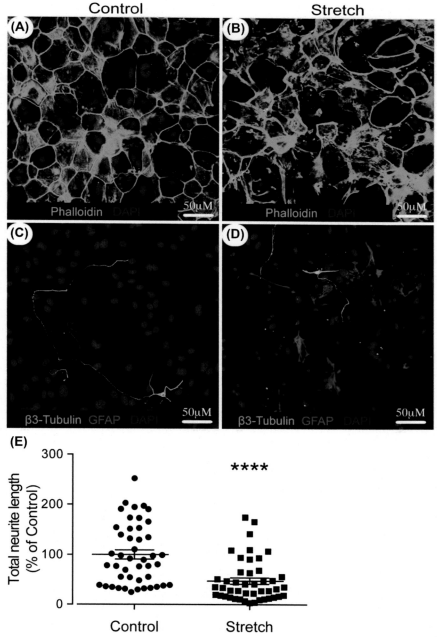

FIG. 1.3.4 Mechanical stretch induces reactive astrogliosis and inhibits neurite outgrowth. **(A, B)** Immunostaining of phalloidin (F-actin; *green*) and DAPI (4′,6-diamidino-2-phenylindole) (cell nuclei; *blue*) for control astrocytes or astrocytes exposed to mechanical stretch for 24 hours. The morphology is changed in the stretch group. **(C, D)** Immunostaining of β3-tubulin (neuron; *green*), glial fibrillary acidic protein (GFAP; reactive astrocyte; *red*), and DAPI (cell nuclei; *blue*) for mouse neurons plating on top of control astrocytes or astrocytes exposed to mechanical stretch for 24 hours. Astrocytes become reactive and neurite outgrowth is reduced in the stretch group. **(E)** Quantification of total neurite length confirms impaired neurite outgrowth in the stretch group. (Image credit Berretta A, Gowing EK, Jasoni CL, Clarkson AN. Sonic hedgehog stimulates neurite outgrowth in a mechanical stretch model of reactive-astrogliosis. *Sci Rep.* 2016;6:21896.)

proximally located.[140,141] Such studies indicate remodeling of the LC may affect the size and location of disc hemorrhages, further linking LC changes to the progression of glaucoma.

## 4.4. Mechanical Stretch

As described in a recent review article, mechanical stress is a measure of the load applied to a tissue, while strain is a measure of the local deformation in a tissue.[142] The material properties refer to the ability to resist deformation under applied load and therefore relate stress to strain.[142] Kirwan et al. have shown that cyclic mechanical stretch of LC cells induces increased MMP activity and changes in *TGFB1* gene expression in vitro.[83,143,144] According to Quill et al.,[145] 20% mechanical stretch significantly increases gene expression of *VEGFA*, *TIMP3*, *MMP2*, *MMP14*, *COL1A1*, *COL5A1*, *COL6A3*, *COL18A1*, *TGFB1*, and *TGFBR1* when compared to 3% or 12.5% stretch in LC cells. In a subsequent study, they exposed LC cell monolayers to cyclic mechanical strain (15% strain at 1 Hz) compared to nothing (0% strain) for 24 hours and found the potential involvement of calcium influx in the initiation of matrix remodeling.[146] In astrocytes, the level of calcium ions increases in intracellular storage during signal transmission, suggesting calcium channel blockers may attenuate disease progression in glaucoma.[16]

To control mechanical stretch, BioFlex culture plates were utilized to mechanically traumatize astrocyte monolayers using triangular stretch for 24 hours and found that mechanical stretch induces reactive astrogliosis and inhibits neurite outgrowth (Fig. 1.3.4).[147] When cyclic mechanical stretch (15% at 1 Hz for 24 hours) was applied to trabecular meshwork cells to mimic

ocular pulse fluctuation, caveolae responded to mechanical stretch in outflow pathway cells, suggesting *CAV1* (gene encoded caveolin-1 protein) may increase the risk of glaucoma pathogenesis.[148] This level of mechanical stretch also induces release of vascular endothelial growth factor to the conditioned medium, affecting the permeability of Schlemm's canal endothelium and modulation of outflow facility in the trabecular meshwork.[149] When increasing the mechanical stretch to 48 hours, *ADAMTS4* (gene encoded aggrecanase-1) gene expression is significantly increased.[150] Both vascular endothelial growth factor and aggrecanase are catabolic markers in matrix remodeling.

## 4.5. Summary

ECM remodeling is initiated by complex cellular and molecular mechanisms. Molecules including TGFβ2, BMP4, gremlin, MMP2, plasminogen, and PAI-1 and their interactions with mechanical stretch all play critical roles in the balance of matrix production and degradation, contributing to the mechanical properties of ONH, as summarized in Fig. 1.3.5. Further details on the cellular and molecular biology of the ONH in glaucoma can also be found in several review articles.[78,98,112,113]

## 5. FUTURE PERSPECTIVE

In this chapter, we have described the structure and function of the ONH and discussed the effects of mechanical variables including IOP, displacement, compliance, and mechanical stretch on the development of glaucoma to provide a current synopsis of the mechanobiology of the ONH. It is clear, however, that the current state of

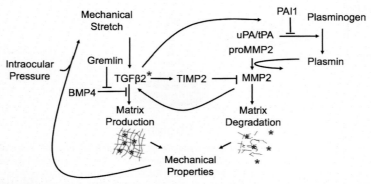

**FIG. 1.3.5** Diagram of how mechanical stretch, transforming growth factor (TGF) β2, matrix metalloproteinase (MMP) 2, gremlin, and plasminogen are involved in matrix production and degradation contributing to different mechanical properties observed in racioethnic groups. *BMP4*, bone morphogenetic protein 4; *PAI-1*: plasminogen activator inhibitor 1; *TIMP2*, tissue inhibitor of metalloproteinase 2; *uPA/tPA*, urokinase-type/tissue plasminogen activator.

knowledge in this field is limited and that future investigations are necessary to further tease out the mechanisms linking mechanics of the ONH to glaucoma initiation and progression.[151–153] For example, several specific questions remain of interest in this area. How does the ONH remodel in response to elevated IOP versus alterations in LC stiffness? What are the extra- and intracellular pathways that drive the mechanosensitivity and mechanotransduction of ONH astrocytes and LC cells? What is the link between ECM maintenance and RGC axonal degeneration? If axonal degeneration is a result of a dysregulated ECM in the ONH, would manipulating this process slow down axonal degeneration? Are there patient-specific differences in IOP fluctuation that play a role in the mechanosensitivity of ONH cells? Answers to these questions may lead to novel therapeutic targets to halt or slow the neurodegeneration in POAG.

## ACKNOWLEDGMENT

The authors would like to thank several eye banks for the donation of the tissues used in the generation of Fig. 1.3.3 (San Diego Eye Bank, Eversight, Alabama Eye Bank, Donor Network of Arizona). All eyes were received using an approved exempt IRB protocol at the University of Arizona. This research is supported by the National Eye Institute of the National Institutes of Health (NIH) under award number R01EY020890 to JPVG.

## REFERENCES

1. Kapetanakis VV, Chan MP, Foster PJ, Cook DG, Owen CG, Rudnicka AR. Global variations and time trends in the prevalence of primary open angle glaucoma (POAG): a systematic review and meta-analysis. *Br J Ophthalmol.* 2016;100(1):86–93.
2. Stamper RL, Wigginton SA, Higginbotham EJ. Primary drug treatment for glaucoma: beta-blockers versus other medications. *Surv Ophthalmol.* 2002;47(1):63–73.
3. Ito T, Ohguro H, Mamiya K, Ohguro I, Nakazawa M. Effects of antiglaucoma drops on MMP and TIMP balance in conjunctival and subconjunctival tissue. *Invest Ophthalmol Vis Sci.* 2006;47(3):823–830.
4. Weinreb RN. IOP and the risk of progression to glaucoma. *Graefes Arch Clin Exp Ophthalmol.* 2005; 243(6):511–512.
5. Hernandez MR, Andrzejewska WM, Neufeld AH. Changes in the extracellular matrix of the human optic nerve head in primary open-angle glaucoma. *Am J Ophthalmol.* 1990;109(2):180–188.
6. Hernandez MR, Ye H. Glaucoma: changes in extracellular matrix in the optic nerve head. *Ann Med.* 1993;25(4): 309–315.
7. Yang H, Ren R, Lockwood H, et al. The connective tissue components of optic nerve head cupping in monkey experimental glaucoma Part 1: global change. *Invest Ophthalmol Vis Sci.* 2015;56(13):7661–7678.
8. Munemasa Y, Kitaoka Y. Molecular mechanisms of retinal ganglion cell degeneration in glaucoma and future prospects for cell body and axonal protection. *Front Cell Neurosci.* 2012;6:60.
9. Leske MC. Factors for glaucoma progression and the effect of treatment. *Arch Ophthalmol.* 2003;121(1).
10. Coleman AL, Miglior S. Risk factors for glaucoma onset and progression. *Surv Ophthalmol.* 2008;53(suppl 1): S3–S10.
11. Zhavoronkov A, Izumchenko E, Kanherkar RR, et al. Profibrotic pathway activation in trabecular meshwork and lamina cribrosa is the main driving force of glaucoma. *Cell Cycle.* 2016;15(12):1643–1652.
12. Tezel G, Wax MB. Hypoxia-inducible factor 1alpha in the glaucomatous retina and optic nerve head. *Arch Ophthalmol.* 2004;122(9):1348–1356.
13. Abe RY, Gracitelli CP, Diniz-Filho A, Tatham AJ, Medeiros FA. Lamina cribrosa in glaucoma: diagnosis and monitoring. *Curr Ophthalmol Rep.* 2015;3(2):74–84.
14. Tovar-Vidales T, Wordinger RJ, Clark AF. Identification and localization of lamina cribrosa cells in the human optic nerve head. *Exp Eye Res.* 2016;147:94–97.
15. Downs JC, Girkin CA. Lamina cribrosa in glaucoma. *Curr Opin Ophthalmol.* 2017;28(2):113–119.
16. Sofroniew MV, Vinters HV. Astrocytes: biology and pathology. *Acta Neuropathol.* 2010;119(1):7–35.
17. Hernandez MR, Miao H, Lukas T. Astrocytes in glaucomatous optic neuropathy. *Prog Brain Res.* 2008;173: 353–373.
18. Yang H, Downs JC, Sigal IA, Roberts MD, Thompson H, Burgoyne CF. Deformation of the normal monkey optic nerve head connective tissue after acute IOP elevation within 3-D histomorphometric reconstructions. *Invest Ophthalmol Vis Sci.* 2009;50(12):5785–5799.
19. Hernandez MR, Igoe F, Neufeld AH. Cell culture of the human lamina cribrosa. *Invest Ophthalmol Vis Sci.* 1988; 29(1):78–89.
20. Lambert W, Agarwal R, Howe W, Clark AF, Wordinger RJ. Neurotrophin and neurotrophin receptor expression by cells of the human lamina cribrosa. *Invest Ophthalmol Vis Sci.* 2001;42(10):2315–2323.
21. Agapova OA, Ricard CS, Salvador-Silva M, Hernandez MR. Expression of matrix metalloproteinases and tissue inhibitors of metalloproteinases in human optic nerve head astrocytes. *Glia.* 2001;33(3):205–216.
22. Hernandez MR, Agapova OA, Yang P, Salvador-Silva M, Ricard CS, Aoi S. Differential gene expression in astrocytes from human normal and glaucomatous optic nerve head analyzed by cDNA microarray. *Glia.* 2002;38(1): 45–64.
23. Nikolskaya T, Nikolsky Y, Serebryiskaya T, et al. Network analysis of human glaucomatous optic nerve head astrocytes. *BMC Med Genomics.* 2009;2:24.

24. Luo C, Yang X, Kain AD, Powell DW, Kuehn MH, Tezel G. Glaucomatous tissue stress and the regulation of immune response through glial Toll-like receptor signaling. *Invest Ophthalmol Vis Sci.* 2010;51(11):5697–5707.

25. Clark AF. The cell and molecular biology of glaucoma: biomechanical factors in glaucoma. *Invest Ophthalmol Vis Sci.* 2012;53(5):2473–2475.

26. Howell GR, Libby RT, Jakobs TC, et al. Axons of retinal ganglion cells are insulted in the optic nerve early in DBA/2J glaucoma. *J Cell Biol.* 2007;179(7):1523–1537.

27. Morrison JC. Integrins in the optic nerve head: potential roles in glaucomatous optic neuropathy (an American Ophthalmological Society thesis). *Trans Am Ophthalmol Soc.* 2006;104:453–477.

28. Li L, Bian A, Cheng G, Zhou Q. Posterior displacement of the lamina cribrosa in normal-tension and high-tension glaucoma. *Acta Ophthalmol.* 2016;94(6):e492–e500.

29. Lee SH, Kim TW, Lee EJ, Girard MJA, Mari JM, Ritch R. Ocular and clinical characteristics associated with the extent of posterior lamina cribrosa curve in normal tension glaucoma. *Sci Rep.* 2018;8(1):961.

30. Kadziauskiene A, Jasinskiene E, Asoklis R, et al. Long-term shape, curvature, and depth changes of the lamina cribrosa after trabeculectomy. *Ophthalmology.* 2018;125(11):1729–1740.

31. Shin JW, Sung KR, Uhm KB, et al. Peripapillary microvascular improvement and lamina cribrosa depth reduction after trabeculectomy in primary open-angle glaucoma. *Invest Ophthalmol Vis Sci.* 2017;58(13):5993–5999.

32. Lee SH, Yu DA, Kim TW, Lee EJ, Girard MJ, Mari JM. Reduction of the lamina cribrosa curvature after trabeculectomy in glaucoma. *Invest Ophthalmol Vis Sci.* 2016;57(11):5006–5014.

33. Esfandiari H, Efatizadeh A, Hassanpour K, Doozandeh A, Yaseri M, Loewen NA. Factors associated with lamina cribrosa displacement after trabeculectomy measured by optical coherence tomography in advanced primary open-angle glaucoma. *Graefes Arch Clin Exp Ophthalmol.* 2018;256(12):2391–2398.

34. Fazio MA, Johnstone JK, Smith B, Wang L, Girkin CA. Displacement of the lamina cribrosa in response to acute intraocular pressure elevation in normal individuals of African and European descent. *Invest Ophthalmol Vis Sci.* 2016;57(7):3331–3339.

35. Miki A, Ikuno Y, Asai T, Usui S, Nishida K. Defects of the lamina cribrosa in high myopia and glaucoma. *PLoS One.* 2015;10(9):e0137909.

36. Albon J, Purslow PP, Karwatowski WS, Easty DL. Age related compliance of the lamina cribrosa in human eyes. *Br J Ophthalmol.* 2000;84(3):318–323.

37. Girard MJ, Suh JK, Bottlang M, Burgoyne CF, Downs JC. Scleral biomechanics in the aging monkey eye. *Invest Ophthalmol Vis Sci.* 2009;50(11):5226–5237.

38. Downs JC, Suh JK, Thomas KA, Bellezza AJ, Hart RT, Burgoyne CF. Viscoelastic material properties of the peripapillary sclera in normal and early-glaucoma monkey eyes. *Invest Ophthalmol Vis Sci.* 2005;46(2):540–546.

39. Girard MJ, Suh JK, Bottlang M, Burgoyne CF, Downs JC. Biomechanical changes in the sclera of monkey eyes exposed to chronic IOP elevations. *Invest Ophthalmol Vis Sci.* 2011;52(8):5656–5669.

40. Downs JC, Burgoyne CF, Seigfried WP, Reynaud JF, Strouthidis NG, Sallee V. 24-hour IOP telemetry in the nonhuman primate: implant system performance and initial characterization of IOP at multiple timescales. *Invest Ophthalmol Vis Sci.* 2011;52(10):7365–7375.

41. Turner DC, Samuels BC, Huisingh C, Girkin CA, Downs JC. The magnitude and time course of IOP change in response to body position change in nonhuman primates measured using continuous IOP telemetry. *Invest Ophthalmol Vis Sci.* 2017;58(14):6232–6240.

42. Markert JE, Jasien JV, Turner DC, Huisingh C, Girkin CA, Downs JC. IOP, IOP transient impulse, ocular perfusion pressure, and mean arterial pressure relationships in nonhuman primates instrumented with telemetry. *Invest Ophthalmol Vis Sci.* 2018;59(11):4496–4505.

43. Sawada Y, Hangai M, Murata K, Ishikawa M, Yoshitomi T. Lamina cribrosa depth variation measured by spectral-domain optical coherence tomography within and between four glaucomatous optic disc phenotypes. *Invest Ophthalmol Vis Sci.* 2015;56(10):5777–5784.

44. Oh BL, Lee EJ, Kim H, Girard MJ, Mari JM, Kim TW. Anterior lamina cribrosa surface depth in open-angle glaucoma: relationship with the position of the central retinal vessel trunk. *PLoS One.* 2016;11(6):e0158443.

45. Jonas JB, Holbach L. Central corneal thickness and thickness of the lamina cribrosa in human eyes. *Invest Ophthalmol Vis Sci.* 2005;46(4):1275–1279.

46. Wu Z, Xu G, Weinreb RN, Yu M, Leung CK. Optic nerve head deformation in glaucoma: a prospective analysis of optic nerve head surface and lamina cribrosa surface displacement. *Ophthalmology.* 2015;122(7):1317–1329.

47. Olsen TW, Edelhauser HF, Lim JI, Geroski DH. Human scleral permeability. Effects of age, cryotherapy, transscleral diode laser, and surgical thinning. *Invest Ophthalmol Vis Sci.* 1995;36(9):1893–1903.

48. Quigley H, Arora K, Idrees S, et al. Biomechanical responses of lamina cribrosa to intraocular pressure change assessed by optical coherence tomography in glaucoma eyes. *Invest Ophthalmol Vis Sci.* 2017;58(5):2566–2577.

49. Mitchell P, Smith W, Attebo K, Healey PR. Prevalence of open-angle glaucoma in Australia. *Ophthalmology.* 1996;103(10):1661–1669.

50. Hulsman CAA. Is open-angle glaucoma associated with early menopause?: the Rotterdam study. *Am J Epidemiol.* 2001;154(2):138–144.

51. Drance S, Anderson DR, Schulzer M. Risk factors for progression of visual field abnormalities in normal-tension glaucoma. *Am J Ophthalmol.* 2001;131(6):699–708.

52. Higginbotham EJ. Does sex matter in glaucoma? *Arch Ophthalmol.* 2004;122(3):374–375.

53. Vajaranant TS, Nayak S, Wilensky JT, Joslin CE. Gender and glaucoma: what we know and what we need to know. *Curr Opin Ophthalmol.* 2010;21(2):91–99.

54. Kim YW, Kim DW, Jeoung JW, Kim DM, Park KH. Peripheral lamina cribrosa depth in primary open-angle glaucoma: a swept-source optical coherence tomography study of lamina cribrosa. *Eye*. 2015;29(10):1368–1374.

55. Park SC, Brumm J, Furlanetto RL, et al. Lamina cribrosa depth in different stages of glaucoma. *Invest Ophthalmol Vis Sci*. 2015;56(3):2059–2064.

56. Tun TA, Thakku SG, Png O, et al. Shape changes of the anterior lamina cribrosa in normal, ocular hypertensive, and glaucomatous eyes following acute intraocular pressure elevation. *Invest Ophthalmol Vis Sci*. 2016;57(11):4869–4877.

57. Villarruel JM, Li XQ, Bach-Holm D, Hamann S. Anterior lamina cribrosa surface position in idiopathic intracranial hypertension and glaucoma. *Eur J Ophthalmol*. 2017;27(1):55–61.

58. Agoumi Y, Sharpe GP, Hutchison DM, Nicolela MT, Artes PH, Chauhan BC. Laminar and prelaminar tissue displacement during intraocular pressure elevation in glaucoma patients and healthy controls. *Ophthalmology*. 2011;118(1):52–59.

59. Strouthidis NG, Fortune B, Yang H, Sigal IA, Burgoyne CF. Effect of acute intraocular pressure elevation on the monkey optic nerve head as detected by spectral domain optical coherence tomography. *Invest Ophthalmol Vis Sci*. 2011;52(13):9431–9437.

60. Fazio MA, Grytz R, Morris JS, Bruno L, Girkin CA, Downs JC. Human scleral structural stiffness increases more rapidly with age in donors of African descent compared to donors of European descent. *Invest Ophthalmol Vis Sci*. 2014;55(11):7189–7198.

61. Tamimi EA, Pyne JD, Muli DK, et al. Racioethnic differences in human posterior scleral and optic nerve stump deformation. *Invest Ophthalmol Vis Sci*. 2017;58(10):4235–4246.

62. Behkam R, Kollech HG, Jana A, et al. Racioethnic differences in the biomechanical response of the lamina cribrosa. *Acta Biomater*. 2019;88:131–140.

63. Leung LK, Ko MW, Lam DC. Effect of age-stiffening tissues and intraocular pressure on optic nerve damages. *Mol Cell Biomech: MCB*. 2012;9(2):157–173.

64. Hommer A, Fuchsjager-Mayrl G, Resch H, Vass C, Garhofer G, Schmetterer L. Estimation of ocular rigidity based on measurement of pulse amplitude using pneumotonometry and fundus pulse using laser interferometry in glaucoma. *Invest Ophthalmol Vis Sci*. 2008;49(9):4046–4050.

65. Vianna JR, Lanoe VR, Quach J, et al. Serial changes in lamina cribrosa depth and neuroretinal parameters in glaucoma: impact of choroidal thickness. *Ophthalmology*. 2017;124(9):1392–1402.

66. Burgoyne CF, Morrison JC. The anatomy and pathophysiology of the optic nerve head in glaucoma. *J Glaucoma*. 2001;10(5 Suppl 1):S16–S18.

67. Burgoyne CF, Downs JC, Bellezza AJ, Suh JK, Hart RT. The optic nerve head as a biomechanical structure: a new paradigm for understanding the role of IOP-related stress and strain in the pathophysiology of glaucomatous optic nerve head damage. *Prog Retin Eye Res*. 2005;24(1):39–73.

68. Burgoyne CF, Downs JC. Premise and prediction-how optic nerve head biomechanics underlies the susceptibility and clinical behavior of the aged optic nerve head. *J Glaucoma*. 2008;17(4):318–328.

69. Burgoyne CF. A biomechanical paradigm for axonal insult within the optic nerve head in aging and glaucoma. *Exp Eye Res*. 2011;93(2):120–132.

70. Liton PB, Challa P, Stinnett S, Luna C, Epstein DL, Gonzalez P. Cellular senescence in the glaucomatous outflow pathway. *Exp Gerontol*. 2005;40(8–9):745–748.

71. Albon J, Karwatowski WS, Avery N, Easty DL, Duance VC. Changes in the collagenous matrix of the aging human lamina cribrosa. *Br J Ophthalmol*. 1995;79(4):368–375.

72. Albon J, Karwatowski WS, Easty DL, Sims TJ, Duance VC. Age related changes in the non-collagenous components of the extracellular matrix of the human lamina cribrosa. *Br J Ophthalmol*. 2000;84(3):311–317.

73. Hernandez MR, Luo XX, Andrzejewska W, Neufeld AH. Age-related changes in the extracellular matrix of the human optic nerve head. *Am J Ophthalmol*. 1989;107(5):476–484.

74. Kotecha A, Izadi S, Jeffery G. Age-related changes in the thickness of the human lamina cribrosa. *Br J Ophthalmol*. 2006;90(12):1531–1534.

75. Tezel G, Luo C, Yang X. Accelerated aging in glaucoma: immunohistochemical assessment of advanced glycation end products in the human retina and optic nerve head. *Invest Ophthalmol Vis Sci*. 2007;48(3):1201–1211.

76. Derhaschnig U, Shehata M, Herkner H, et al. Increased levels of transforming growth factor-beta1 in essential hypertension. *Am J Hypertens*. 2002;15(3):207–211.

77. Schlotzer-Schrehardt U, Zenkel M, Kuchle M, Sakai LY, Naumann GO. Role of transforming growth factor-beta1 and its latent form binding protein in pseudoexfoliation syndrome. *Exp Eye Res*. 2001;73(6):765–780.

78. Fuchshofer R, Tamm ER. The role of TGF-beta in the pathogenesis of primary open-angle glaucoma. *Cell Tissue Res*. 2012;347(1):279–290.

79. Zode GS, Sethi A, Brun-Zinkernagel AM, Chang IF, Clark AF, Wordinger RJ. Transforming growth factor-beta2 increases extracellular matrix proteins in optic nerve head cells via activation of the Smad signaling pathway. *Mol Vis*. 2011;17:1745–1758.

80. Tovar-Vidales T, Clark AF, Wordinger RJ. Transforming growth factor-beta2 utilizes the canonical Smad-signaling pathway to regulate tissue transglutaminase expression in human trabecular meshwork cells. *Exp Eye Res*. 2011;93(4):442–451.

81. Yoneda K, Nakano M, Mori K, Kinoshita S, Tashiro K. Disease-related quantitation of TGF-beta3 in human aqueous humor. *Growth Factors*. 2007;25(3):160–167.

82. Kirwan RP, Leonard MO, Murphy M, Clark AF, O'Brien CJ. Transforming growth factor-beta-regulated gene transcription and protein expression in human GFAP-negative lamina cribrosa cells. *Glia*. 2005;52(4):309–324.

83. Kirwan RP, Fenerty CH, Crean J, Wordinger RJ, Clark AF, O'Brien CJ. Influence of cyclical mechanical strain on extracellular matrix gene expression in human lamina cribrosa cells in vitro. *Mol Vis.* 2005;11:798–810.

84. Kirwan RP, Felice L, Clark AF, O'Brien CJ, Leonard MO. Hypoxia regulated gene transcription in human optic nerve lamina cribrosa cells in culture. *Invest Ophthalmol Vis Sci.* 2012;53(4):2243–2255.

85. Wallace MP, Stewart CE, Moseley MJ, Stephens DA, Fielder AR. Monitored Occlusion Treatment Amblyopia Study C, et al. Compliance with occlusion therapy for childhood amblyopia. *Invest Ophthalmol Vis Sci.* 2013; 54(9):6158–6166.

86. Ferris 3rd FL, Tielsch JM. Blindness and visual impairment: a public health issue for the future as well as today. *Arch Ophthalmol.* 2004;122(4):451–452.

87. Friedman DS, Wolfs RC, O'Colmain BJ, et al. Prevalence of open-angle glaucoma among adults in the United States. *Arch Ophthalmol.* 2004;122(4):532–538.

88. Lukas TJ, Miao H, Chen L, et al. Susceptibility to glaucoma: differential comparison of the astrocyte transcriptome from glaucomatous African American and Caucasian American donors. *Genome Biol.* 2008;9(7): R111.

89. Chen L, Lukas TJ, Hernandez MR. Hydrostatic pressure-dependent changes in cyclic AMP signaling in optic nerve head astrocytes from Caucasian and African American donors. *Mol Vis.* 2009;15:1664–1672.

90. Suthanthiran M, Khanna A, Cukran D, et al. Transforming growth factor-beta 1 hyperexpression in African American end-stage renal disease patients. *Kidney Int.* 1998;53(3):639–644.

91. Jester JV, Petroll WM, Cavanagh HD. Corneal stromal wound healing in refractive surgery: the role of myofibroblasts. *Prog Retin Eye Res.* 1999;18(3):311–356.

92. Young GD, Murphy-Ullrich JE. The tryptophan-rich motifs of the thrombospondin type 1 repeats bind VLAL motifs in the latent transforming growth factor-beta complex. *J Biol Chem.* 2004;279(46):47633–47642.

93. Bornstein P. Thrombospondins function as regulators of angiogenesis. *J Cell Commun Signal.* 2009;3(3–4): 189–200.

94. Wallace DM, O'Brien CJ. The role of lamina cribrosa cells in optic nerve head fibrosis in glaucoma. *Exp Eye Res.* 2016;142:102–109.

95. Fuchshofer R, Welge-Lussen U, Lutjen-Drecoll E. The effect of TGF-beta2 on human trabecular meshwork extracellular proteolytic system. *Exp Eye Res.* 2003;77(6): 757–765.

96. Steely Jr HT, English-Wright SL, Clark AF. The similarity of protein expression in trabecular meshwork and lamina cribrosa: implications for glaucoma. *Exp Eye Res.* 2000; 70(1):17–30.

97. Guo L, Moss SE, Alexander RA, Ali RR, Fitzke FW, Cordeiro MF. Retinal ganglion cell apoptosis in glaucoma is related to intraocular pressure and IOP-induced effects on extracellular matrix. *Invest Ophthalmol Vis Sci.* 2005;46(1):175–182.

98. Fuchshofer R. The pathogenic role of transforming growth factor-beta2 in glaucomatous damage to the optic nerve head. *Exp Eye Res.* 2011;93(2):165–169.

99. Tsai CC, Wu SB, Kau HC, Wei YH. Essential role of connective tissue growth factor (CTGF) in transforming growth factor-beta1 (TGF-beta1)-induced myofibroblast transdifferentiation from Graves' orbital fibroblasts. *Sci Rep.* 2018;8(1):7276.

100. Fuchshofer R, Stephan DA, Russell P, Tamm ER. Gene expression profiling of TGFbeta2- and/or BMP7-treated trabecular meshwork cells: identification of Smad7 as a critical inhibitor of TGF-beta2 signaling. *Exp Eye Res.* 2009;88(6):1020–1032.

101. Wordinger RJ, Zode G, Clark AF. Focus on molecules: gremlin. *Exp Eye Res.* 2008;87(2):78–79.

102. Wordinger RJ, Fleenor DL, Hellberg PE, et al. Effects of TGF-beta2, BMP-4, and gremlin in the trabecular meshwork: implications for glaucoma. *Invest Ophthalmol Vis Sci.* 2007;48(3):1191–1200.

103. Bradham DM, Igarashi A, Potter RL, Grotendorst GR. Connective tissue growth factor: a cysteine-rich mitogen secreted by human vascular endothelial cells is related to the SRC-induced immediate early gene product CEF-10. *J Cell Biol.* 1991;114(6):1285–1294.

104. Ho SL, Dogar GF, Wang J, et al. Elevated aqueous humour tissue inhibitor of matrix metalloproteinase-1 and connective tissue growth factor in pseudoexfoliation syndrome. *Br J Ophthalmol.* 2005;89(2):169–173.

105. Browne JG, Ho SL, Kane R, et al. Connective tissue growth factor is increased in pseudoexfoliation glaucoma. *Invest Ophthalmol Vis Sci.* 2011;52(6):3660–3666.

106. Fuchshofer R, Yu AH, Welge-Lussen U, Tamm ER. Bone morphogenetic protein-7 is an antagonist of transforming growth factor-beta2 in human trabecular meshwork cells. *Invest Ophthalmol Vis Sci.* 2007;48(2):715–726.

107. Bollinger KE, Crabb JS, Yuan X, Putliwala T, Clark AF, Crabb JW. Quantitative proteomics: TGFbeta(2) signaling in trabecular meshwork cells. *Invest Ophthalmol Vis Sci.* 2011;52(11):8287–8294.

108. Junglas B, Yu AH, Welge-Lussen U, Tamm ER, Fuchshofer R. Connective tissue growth factor induces extracellular matrix deposition in human trabecular meshwork cells. *Exp Eye Res.* 2009;88(6):1065–1075.

109. Folk JE, Finlayson JS. The epsilon-(gamma-glutamyl) lysine crosslink and the catalytic role of transglutaminases. *Adv Protein Chem.* 1977;31:1–133.

110. Tovar-Vidales T, Roque R, Clark AF, Wordinger RJ. Tissue transglutaminase expression and activity in normal and glaucomatous human trabecular meshwork cells and tissues. *Invest Ophthalmol Vis Sci.* 2008;49(2):622–628.

111. Sethi A, Mao W, Wordinger RJ, Clark AF. Transforming growth factor-beta induces extracellular matrix protein cross-linking lysyl oxidase (LOX) genes in human trabecular meshwork cells. *Invest Ophthalmol Vis Sci.* 2011; 52(8):5240–5250.

112. Paula JS, O'Brien C, Stamer WD. Life under pressure: the role of ocular cribriform cells in preventing glaucoma. *Exp Eye Res.* 2016;151:150–159.

113. Schneider M, Fuchshofer R. The role of astrocytes in optic nerve head fibrosis in glaucoma. *Exp Eye Res.* 2016;142: 49–55.

114. Yuan L, Neufeld AH. Activated microglia in the human glaucomatous optic nerve head. *J Neurosci Res.* 2001; 64(5):523–532.

115. Bradley JM, Kelley MJ, Zhu X, Anderssohn AM, Alexander JP, Acott TS. Effects of mechanical stretching on trabecular matrix metalloproteinases. *Invest Ophthalmol Vis Sci.* 2001;42(7):1505–1513.

116. Bradley JM, Kelley MJ, Rose A, Acott TS. Signaling pathways used in trabecular matrix metalloproteinase response to mechanical stretch. *Invest Ophthalmol Vis Sci.* 2003;44(12):5174–5181.

117. Agapova OA, Kaufman PL, Lucarelli MJ, Gabelt BT, Hernandez MR. Differential expression of matrix metalloproteinases in monkey eyes with experimental glaucoma or optic nerve transection. *Brain Res.* 2003; 967(1–2):132–143.

118. Schlotzer-Schrehardt U, Lommatzsch J, Kuchle M, Konstas AG, Naumann GO. Matrix metalloproteinases and their inhibitors in aqueous humor of patients with pseudoexfoliation syndrome/glaucoma and primary open-angle glaucoma. *Invest Ophthalmol Vis Sci.* 2003; 44(3):1117–1125.

119. Ashworth Briggs EL, Toh T, Eri R, Hewitt AW, Cook AL. TIMP1, TIMP2, and TIMP4 are increased in aqueous humor from primary open angle glaucoma patients. *Mol Vis.* 2015;21:1162–1172.

120. Maatta M, Tervahartiala T, Harju M, Airaksinen J, Autio-Harmainen H, Sorsa T. Matrix metalloproteinases and their tissue inhibitors in aqueous humor of patients with primary open-angle glaucoma, exfoliation syndrome, and exfoliation glaucoma. *J Glaucoma.* 2005; 14(1):64–69.

121. Nga AD, Yap SL, Samsudin A, Abdul-Rahman PS, Hashim OH, Mimiwati Z. Matrix metalloproteinases and tissue inhibitors of metalloproteinases in the aqueous humour of patients with primary angle closure glaucoma - a quantitative study. *BMC Ophthalmol.* 2014;14:33.

122. Zalewska R, Reszec J, Kisielewski W, Mariak Z. Metalloproteinase 9 and TIMP-1 expression in retina and optic nerve in absolute angle closure glaucoma. *Adv Med Sci.* 2016;61(1):6–10.

123. Fountoulakis N, Labiris G, Aristeidou A, et al. Tissue inhibitor of metalloproteinase 4 in aqueous humor of patients with primary open angle glaucoma, pseudoexfoliation syndrome and pseudoexfoliative glaucoma and its role in proteolysis imbalance. *BMC Ophthalmol.* 2013;13:69.

124. Helin-Toiviainen M, Ronkko S, Puustjarvi T, Rekonen P, Ollikainen M, Uusitalo H. Conjunctival matrix metalloproteinases and their inhibitors in glaucoma patients. *Acta Ophthalmol.* 2015;93(2):165–171.

125. Ghosh AK, Vaughan DE. PAI-1 in tissue fibrosis. *J Cell Physiol.* 2012;227(2):493–507.

126. Yang Y, Duan JZ, Di Y, Gui DM, Gao DW. Bioinformatics analysis of potential essential genes that response to the high intraocular pressure on astrocyte due to glaucoma. *Int J Ophthalmol.* 2015;8(2):395–398.

127. Ford ES, Giles WH, Dietz WH. Prevalence of the metabolic syndrome among US adults. *JAMA.* 2002;287(3): 356.

128. Zhou G, Liu B. Single nucleotide polymorphisms of metabolic syndrome-related genes in primary open angle glaucoma. *Int J Ophthalmol.* 2010;3(1):36–42.

129. Zode GS, Clark AF, Wordinger RJ. Bone morphogenetic protein 4 inhibits TGF-beta2 stimulation of extracellular matrix proteins in optic nerve head cells: role of gremlin in ECM modulation. *Glia.* 2009;57(7):755–766.

130. Sethi A, Jain A, Zode GS, Wordinger RJ, Clark AF. Role of TGFbeta/Smad signaling in gremlin induction of human trabecular meshwork extracellular matrix proteins. *Invest Ophthalmol Vis Sci.* 2011;52(8):5251–5259.

131. Yan X, Tezel G, Wax MB, Edward DP. Matrix metalloproteinases and tumor necrosis factor alpha in glaucomatous optic nerve head. *Arch Ophthalmol.* 2000;118(5): 666–673.

132. Markiewicz L, Pytel D, Mucha B, et al. Altered expression levels of MMP1, MMP9, MMP12, TIMP1, and IL-1beta as a risk factor for the elevated IOP and optic nerve head damage in the primary open-angle glaucoma patients. *BioMed Res Int.* 2015;2015:812503.

133. Sahay P, Rao A, Padhy D, et al. Functional activity of matrix metalloproteinases 2 and 9 in tears of patients with glaucoma. *Invest Ophthalmol Vis Sci.* 2017;58(6): BIO106–BIO113.

134. Susarla BT, Laing ED, Yu P, Katagiri Y, Geller HM, Symes AJ. Smad proteins differentially regulate transforming growth factor-beta-mediated induction of chondroitin sulfate proteoglycans. *J Neurochem.* 2011;119(4): 868–878.

135. Wallace DM, Clark AF, Lipson KE, Andrews D, Crean JK, O'Brien CJ. Anti-connective tissue growth factor antibody treatment reduces extracellular matrix production in trabecular meshwork and lamina cribrosa cells. *Invest Ophthalmol Vis Sci.* 2013;54(13):7836–7848.

136. Yu AL, Fuchshofer R, Birke M, Kampik A, Bloemendal H, Welge-Lussen U. Oxidative stress and TGF-beta2 increase heat shock protein 27 expression in human optic nerve head astrocytes. *Invest Ophthalmol Vis Sci.* 2008;49(12): 5403–5411.

137. McElnea EM, Quill B, Docherty NG, et al. Oxidative stress, mitochondrial dysfunction and calcium overload in human lamina cribrosa cells from glaucoma donors. *Mol Vis.* 2011;17:1182–1191.

138. Kortuem K, Geiger LK, Levin LA. Differential susceptibility of retinal ganglion cells to reactive oxygen species. *Invest Ophthalmol Vis Sci.* 2000;41(10):3176–3182.

139. Tezel G, Yang X, Luo C, Peng Y, Sun SL, Sun D. Mechanisms of immune system activation in glaucoma: oxidative stress-stimulated antigen presentation by the retina and optic nerve head glia. *Invest Ophthalmol Vis Sci.* 2007;48(2):705–714.

140. Chidlow G, Wood JPM, Casson RJ. Investigations into hypoxia and oxidative stress at the optic nerve head in a rat model of glaucoma. *Front Neurosci.* 2017;11:478.

141. Kerr J, Nelson P, O'Brien C. A comparison of ocular blood flow in untreated primary open-angle glaucoma and ocular hypertension. *Am J Ophthalmol.* 1998; 126(1):42—51.

142. Drance SM. Disc hemorrhages in the glaucomas. *Surv Ophthalmol.* 1989;33(5):331—337.

143. Kim YK, Jeoung JW, Park KH. Effect of focal lamina cribrosa defect on disc hemorrhage area in glaucoma. *Invest Ophthalmol Vis Sci.* 2016;57(3):899—907.

144. Lee EJ, Kim TW, Kim M, Girard MJ, Mari JM, Weinreb RN. Recent structural alteration of the peripheral lamina cribrosa near the location of disc hemorrhage in glaucoma. *Invest Ophthalmol Vis Sci.* 2014;55(4): 2805—2815.

145. Downs JC. Optic nerve head biomechanics in aging and disease. *Exp Eye Res.* 2015;133:19—29.

146. Kirwan RP, Crean JK, Fenerty CH, Clark AF, O'Brien CJ. Effect of cyclical mechanical stretch and exogenous transforming growth factor-beta1 on matrix metalloproteinase-2 activity in lamina cribrosa cells from the human optic nerve head. *J Glaucoma.* 2004;13(4):327—334.

147. Rogers R, Dharsee M, Ackloo S, Flanagan JG. Proteomics analyses of activated human optic nerve head lamina cribrosa cells following biomechanical strain. *Invest Ophthalmol Vis Sci.* 2012;53(7):3806—3816.

148. Quill B, Docherty NG, Clark AF, O'Brien CJ. The effect of graded cyclic stretching on extracellular matrix-related gene expression profiles in cultured primary human lamina cribrosa cells. *Invest Ophthalmol Vis Sci.* 2011; 52(3):1908—1915.

149. Quill B, Irnaten M, Docherty NG, et al. Calcium channel blockade reduces mechanical strain-induced extracellular matrix gene response in lamina cribrosa cells. *Br J Ophthalmol.* 2015;99(7):1009—1014.

150. Berretta A, Gowing EK, Jasoni CL, Clarkson AN. Sonic hedgehog stimulates neurite outgrowth in a mechanical stretch model of reactive-astrogliosis. *Sci Rep.* 2016;6: 21896.

151. Elliott MH, Ashpole NE, Gu X, et al. Caveolin-1 modulates intraocular pressure: implications for caveolae mechanoprotection in glaucoma. *Sci Rep.* 2016;6:37127.

152. Reina-Torres E, Wen JC, Liu KC, et al. VEGF as a paracrine regulator of conventional outflow facility. *Invest Ophthalmol Vis Sci.* 2017;58(3):1899—1908.

153. Keller KE, Bradley JM, Acott TS. Differential effects of ADAMTS-1, -4, and -5 in the trabecular meshwork. *Invest Ophthalmol Vis Sci.* 2009;50(12):5769—5777.

154. Stowell C, Burgoyne CF, Tamm ER, Ethier CR, Lasker/ IRRF Initiative on Astrocytes and Glaucomatous Neurodegeneration Participants. Biomechanical aspects of axonal damage in glaucoma: a brief review. *Exp Eye Res.* 2017;157:13—19.

155. Tamm ER, Ethier CR, Lasker/IRRF Initiative on Astrocytes and Glaucomatous Neurodegeneration Participants. Biological aspects of axonal damage in glaucoma: a brief review. *Exp Eye Res.* 2017;157:5—12.

156. Albrecht May C. Comparative anatomy of the optic nerve head and inner retina in non-primate animal models used for glaucoma research. *Open Ophthalmol J.* 2008;2: 94—101.

# The Role of Mechanobiology in Cancer Metastasis

MAUREEN E. LYNCH, PHD • COREY P. NEU, PHD • BENJAMIN SEELBINDER, MS • KAITLIN P. MCCREERY, BS

## 1 INTRODUCTION

The skeleton is the preferred site of metastasis for several of the most common cancers, including breast, prostate, and multiple myeloma.[1,2] Metastasis to bone is particularly devastating, as it is incurable and is associated with significant comorbidities. In particular, roughly three in four patients with advanced breast and prostate cancers develop incurable bone metastases,[2] with survival rates of approximately 30%.[1] Once tumor cells localize the skeletal environment, they dysregulate normal tissue homeostatic processes and cause heightened risk for skeletal-related events, such as bone fragility, pain, and fracture and hypercalcemia of malignancy.[3] Mechanical cues are the primary regulators of bone tissue homeostasis, and, in the primary tumor, they are well recognized to modulate tumor progression and metastasis. Therefore mechanical cues likely continue to play a role in tumor progression once disseminated tumor cells have localized to the skeleton. However, the relationship between the skeletal mechanical environment and metastatic bone cancer is poorly characterized.

Physical changes in the primary tumor microenvironment arise as a result of tumor growth and progression and feedback to promote tumor malignancy and migration.[4] As tumor cells grow uncontrollably in a confined environment, solid stresses arise in both the surrounding healthy tissue and in the tumor itself. These stresses are exacerbated by the concomitant rise in extracellular matrix (ECM) stiffness, which develops due to increased deposition of matrix proteins as well as increased cross-linking. Finally, increased formation of leaky vessels within the tumor raises the interstitial fluid pressure and causes net flux of fluid from the tumor into the surrounding environment, exposing tumor cells to interstitial fluid flow. At the cellular level, these mechanical cues are first sensed via structures in the plasma membrane (e.g., integrins) and are transduced to the cell interior via linkages with the cytoskeleton and nucleus, ultimately impacting chromatin reorganization. Thus mechanical stress activates cellular machinery to ultimately impact a variety of tumor cell behaviors, including proliferation, migration, malignancy, and extravasation. Evidence is emerging that, in addition to those in the primary site, mechanical cues in the skeletal microenvironment are important modulators of bone metastatic progression. In preclinical models, increased mechanical loading, which typically results in net bone formation in healthy subjects, inhibited or slowed cancer-induced bone loss.[5,6] Furthermore, tumor-induced rise in intramedullary pressure interfered with normal bone cell signaling.[7] Results from in vitro studies further support an intersection between mechanical loading and bone cancer and are helping to elucidate the underlying mechanisms, but more work is needed before this intersection can be leveraged for improving patient outcomes.

Mechanical cues arising from daily physical activity are ubiquitous and are the primary regulator of bone tissue homeostasis, including management of bone size and strength as well as mineral metabolism.[8] Intuitively, they likely play a role in metastatic processes. As they are well documented to be mechanosensitive at the primary site, as discussed earlier, they likely have the capability to sense skeletal mechanical cues directly. For example, upon applied cyclic compression, breast cancer cells decreased their expression of osteolytic genes[5] and Ewing sarcoma cells increased their drug resistance.[9] Additionally, tumor cells may be indirectly impacted by the downstream effects of mechanical cues applied to neighboring cells. For example, mechanical cues may be transduced to tumor cells via resident bone cells such as osteoclasts or endothelial cells.[10,11] Furthermore, increased mechanical loading

Mechanobiology. https://doi.org/10.1016/B978-0-12-817931-4.00004-2

results in increased osteogenesis, and metastatic tumor cells tend to colonize in endosteal niches that are osteogenic.[12] Finally, metastatic cancer cells may interfere with normal bone cell mechanosensing and mechanotransducing capabilities. For example, conditioned media from cyclically loaded breast cancer cells interfered with mesenchymal stem cell (MSC) loading-induced osteogenic differentiation, resulting in greater deposition of bone matrix proteins favorable to tumor cell adhesion.[13] Taken together, these data support that a relationship between mechanical cues exists in the skeleton and bone metastasis. Here, we review what is currently known about the relationship between bone mechanical cues and bone metastatic cancer, provide an overview of mechanical cues in the skeletal microenvironment, and discuss possible mechanisms by which tumor cells may sense these cues.

## 2 MECHANICAL CUES IN THE BONE MICROENVIRONMENT

### 2.1 Overview

During daily physical activity, forces are applied to the whole bone, causing tissue deformations to arise in both the bone marrow and the mineralized tissue, which are both fundamentally hydrated, porous tissue compartments. The deforming tissue pressurizes the interstitial fluid, resulting in local matrix deformations, fluid pressure gradients, and net fluid flow from high to low pressure. Cells attached to the matrix, including metastasized tumor cells, are exposed to these mechanical signals and alter their behavior in response.[14] Collectively, matrix deformation and the movement of fluid transmits chemical and mechanical signals,[15] which functionally couple loading-induced mechanical forces and cell signaling (i.e., mechanotransduction).

Key differences exist between the mechanical microenvironment of the primary site and that of the skeleton, calling to question whether what we know of mechanobiology at the primary site translates to the secondary site. The most notable difference is that healthy bone tissue remodeling requires intermittent forces (e.g., physical activity is in the range of 1−5 Hz), whereas cues in the primary tumor can be driven by either relatively static (e.g., ECM stiffness) or steady (e.g., fluid flow) cues. Also, the magnitude of some cues differs between sites. For example, estimated flow velocities at the primary site are on the order of 1−10 μm/second,[16] whereas velocities in the bone are ∼2 μm/second during walking[17] and as high as ∼60 μm/second during anabolic loading.[18] Whether these differences in mechanical cues are physiologically relevant in metastasis progression remains an open question.

### 2.2 Bone Marrow Mechanics and Relevance to Metastatic Bone Cancer

Bone marrow, where metastatic tumors first arrive, is a mechanosensitive tissue that participates in bone remodeling (Fig. 1.4.1A). Thus understanding the connections between bone marrow mechanics and bone metastasis is critical. When forces are applied to the whole bone, the marrow space, enclosed by a deforming mineralized cortex, experiences heightened intramedullary pressure that results in marrow tissue deformation and interstitial fluid flow. In fact, marrow forces in the absence of bone matrix strains are anabolic, resulting in bone formation[19,20] or inhibition of bone loss.[21,22] For example, intramedullary pressurization resulted in anabolic interstitial fluid flow throughout the whole bone, including the lacunocanaliculi system, and inhibited bone loss.[21,22] Thus the marrow mechanical environment is coupled to the mineralized tissue and is an important source of mechanical cues for remodeling of the mineralized tissue. Indeed, strain energy density gradients in the bone marrow predicted locations of bone formation and resorption better than those in the mineralized tissue itself.[23]

The marrow as a material is quite distinct from the mineralized tissue. The marrow is often (incorrectly) treated as a fluid, but in reality, it has an ECM and is a relatively compliant, viscoelastic material compared to bone.[24,25] The stiffness of bone is on the order of ∼10 GPa, whereas that of marrow is ∼1−100 kPa.[24] Its viscosity ranges between 10 and 100 Pa-s at a shear rate of 1 $s^{-1}$ and decreases with increasing shear rate according to a power law.[24,25] Despite these recent advances, the bone marrow mechanical properties are still relatively poorly characterized and depend on the composition,[26] anatomic site,[27] age,[28,29] and bone density.[30]

Marrow tissue is host to a variety of cell types important to bone homeostasis that are also implicated in many tumor processes: cells of hematopoietic origin (hematopoietic stem cells, osteoclasts, macrophages, etc.), cells of mesenchymal origin (MSCs, osteoblasts, adipocytes, etc.), endothelial cells, and nerve cells (Fig. 1.4.1A). Stresses and strains in the marrow tissue are imparted to resident cells, many of which are mechanoresponsive and are integral to bone homeostasis. Megakaryocytes in the mouse ulna altered their gene expression due to mechanical loading,[31] and loading

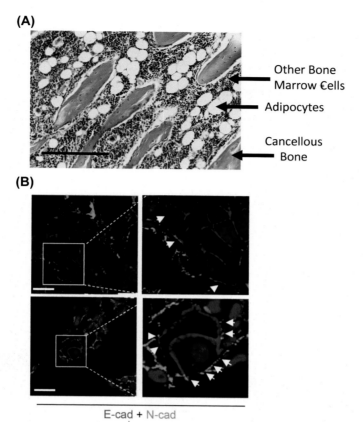

FIG. 1.4.1 Bone marrow tissue is the site for disseminated tumor cells. **(A)** Bone marrow tissue is enclosed by a mineralized cortex and interposed with mineralized cancellous bone. The tissue is a viscoelastic material that is host to numerous mechanosensitive cells. **(B)** Osteogenic niches at the interface between bone marrow and mineralized tissue promote early-stage micrometastases via activation of adherens junctions between breast cancer-derived E-cadherin and osteogenic cell N-cadherin. Immunofluorescent costaining of E-cadherin (*red*) and N-cadherin (*green*) on bone lesions of different sizes is shown.[12] (Reproduced from Wang H, Yu C, Gao X, et al. The osteogenic niche promotes early-stage bone colonization of disseminated breast cancer cells. *Cancer Cell*. 2015;27:193–210.)

applied to trabecular bone cores with intact marrow caused cellular deformations as high as 10,000 με at cell-cell junctions, more than sufficient to elicit a mechanoresponse in resident cells.[32,33] Further supporting this, loading applied to bone marrow explants stimulated ossification in the marrow itself.[34,35] Thus the marrow mechanical environment is a rich source of cues that modulate resident cell behavior and is thus likely to elicit responses from metastatic cancer cells as well. For example, osteogenic niches at the endosteal surface promote early-stage micrometastases via the activation of adherens junctions between cancer-derived E-cadherin and osteogenic N-cadherin[12] (Fig. 1.4.1B) and the endosteal surface experiences very high shear stresses that are correlated with bone formation,[36–38] further suggesting a link between tumor function and marrow biomechanical cues. Taken together, these data show that the bone marrow tissue and its cell populations experience significant mechanical cues; these cues regulate bone homeostatic processes and they likely affect any local tumor cells. Next, we describe relevant mechanical cues in the marrow in greater detail and make connections with tumor cell behavior.

### 2.2.1 Interstitial fluid flow

In the bone marrow, fluid velocity and shear stress values are highest at the marrow-bone surface, with peak values estimated to be ~50 μm/second and 5–10 Pa, respectively[36–38] (Fig. 1.4.2A). These values, generally derived from computational modeling, are highly dependent on the marrow mechanical properties such as assigned viscosity,[38,39] which in part depend on the cellular makeup. Wall shear stresses decreased with increasing adipocyte content because adipocytes are more compliant than other marrow cells and act to "stress-shield" the neighboring cells.[32,33] For example,

maximum shear stress decreased by 20% when the adipocyte volume fraction increased from 30% (young healthy marrow) to 60% (osteoporosis).[32] A similar result may occur in the presence of tumor cells, which are generally more compliant than healthy cells,[40,41] and may cause a rise in net bone loss signals due to the loss of mechanical stimulus. However, reduced trabecular bone volume, as would occur in the case of osteolytic metastases, increased fluid velocity and pore

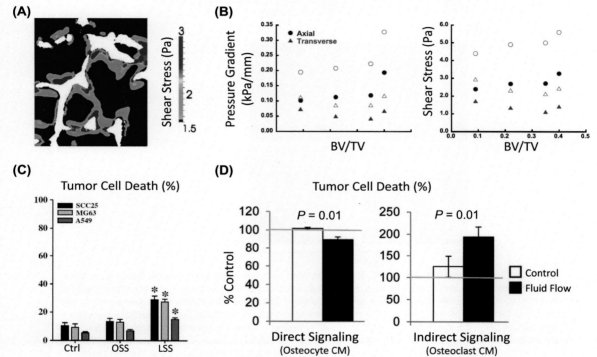

FIG. 1.4.2 Effects of interstitial fluid flow on bone marrow tissue and on tumor cells. (A) The maximum fluid-induced shear stresses in bone marrow can reach as high as 5 Pa under applied loading and occur at the bone-marrow interface.[37] (B) Estimates of fluid pressure gradient and flow-induced shear stress arising from applied compression increased with increasing bone volume fraction (bone volume/total volume [BV/TV]) and with increasing strain rate (open circles/triangles = high strain rate, 0.32% s[−1]; solid circles/triangles = low strain rate, 0.16% s[−1]).[32] (C) Two-dimensional (2D) laminar shear stress (LSS) at 0.05 Pa increased tumor cell death in the human tumor cell lines SCC25 (oral squamous carcinoma cells), MG63 (osteosarcoma cells), and A549 (carcinomic alveolar basal epithelial cells), but oscillatory shear stress (OSS) at 0.05 ± 4 Pa did not alter cell viability.[46] (D) Breast cancer cell death decreased when the cells were directly exposed to conditioned media (CMs) from osteocytes that had been flowed at shear stresses of ~1.6 Pa (2D flow), but tumor cell death increased when they were indirectly exposed to osteocytes' flowed CMs (i.e., when tumor cells were cultured with CMs from osteoclasts that had been conditioned in flowed osteocytes' CM).[10] ((A) Reproduced from Coughlin TR, Niebur GL Fluid shear stress in trabecular bone marrow due to low-magnitude high-frequency vibration. *J Biomech.* 2012;45:2222–2229. (B) Reproduced from Metzger TA, Vaughan TJ, McNamara LM, Niebur GL. Altered architecture and cell populations affect bone marrow mechanobiology in the osteoporotic human femur. *Biomech Model Mechanobiol.* 2017;16:841–850. (C) Reproduced from Lien SC, Chang SF, Lee PL, et al. Mechanical regulation of cancer cell apoptosis and autophagy: roles of bone morphogenetic protein receptor, Smad1/5, and p38 MAPK. *Biochim Biophys Acta.* 2013;1833:3124–3133. (D) Modified from Ma YV, Lam C, Dalmia S, et al. Mechanical regulation of breast cancer migration and apoptosis via direct and indirect osteocyte signaling. *J Cell Biochem.* 2018;119:5665–5675.)

pressure in the marrow compartment, thus leading to increased shear stress at the cell level[32,33,37,39] (Fig. 1.4.2B). But the integrated effects of bone loss and increased marrow compliance on overall bone remodeling are relatively unknown.

Interstitial fluid flow on primary tumor cells is generally considered to be a driver of metastasis.[16] For example, the percentage of breast cancer cells that were migratory as well as their average migration speed increased in response to ~1−5 μm/second flow, compared with static culture,[42,43] and flow velocity influenced the overall directional bias.[42] When Ewing sarcoma cells underwent in vitro interstitial fluid flow velocities of 0.5−4 μm/second, their resistance to chemotherapies increased.[9] However, high fluid flow values that are more in line with those in the bone microenvironment may have antitumorigenic effects. In human osteosarcoma (OS) cells, 1.2 Pa shear stress induced cell cycle blockage via integrin activation.[44] Similarly, high levels of shear stress (6 Pa) induced apoptosis in circulating tumor cells.[45] In contrast, oscillatory shear stress (0.05 ± 4 Pa) failed to alter MG63 OS cell viability, while low laminar shear (0.05 Pa) increased apoptosis[46] (Fig. 1.4.2C), suggesting that the intermittent nature of fluid flow in the bone mechanical environment should be considered when studying metastatic cancer. Additionally, when prostate cancer cells were repeatedly exposed to high shear stress values, they exhibited increased survival relative to normal prostate epithelial cells,[40] thus disseminated tumor cells that ultimately survive in the bone environment may have already adapted to higher shear stresses. More work is needed to understand the impact of bone-specific mechanical cues. Finally, the influence of fluid flow in the bone microenvironment is likely dependent on the particular type of cancer (e.g., primary vs. metastatic, osteolytic vs. osteoblastic) and the cell types (e.g., osteocytes, osteoclasts, endothelial cells) responsible for transducing the impacts of flow. For example, breast cancer cell migration was increased and apoptosis was reduced when the cells were directly exposed to media conditioned by flowed osteocytes; however, their response was the opposite when they received conditioned medium from osteoclasts conditioned in flowed osteocytes' conditioned medium[10] (Fig. 1.4.2D).

### 2.2.2 Tissue strain

Although marrow is often considered to be a fluid in computational models, a few studies have provided evidence of marrow ECM deformation under loading. Fluid-structure models of bone marrow cells exposed to loading-induced fluid flow have shown that loading applied to bone cores with intact marrow caused cellular deformations as high as 10,000 μe at cell-cell junctions[32,33] (see Fig. 1.4.1). Furthermore, strain energy density gradients (peak values ~0.8 mJ/mm$^3$) in bone marrow arising under the compression of vertebrae predicted sites of bone formation and resorption better than the gradients in the mineralized tissue[23] (Fig. 1.4.3A). Taken together, bone marrow ECM strains modulate the behavior of local marrow constituents with downstream effects on the mineralized tissue, and their effects on localized tumor cells may similarly impact remodeling.

Primary tumors and their microenvironment experience both stretching and compression during tumor growth. The outer periphery of primary tumors becomes stretched due to heightened interstitial fluid pressure and compressive forces from proliferating cancer cells, which may increase cancer cell proliferation.[47,48] Additionally, static stretching stimulated growth-arrested human mammary cells to proliferate, mediated by F-actin activation,[49] thus stretching may serve as an initiator of tumor growth and proliferation. Similarly, uniaxial stretching at 1 Hz increased tumor proliferation in multiple, though not all, OS cell lines[50] (Fig. 1.4.3B). Thus the effects of stretching may be preserved whether it is static (primary ECM) or intermittent (marrow ECM), but preservation may depend on the specific cancer cell type. The effects of compression are less clear. While the periphery is stretched, uncontrolled proliferation and growth of cancer cells in the tumor exert pressure on the ECM and neighboring tissue, which in turn exerts compressive forces/stresses on the tumor. Compressive stresses may not only inhibit growth in mammary carcinoma cells[51,52] but also trigger the formation of leader cells (tumor cells with a more invasive phenotype that lead other tumor cells during collective invasion and migration) and increase cell-matrix adhesions, leading to increased migration independent of changes to cell proliferation[53] (Fig. 1.4.3C). When dynamic compression was applied to breast cancer cells in a bone mineral-containing scaffold, their expression of *RUNX2*, which regulates downstream remodeling in bone cells, was reduced[5] (Fig. 1.4.3D). In contrast, when Ewing sarcoma cells in a collagen-based scaffold underwent dynamic compression, their *RUNX2* expression increased.[9] These studies highlight the need for additional investigations of how bone matrix strains regulate bone cancer cell function. In fact, computational models of metastatically involved vertebrae, a common site of metastasis, estimated elevated whole bone displacements relative to nonmetastatic controls. Thus a heightened risk for

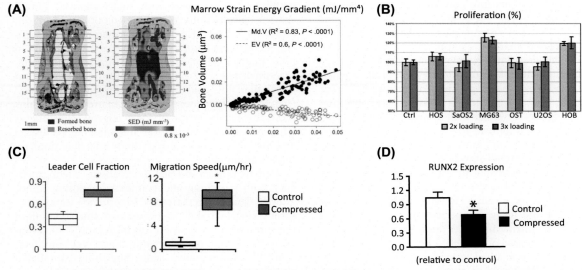

**FIG. 1.4.3** Matrix strain on bone marrow and on tumor cells. **(A)** Strain energy density (SED) gradients in bone marrow arising from the compression of vertebrae predicted sites of bone formation and resorption better than gradients in the mineralized tissue.[23] **(B)** Uniaxial stretching at 1 Hz increased tumor proliferation in several osteosarcoma and osteoblast cell lines.[50] **(C)** Compressive matrix strains applied to a monolayer triggered the formation of leader cells among breast cancer cells and increased cell-matrix adhesions, leading to increased tumor cell migration without changes to proliferation.[53] **(D)** Applied compression in a three-dimensional bone-mimicking scaffold reduced RUNX2 expression in breast cancer cells.[5] (**(A)** Modified from Webster D, Schulte FA, Lambers FM, Kuhn G, Muller R. Strain energy density gradients in bone marrow predict osteoblast and osteoclast activity: a finite element study. *J Biomech.* 2015;48:866–874. **(B)** Reproduced from Hoberg M, Gratz HH, Noll M, Jones DB. Mechanosensitivity of human osteosarcoma cells and phospholipase C β2 expression. *Biochem Biophys Res Commun.* 2005;333:142–149. **(C)** Modified from Tse JM, Cheng G, Tyrrell JA, et al. Mechanical compression drives cancer cells toward invasive phenotype. *Proc Natl Acad Sci U S A.* 2012;109:911–916. **(D)** Modified from Lynch ME, Brooks D, Mohanan S, et al. In vivo tibial compression decreases osteolysis and tumor formation in a human metastatic breast cancer model. *J Bone Min Res.* 2013; 28:2357–2367.)

fracture was predicted for increasing tumor size and reduced trabecular bone density.[54,55] The integrated effects of bone loss and increased marrow compliance, however, on overall bone remodeling during metastasis are relatively unknown.

### 2.2.3 Extracellular matrix stiffness

The stiffness of the ECM is an important factor in tumor growth and metastasis that is well documented to increase malignancy and migration. Simply increasing ECM stiffness is sufficient to induce the malignant transformation of normal mammary epithelial cells, a process that is mediated by integrin signaling.[56–58] Furthermore, the stiffness of tumor ECM is much higher than that of healthy tissues (tens of kilopascals vs. hundreds of pascals),[57] which correlates with changes in the metastatic potential of cells within the tumor.[59] For example, the higher stiffness of breast tumor tissue promotes a malignant phenotype of the mammary

epithelium by controlling integrin expression and signaling.[57,60]

The stiffness of the mineralized bone tissue is on the order of ~10 GPa, whereas that of bone marrow is ~1–100 kPa.[24,25] Matrix stiffness, similar to bone tissue, induced osteolytic gene expression in metastatic breast cancer cells via actomyosin contractility[61] (Fig. 1.4.4A); however, whether migration from the marrow to the endosteal surface is required for this to occur in vivo is unclear. Besides, tissue tropism to metastatic sites may be governed by intrinsic cellular contractility of the cytoskeleton. Expression of genes associated with cytoskeletal tension and contractility was higher in cancer cells that preferred stiff environments in terms of increased malignancy and proliferation, both when comparing MDA-MB-231 breast cancer cells to SKOV3 ovarian cells and when comparing bone-metastatic to lung-metastatic MDA-MB-231 subclones[62] (Fig. 1.4.4B). Although clearly

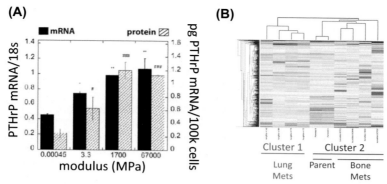

FIG. 1.4.4 Matrix stiffness is pro-osteolytic and mediated by cytoskeleton contractility in breast cancer cells. **(A)** Matrix stiffness, similar to bone tissue, increased expression and secretion of parathyroid-hormone-related protein (PTHrP), a cancer-derived protein that induces bone resorption, in metastatic breast cancer cells via actomyosin contractility.[61] **(B)** Unsupervised hierarchic clustering of lung metastatic, bone metastatic, and parental MDAMB231 human breast cancer cells shows that parental cell clusters with the bone metastatic subclones, both of which prefer stiffer matrices, indicating they are genetically similar, particularly among genes related to actomyosin contractility.[62] *mRNA*, messenger RNA. ((A) Reproduced from Ruppender NS, Merkel AR, Martin TJ, et al. Matrix rigidity induces osteolytic gene expression of metastatic breast cancer cells. *PLoS One.* 2010;5:e15451. **(B)** Reproduced from McGrail DJ, Kieu QM, Iandoli JA, Dawson MR. Actomyosin tension as a determinant of metastatic cancer mechanical tropism. *Phys Biol.* 2015;12:026001.)

important in the primary site, the extent to which matrix stiffness is a factor relative to other physical cues is still uncertain in the context of bone metastatic cancer.

### 2.2.4 Hydraulic pressure

The role of hydraulic pressure in bone is less documented. Changes to fluid pressure are obviously coupled with matrix deformation and interstitial fluid flow, but few studies exist in which the changes are explicitly studied in bone remodeling. In bones, the peak pressure in human femora subjected to physiologic loading ex vivo was measured at ~4 kPa.[63] Marrow pore pressure in lumbar vertebrae was predicted via computational modeling to be ~700 kPa under uniaxial compression, but it increased to above 1.1 MPa with a tumor.[54] The applied intramedullary pressurization of mouse tibiae resulted in peak values of 15.5 kPa (50 mmHg) and stimulated bone formation in the absence of bone matrix strains measured via ex vivo strain gauging.[22]

Tumors are well documented to exhibit greater intratumoral fluid pressure, leading to net fluid efflux, as described previously.[4,16] Applied pressure gradients, resulting in fluid flow, have resulted in more migratory cancer cells that are faster and more persistent,[42,43] which was mediated by a transcellular gradient in β1-integrin activation with F-actin-dependent protrusion formation localizing to the upstream side of the cell,

where matrix adhesions are under maximum tension.[43] For those studies, the pressure gradients were lower than those in bone, ranging from 40 Pa (3 μm/second, ~1.3 Pa shear stresses)[64] to ~900 Pa (~10 μm/second),[42] and much lower than what was predicted in tumor-bearing vertebrae, which ranged from 0.7 to 1.25 MPa.[54] The effects of the higher pressure values found in bone marrow on metastatic tumor cells is unknown.

## 3 POSSIBLE MECHANISMS OF CANCER CELL MECHANOSENSING OF BONE CUES
### 3.1 Overview

Tumor cell progression involves interaction with the local microenvironment, including physical aspects such as stiffness and fluid flow. Changes to these exogenous mechanical cues can first be detected by the cell via plasma membrane structures such as integrins and ion channels. These membrane structures are linked to cytoskeletal elements, such as actin and microtubules, that rearrange in response to mechanical cues. Finally, the cytoskeleton is connected to the nuclear envelope surrounding the nucleus, similar to the plasma membrane, with the linker of nucleoskeleton and cytoskeleton (LINC) complex providing a direct physical connection between the cytoplasm and the nucleoplasm, which could facilitate transduction of mechanical cues from the microenvironment. Here, we will provide an overview of several major cellular structures that are involved in mechanosensing and mechanotransduction and that have been implicated in cancer behavior.

## 3.2 Cell Surface/Plasma Membrane Structures

### 3.2.1 Extracellular matrix linkers to actin cytoskeleton

**3.2.1.1 Integrins and focal adhesions.** Integrins represent the main mechanoreceptor that connects the ECM and the cytoskeleton.[65] Integrins are heterodimeric cell adhesion receptors that transmit signals across the plasma membrane after binding ECM ligands. They contain a large extracellular domain that binds to substrates in the ECM (α-chain), a single transmembrane domain (β-chain), and a cytoplasmic domain (β-chain). Upon ECM binding, the heterodimeric receptor undergoes a conformational change, which activates the intracellular portion of the β-chain. Activated integrins have improved binding affinity to the F-actin cytoskeleton and activate more integrins to form clusters, recruiting proteins to form focal adhesions (FAs). FAs serve not only as the main link between the ECM and cytoskeleton inside the cell but also as the basis for large signaling complexes, and therefore regulate many cellular functions, including cell attachment, proliferation, migration, and differentiation.[66] When integrins are exposed to exogenous mechanical cues, the cell responds by recruiting FA proteins and mechanically strengthening itself against additional stress.[67–69] In breast cancer cells, drag-induced tension at matrix adhesion sites induced by interstitial fluid flow (4.6 μm/second due to a 60-Pa pressure gradient) resulted in activated β1-integrin adhesion complexes and FA proteins to counteract the tension.[43] In human OS cells, αvβ3 and β1 integrins were activated by fluid-flow-induced shear stress, resulting in cell cycle arrest.[44]

**3.2.1.2 Ion channels.** Ion channels are also mechanosensing and mechanotransducing structures.[70,71] These channels are able to sense changes in plasma membrane tension by changing their permeability to ions, such as $Na^+$, $K^+$, and $Ca^{2+}$, that can act as secondary messengers to regulate gene expression and modulate cell behavior, including migration and cytoskeletal organization.[70,72,73] The influx of $Ca^{2+}$, in particular, increases cellular migration[32,74] as well as regulates cell growth, contraction, apoptosis, and differentiation.[75] Members of the transient receptor potential (TRP) family of cation channels are mechanically gated and are

important structures in mechanotransduction, especially for shear stresses,[32,74] and they have also been implicated in cancer.[74,76,77] For example, in patients with breast cancer, high TRPM7 expression independently predicted poor outcomes, and in a mouse model, TRPM7 was required for metastasis formation by modifying cytoskeletal tension.[74] Furthermore, changes to cell function occur upon mechanical stimulation of ion channels because of their linkage to the cytoskeleton. For example, mechanical cues can activate TRPV4 to permit $Ca^{2+}$ influx and actin reorganization.[78]

**3.2.1.3 Cadherins.** Cell-cell adhesions contribute to maintaining the integrity of a tissue, and they relay tension across groups of cells to regulate collective cell behaviors. A major class of cell-cell adhesion molecules is the cadherin superfamily of transmembrane structures composed of a long extracellular domain, a single-pass transmembrane domain, and an intracellular tail that anchor cadherin to the cytoskeleton by associating with multiprotein complexes that include vinculin, α-catenin, and β-catenin.[79,80] Cadherin complexes are force sensors that transduce changes in cytoskeletal tension into intracellular signals that regulate tissue functions such as remodeling.[81,82] They connect the actomyosin cytoskeletons of adjacent cells and are thus part of a mechanical chain that experiences cytoskeletal deformations. In this way, they convey mechanical information to the cells by resisting the exogenous forces, for example, fluid shear stress, tissue rigidity, or compressive and extensive forces.[79] For example, long-distance mechanical communication between two premalignant mammary acini increases the transition to a more invasive morphology.[83] Its prototypic member, E-cadherin, displays strong anti-invasive and antimetastatic roles.[84,85] In the bone marrow, metastatic tumor cells engage with resident cells via cadherin-based adhesions. For example, activation of junctions between cancer-derived E-cadherin and osteogenic N-cadherin promotes early-stage micrometastases at the endosteal surface in bone marrow.[12] Similarly, myeloma cells with aberrant N-cadherin expression are more strongly retained in the bone marrow, where they interact with N-cadherin-positive osteoblasts, facilitating the onset and progression of metastatic lesions.[86] However, how these cadherin-based mechanical cues may impact metastatic progression is unknown.

Cadherins are integral to the epithelial-mesenchymal transition (EMT), a process in which epithelial cells lose cell-cell adhesion properties, gain migratory and invasive properties, and become more mesenchyma-like. Mechanotransduction plays a role in this process. For example, stiffer ECM drives mammary epithelial cells to undergo EMT.[87,88] Epithelial tumors often lose E-cadherin partially or completely as they progress toward malignancy, and reduction in E-cadherin expression is a hallmark of advanced carcinoma.[89] N-cadherin expression is increased concomitantly. The occurrence of cadherin switching (from E-cadherin to N-cadherin) is accompanied by redistribution of forces from junctions to cell-ECM adhesions.[90] For example, induction of N-cadherin expression in several cancer cell lines enhanced cell migration, invasion, and metastasis in mice even in the presence of E-cadherin.[91]

### 3.2.2 Other receptors/channels

Several other cell surface receptors such as G-protein-coupled receptors, primary cilia, and the cellular glycocalyx have emerged as possible mechanisms of mechanosensing in cancer progression.[92] G proteins are a family of membrane proteins, localized at FAs, that participate in mechanotransduction. Mechanical forces, such as shear stress or stretching, cause conformational changes to G proteins, which initiate signaling cascades, thus leading to cell growth. Primary cilia are mechanosensory organelles projecting outward from the plasma membrane and have been implicated in some cancers. For example, overexpression of polycystin 1 and 2, which are proteins present in the plasma membrane and cilia, is correlated with negative clinical outcomes and invasiveness in colorectal cancer.[93] They have also been found on the MG63 human OS cell line,[94] suggesting a mechanism by which cancer cells sense fluid flow. Finally, cells, including tumor cells, are covered with a surface proteoglycan (glycocalyx) layer that contributes to sensing fluid flow, and interstitial flow-driven motility in tumor cells is mediated by the cell surface glycocalyx.[95] Furthermore, a bulky glycocalyx facilitated integrin clustering by funneling active integrins into adhesions and applied tension to matrix-bound integrins independent of actomyosin contractility, thereby altering their activation state.[96]

### 3.3 Cytoskeleton

The cytoskeleton is an integrated network of microfilaments, microtubules, and intermediate filaments.[97] Microfilaments are semiflexible, composed of actin monomers attached with small binding proteins (e.g., talin, vinculin), which act to cross-link actin filaments to form contractile stress fibers in response to physical cues. In most cases, stress fibers connect to FAs, and hence are crucial in mechanotransduction. The presence of motor proteins in stress fibers facilitates contractility, underlying the cell response to its environment, for example, by regulating motility.[98] Additionally, cells cultured on ECM substrates of variable moduli displayed different stress fiber organization.[99] Microtubules are stiffer and rodlike polymers, whereas intermediate filaments are very flexible.[100] The cytoskeleton, being overall elastic and flexible, stabilizes the cell against shape distortion, especially in the presence of mechanical stresses. It also governs the mechanical properties of the cell itself, dictating how cells deform under stress,[101] whereby mechanical cues transferred across integrins and FAs can be transduced into a biochemical response via changes in the cytoskeleton at the site of binding or at other intracellular locations.[102] Mechanical stresses applied to integrins can alter the cytoskeletal structure (e.g., stress fiber formation) and activate signal transduction and gene expression in a stress-dependent manner.[69,103,104] Additionally, intrinsic cell contractility of the cytoskeleton may govern tissue tropism. For example, tumor cells that preferred a stiff ECM exhibited higher expression of genes associated with cytoskeletal tension and contractility.[62]

F-actin, composed of intertwining actin filament chains, is recognized as the major regulator and transducer of mechanical stimuli to cells.[105–107] In the case of increasing ECM stiffness, actin fibers become more ordered and numerous.[105,108] F-actin can be further organized into multiple ultrastructures, such as filopodia, which are used by cancer cells to sense physical cues.[109] For example, actin-enriched protrusions in breast cancer cells were utilized to sense the topographic features of an in vitro microenvironment mimicking the porosity of a bone, resulting in "mechanically induced dormancy" and reduced proliferation.[110] The interaction between the ECM and FAs, which are bound to the F-actin network and experience tension between the network and the ECM, drives the force-sensing mechanism.[111]

### 3.4 Nucleus

The nucleus is compartmentalized by two lipid bilayer membranes. The outer nuclear membrane has large outward bulges that form the endoplasmic reticulum in which all membrane proteins are synthesized. The inner nuclear membrane is lined with a fibrous protein meshwork called the nuclear lamina, which provides mechanical stability and connects the nucleus to other parts of the cell. The nuclear lamina is composed of two separate, but overlapping, protein networks that in turn consist of two different intermediate filament proteins: A-type and B-type lamins. A-type lamins form a thick meshwork with viscoelastic properties, whereas B-type lamins form a thin meshwork with elastic mechanical properties that is connected to the inner nuclear membrane.[112] Both networks together provide flexibility and resilience to the nucleus to maintain shape and integrity against external forces. As a result, the nucleus is approximately 5- to 10-fold stiffer than the cell body.[113,114]

The nucleus is interconnected with the cytoskeleton and, by extension, to the ECM via LINC complexes.[115] LINC complexes are composed of structural proteins that span the outer (nesprins) and inner (SUNs) nuclear membranes. On the outer side of the nucleus, different nesprin isotypes bind to specific parts of the cytoskeleton: nesprin-1 and nesprin-2 to actin, nesprin-3 to intermediary filaments, and nesprin-4 to microtubules.[116] On the inner side of the nucleus, the tetrameric SUN complex connects to the nuclear lamina.[117] As the nucleus is interconnected via LINC complexes, it might be an integrator of mechanical stimuli. For example, disruption of the LINC complex in experiments abrogated mechanosensitive differentiation[118] and stretch-induced chromatin condensation in MSCs.[119] It is further interesting to note that lamin A expression, the main contributor of mechanical resistance in the nuclear envelope, scales with tissue stiffness up to 30-fold, showing that there is a relationship between nuclear stiffness and tissue stiffness.[112] The most direct evidence, however, comes from experiments in which isolated nuclei were attached to magnetic beads and cyclically stretched.[120] Nuclei showed an increased mechanical resistance, measured by the shortening of the bead displacement, shortly after start of the stimulation. Interestingly, this response was only observed when beads were attached to the LINC complex protein nesprin-1, but not if beads were attached to nuclear pore complexes or nonspecifically bound to the nucleus.

Although extensive research has demonstrated that mechanotransduction pathways result in diseases, only a few pathways have specifically been linked to bone disease and osteosarcomas. Alterations in matrix material properties via a variation in matrix stiffness may lead to a cascade in mechanotransduction signaling, which has been shown to result in the dysregulation of bone remodeling, leading to disease phenotypes such as bone cancer.[121] Detrimental dysregulations of mechanotransduction pathways are caused by altered bone matrix mechanical properties,

which are transferred to the cell nucleus.[122] OS cell migration may occur under tension, leading to the deformation of the nucleus and chromatin reorganization that ultimately influence gene expression. Along with influencing gene expression heterogeneity within tissues, nucleus deformation results in the loss of nuclear envelope integrity and DNA damage, both of which can lead to cancer metastasis.[123] Finally, cancer genome mutations seem to increase with tissue stiffness, and stiff tissues such as bones produce cancer that exhibits more genomic changes than tumors originating from soft tissues.[124]

## 4 FUTURE CONSIDERATIONS

In summary, mechanical cues are well documented to drive cellular processes both in the skeleton and in primary tumors. However, their impacts on tumor cells that have metastasized to the skeleton are less understood. Here, we focused on mechanical cues in the bone marrow compartment, where disseminated tumors first localize, and attempted to provide connections with tumor cell mechanobiology. However, much of what we currently understand about tumor cell mechanosensing has been determined in the context of the primary tumor (e.g., extravasation) or circulation. Future studies that not only include the appropriate mechanical microenvironment but also incorporate both the mineralized tissue and marrow tissue compartments will help shed light on mechanically driven metastatic tumor progression in the bone.

## ACKNOWLEDGMENTS
Funding was provided by NSF CBET 1605060 (MEL) and NSF CAREER 1349735 (CPN).

## REFERENCES
1. Siegel RL, Miller KD, Jemal A. Cancer statistics, 2018. *CA Cancer J Clin.* 2018;68(1):7−30.
2. American Cancer Society. *Cancer Facts & Figures 2017.* Atlanta: American Cancer Society; 2017.
3. Kozlow W, Guise TA. Breast cancer metastasis to bone: mechanisms of osteolysis and implications for therapy. *J Mammary Gland Biol Neoplasia.* 2005;10(2):169−180.
4. Shieh AC. Biomechanical forces shape the tumor microenvironment. *Ann Biomed Eng.* 2011;39(5): 1379−1389.
5. Lynch ME, Brooks D, Mohanan S, et al. In vivo tibial compression decreases osteolysis and tumor formation in a human metastatic breast cancer model. *J Bone Miner Res.* 2013;28(11):2357−2367.

6. Pagnotti GM, Chan ME, Adler BJ, et al. Low intensity vibration mitigates tumor progression and protects bone quantity and quality in a murine model of myeloma. *Bone.* 2016;90:69−79.

7. Sottnik JL, Dai J, Zhang H, Campbell B, Keller ET. Tumor-induced pressure in the bone microenvironment causes osteocytes to promote the growth of prostate cancer bone metastases. *Cancer Res.* 2015;75(11):2151−2158.

8. Robling AG, Castillo AB, Turner CH. Biomechanical and molecular regulation of bone remodeling. *Annu Rev Biomed Eng.* 2006;8:455−498.

9. Marturano-Kruik A, Villasante A, Yaeger K, et al. Biomechanical regulation of drug sensitivity in an engineered model of human tumor. *Biomaterials.* 2018;150:150−161.

10. Ma YV, Lam C, Dalmia S, et al. Mechanical regulation of breast cancer migration and apoptosis via direct and indirect osteocyte signaling. *J Cell Biochem.* 2018;119(7):5665−5675.

11. Ma YV, Xu L, Mei X, Middleton K, You L. Mechanically stimulated osteocytes reduce the bone-metastatic potential of breast cancer cells in vitro by signaling through endothelial cells. *J Cell Biochem.* 2019;120(5):7590−7601.

12. Wang H, Yu C, Gao X, et al. The osteogenic niche promotes early-stage bone colonization of disseminated breast cancer cells. *Cancer Cell.* 2015;27(2):193−210.

13. Lynch ME, Chiou AE, Lee MJ, et al. Three-dimensional mechanical loading modulates the osteogenic response of mesenchymal stem cells to tumor-derived soluble signals. *Tissue Eng A.* 2016;22(15−16):1006−1015.

14. Thompson WR, Rubin CT, Rubin J. Mechanical regulation of signaling pathways in bone. *Gene.* 2012.

15. Fritton SP, Weinbaum S. Fluid and solute transport in bone: flow-induced mechanotransduction. *Annu Rev Fluid Mech.* 2009;41:347−374.

16. Munson JM, Shieh AC. Interstitial fluid flow in cancer: implications for disease progression and treatment. *Cancer Manag Res.* 2014;6:317−328.

17. Fan L, Pei S, Lucas Lu X, Wang L. A multiscale 3D finite element analysis of fluid/solute transport in mechanically loaded bone. *Bone Res.* 2016;4:16032.

18. Price C, Zhou X, Li W, Wang L. Real-time measurement of solute transport within the lacunar-canalicular system of mechanically loaded bone: direct evidence for load-induced fluid flow. *J Bone Miner Res.* 2011;26(2):277−285.

19. Guldberg RE, Richards M, Caldwell NJ, Kuelske CL, Goldstein SA. Trabecular bone adaptation to variations in porous-coated implant technology. *J Biomech.* 1997;30(2):147−153.

20. Metzger TA, Schwaner SA, LaNeve AJ, Kreipke TC, Niebur GL. Pressure and shear stress in trabecular bone marrow during whole bone loading. *J Biomech.* 2015;48(12):3035−3043.

21. Kwon RY, Meays DR, Meilan AS, et al. Skeletal adaptation to intramedullary pressure-induced interstitial fluid flow is enhanced in mice subjected to targeted osteocyte ablation. *PLoS One.* 2012;7(3):e33336.

22. Kwon RY, Meays DR, Tang WJ, Frangos JA. Microfluidic enhancement of intramedullary pressure increases interstitial fluid flow and inhibits bone loss in hindlimb suspended mice. *J Bone Miner Res.* 2010;25(8):1798−1807.

23. Webster D, Schulte FA, Lambers FM, Kuhn G, Muller R. Strain energy density gradients in bone marrow predict osteoblast and osteoclast activity: a finite element study. *J Biomech.* 2015;48(5):866−874.

24. Jansen LE, Birch NP, Schiffman JD, Crosby AJ, Peyton SR. Mechanics of intact bone marrow. *J Mech Behav Biomed Mater.* 2015;50:299−307.

25. Metzger TA, Shudick JM, Seekell R, Zhu Y, Niebur GL. Rheological behavior of fresh bone marrow and the effects of storage. *J Mech Behav Biomed Mater.* 2014;40:307−313.

26. Bryant JD, David T, Gaskell PH, King S, Lond G. Rheology of bovine bone marrow. *Proc Inst Mech Eng H.* 1989;203(2):71−75.

27. Liney GP, Bernard CP, Manton DJ, Turnbull LW, Langton CM. Age, gender, and skeletal variation in bone marrow composition: a preliminary study at 3.0 Tesla. *J Magn Reson Imaging.* 2007;26(3):787−793.

28. Moore SG, Dawson KL. Red and yellow marrow in the femur: age-related changes in appearance at MR imaging. *Radiology.* 1990;175(1):219−223.

29. Ricci C, Cova M, Kang YS, et al. Normal age-related patterns of cellular and fatty bone marrow distribution in the axial skeleton: MR imaging study. *Radiology.* 1990;177(1):83−88.

30. Yeung DK, Griffith JF, Antonio GE, et al. Osteoporosis is associated with increased marrow fat content and decreased marrow fat unsaturation: a proton MR spectroscopy study. *J Magn Reson Imaging.* 2005;22(2):279−285.

31. Soves CP, Miller JD, Begun DL, et al. Megakaryocytes are mechanically responsive and influence osteoblast proliferation and differentiation. *Bone.* 2014;66:111−120.

32. Metzger TA, Vaughan TJ, McNamara LM, Niebur GL. Altered architecture and cell populations affect bone marrow mechanobiology in the osteoporotic human femur. *Biomech Model Mechanobiol.* 2017;16(3):841−850.

33. Vaughan TJ, Voisin M, Niebur GL, McNamara LM. Multiscale modeling of trabecular bone marrow: understanding the micromechanical environment of mesenchymal stem cells during osteoporosis. *J Biomech Eng.* 2015;137(1).

34. Gurkan UA, Kishore V, Condon KW, Bellido TM, Akkus O. A scaffold-free multicellular three-dimensional in vitro model of osteogenesis. *Calcif Tissue Int.* 2011;88(5):388−401.

35. Gurkan UA, Krueger A, Akkus O. Ossifying bone marrow explant culture as a three-dimensional mechanoresponsive in vitro model of osteogenesis. *Tissue Eng A.* 2011;17(3−4):417−428.

36. Birmingham E, Kreipke TC, Dolan EB, et al. Mechanical stimulation of bone marrow in situ induces bone formation in trabecular explants. *Ann Biomed Eng.* 2015;43(4):1036−1050.

37. Coughlin TR, Niebur GL. Fluid shear stress in trabecular bone marrow due to low-magnitude high-frequency vibration. *J Biomech.* 2012;45(13):2222−2229.

38. Metzger TA, Kreipke TC, Vaughan TJ, McNamara LM, Niebur GL. The in situ mechanics of trabecular bone marrow: the potential for mechanobiological response. *J Biomech Eng.* 2015;137(1).

39. Birmingham E, Grogan JA, Niebur GL, McNamara LM, McHugh PE. Computational modelling of the mechanics of trabecular bone and marrow using fluid structure interaction techniques. *Ann Biomed Eng.* 2013;41(4):814–826.

40. Chivukula VK, Krog BL, Nauseef JT, Henry MD, Vigmostad SC. Alterations in cancer cell mechanical properties after fluid shear stress exposure: a micropipette aspiration study. *Cell Health Cytoskelet.* 2015;7:25–35.

41. Guck J, Schinkinger S, Lincoln B, et al. Optical deformability as an inherent cell marker for testing malignant transformation and metastatic competence. *Biophys J.* 2005;88(5):3689–3698.

42. Haessler U, Teo JC, Foretay D, Renaud P, Swartz MA. Migration dynamics of breast cancer cells in a tunable 3D interstitial flow chamber. *Integr Biol (Camb).* 2012; 4(4):401–409.

43. Polacheck WJ, German AE, Mammoto A, Ingber DE, Kamm RD. Mechanotransduction of fluid stresses governs 3D cell migration. *Proc Natl Acad Sci U S A.* 2014; 111(7):2447–2452.

44. Chang SF, Chang CA, Lee DY, et al. Tumor cell cycle arrest induced by shear stress: roles of integrins and Smad. *Proc Natl Acad Sci U S A.* 2008;105(10):3927–3932.

45. Regmi S, Fu A, Luo KQ. High shear stresses under exercise condition destroy circulating tumor cells in a microfluidic system. *Sci Rep.* 2017;7:39975.

46. Lien SC, Chang SF, Lee PL, et al. Mechanical regulation of cancer cell apoptosis and autophagy: roles of bone morphogenetic protein receptor, Smad1/5, and p38 MAPK. *Biochim Biophys Acta.* 2013;1833(12):3124–3133.

47. Hofmann M, Guschel M, Bernd A, et al. Lowering of tumor interstitial fluid pressure reduces tumor cell proliferation in a xenograft tumor model. *Neoplasia.* 2006;8(2):89–95.

48. Padera TP, Stoll BR, Tooredman JB, et al. Pathology: cancer cells compress intratumour vessels. *Nature.* 2004; 427(6976):695.

49. Aragona M, Panciera T, Manfrin A, et al. A mechanical checkpoint controls multicellular growth through YAP/TAZ regulation by actin-processing factors. *Cell.* 2013; 154(5):1047–1059.

50. Hoberg M, Gratz HH, Noll M, Jones DB. Mechanosensitivity of human osteosarcoma cells and phospholipase C beta2 expression. *Biochem Biophys Res Commun.* 2005; 333(1):142–149.

51. Cheng G, Tse J, Jain RK, Munn LL. Micro-environmental mechanical stress controls tumor spheroid size and morphology by suppressing proliferation and inducing apoptosis in cancer cells. *PLoS One.* 2009;4(2):e4632.

52. Helmlinger G, Netti PA, Lichtenbeld HC, Melder RJ, Jain RK. Solid stress inhibits the growth of multicellular tumor spheroids. *Nat Biotechnol.* 1997;15(8):778–783.

53. Tse JM, Cheng G, Tyrrell JA, et al. Mechanical compression drives cancer cells toward invasive phenotype. *Proc Natl Acad Sci U S A.* 2012;109(3):911–916.

54. Whyne CM, Hu SS, Lotz JC. Parametric finite element analysis of vertebral bodies affected by tumors. *J Biomech.* 2001;34(10):1317–1324.

55. Salvatore G, Berton A, Giambini H, et al. Biomechanical effects of metastasis in the osteoporotic lumbar spine: a Finite Element Analysis. *BMC Muscoskelet Disord.* 2018; 19(1):38.

56. Chaudhuri O, Koshy ST, Branco da Cunha C, et al. Extracellular matrix stiffness and composition jointly regulate the induction of malignant phenotypes in mammary epithelium. *Nat Mater.* 2014;13(10):970–978.

57. Paszek MJ, Zahir N, Johnson KR, et al. Tensional homeostasis and the malignant phenotype. *Cancer Cell.* 2005; 8(3):241–254.

58. Shaw LM, Rabinovitz I, Wang HH, Toker A, Mercurio AM. Activation of phosphoinositide 3-OH kinase by the alpha6beta4 integrin promotes carcinoma invasion. *Cell.* 1997;91(7):949–960.

59. Provenzano PP, Eliceiri KW, Campbell JM, et al. Collagen reorganization at the tumor-stromal interface facilitates local invasion. *BMC Med.* 2006;4(1):38.

60. Levental KR, Yu H, Kass L, et al. Matrix crosslinking forces tumor progression by enhancing integrin signaling. *Cell.* 2009;139(5):891–906.

61. Ruppender NS, Merkel AR, Martin TJ, et al. Matrix rigidity induces osteolytic gene expression of metastatic breast cancer cells. *PLoS One.* 2010;5(11):e15451.

62. McGrail DJ, Kieu QM, Iandoli JA, Dawson MR. Actomyosin tension as a determinant of metastatic cancer mechanical tropism. *Phys Biol.* 2015;12(2):026001.

63. Downey DJ, Simkin PA, Taggart R. The effect of compressive loading on intraosseous pressure in the femoral head in vitro. *J Bone Joint Surg Am.* 1988;70(6):871–877.

64. Polacheck WJ, Charest JL, Kamm RD. Interstitial flow influences direction of tumor cell migration through competing mechanisms. *Proc Natl Acad Sci U S A.* 2011; 108(27):11115–11120.

65. Ross TD, Coon BG, Yun S, et al. Integrins in mechanotransduction. *Curr Opin Cell Biol.* 2013;25(5): 613–618.

66. Coppolino MG, Dedhar S. Bi-directional signal transduction by integrin receptors. *Int J Biochem Cell Biol.* 2000; 32(2):171–188.

67. Choquet D, Felsenfeld DP, Sheetz MP. Extracellular matrix rigidity causes strengthening of integrin-cytoskeleton linkages. *Cell.* 1997;88(1):39–48.

68. Riveline D, Zamir E, Balaban NQ, et al. Focal contacts as mechanosensors: externally applied local mechanical force induces growth of focal contacts by an mDia1-dependent and ROCK-independent mechanism. *J Cell Biol.* 2001;153(6):1175–1186.

69. Wang N, Butler JP, Ingber DE. Mechanotransduction across the cell surface and through the cytoskeleton. *Science.* 1993;260(5111):1124–1127.

70. Hamill OP, Martinac B. Molecular basis of mechanotransduction in living cells. *Physiol Rev.* 2001;81(2):685–740.

71. Sachs F. Mechanical transduction by ion channels: a cautionary tale. *World J Neurol.* 2015;5(3):74–87.

72. Ruknudin A, Sachs F, Bustamante JO. Stretch-activated ion channels in tissue-cultured chick heart. *Am J Physiol.* 1993;264(3 Pt 2):H960–H972.

73. Sachs F. Stretch-sensitive ion channels: an update. *Soc Gen Physiol Ser.* 1992;47:241–260.

74. Middelbeek J, Kuipers AJ, Henneman L, et al. TRPM7 is required for breast tumor cell metastasis. *Cancer Res.* 2012;72(16):4250–4261.

75. Wang JH, Thampatty BP. An introductory review of cell mechanobiology. *Biomech Model Mechanobiol.* 2006;5(1):1–16.

76. Lehen'kyi V, Prevarskaya N. Oncogenic TRP channels. *Adv Exp Med Biol.* 2011;704:929–945.

77. Prevarskaya N, Zhang L, Barritt G. TRP channels in cancer. *Biochim Biophys Acta.* 2007;1772(8):937–946.

78. Kuipers AJ, Middelbeek J, van Leeuwen FN. Mechanoregulation of cytoskeletal dynamics by TRP channels. *Eur J Cell Biol.* 2012;91(11–12):834–846.

79. Leckband DE, de Rooij J. Cadherin adhesion and mechanotransduction. *Annu Rev Cell Dev Biol.* 2014;30:291–315.

80. Nelson WJ, Nusse R. Convergence of Wnt, beta-catenin, and cadherin pathways. *Science.* 2004;303(5663):1483–1487.

81. le Duc Q, Shi Q, Blonk I, et al. Vinculin potentiates E-cadherin mechanosensing and is recruited to actin-anchored sites within adherens junctions in a myosin II-dependent manner. *J Cell Biol.* 2010;189(7):1107–1115.

82. Liu Z, Tan JL, Cohen DM, et al. Mechanical tugging force regulates the size of cell-cell junctions. *Proc Natl Acad Sci U S A.* 2010;107(22):9944–9949.

83. Shi Q, Ghosh RP, Engelke H, et al. Rapid disorganization of mechanically interacting systems of mammary acini. *Proc Natl Acad Sci U S A.* 2014;111(2):658–663.

84. Frixen UH, Behrens J, Sachs M, et al. E-cadherin-mediated cell-cell adhesion prevents invasiveness of human carcinoma cells. *J Cell Biol.* 1991;113(1):173–185.

85. Vleminckx K, Vakaet Jr L, Mareel M, Fiers W, van Roy F. Genetic manipulation of E-cadherin expression by epithelial tumor cells reveals an invasion suppressor role. *Cell.* 1991;66(1):107–119.

86. Groen RW, de Rooij MF, Kocemba KA, et al. N-cadherin-mediated interaction with multiple myeloma cells inhibits osteoblast differentiation. *Haematologica.* 2011;96(11):1653–1661.

87. Wei SC, Fattet L, Tsai JH, et al. Matrix stiffness drives epithelial-mesenchymal transition and tumour metastasis through a TWIST1-G3BP2 mechanotransduction pathway. *Nat Cell Biol.* 2015;17(5):678–688.

88. Wei SC, Yang J. Forcing through tumor metastasis: the interplay between tissue rigidity and epithelial-mesenchymal transition. *Trends Cell Biol.* 2016;26(2):111–120.

89. Hanahan D, Weinberg RA. Hallmarks of cancer: the next generation. *Cell.* 2011;144(5):646–674.

90. Scarpa E, Szabo A, Bibonne A, et al. Cadherin switch during EMT in neural crest cells leads to contact inhibition of Locomotion via Repolarization of forces. *Dev Cell.* 2015;34(4):421–434.

91. Hazan RB, Phillips GR, Qiao RF, Norton L, Aaronson SA. Exogenous expression of N-cadherin in breast cancer cells induces cell migration, invasion, and metastasis. *J Cell Biol.* 2000;148(4):779–790.

92. Gasparski AN, Beningo KA. Mechanoreception at the cell membrane: more than the integrins. *Arch Biochem Biophys.* 2015;586:20–26.

93. Gargalionis AN, Korkolopoulou P, Farmaki E, et al. Polycystin-1 and polycystin-2 are involved in the acquisition of aggressive phenotypes in colorectal cancer. *Int J Cancer.* 2015;136(7):1515–1527.

94. Kowal TJ, Falk MM. Primary cilia found on HeLa and other cancer cells. *Cell Biol Int.* 2015;39(11):1341–1347.

95. Qazi H, Palomino R, Shi ZD, Munn LL, Tarbell JM. Cancer cell glycocalyx mediates mechanotransduction and flow-regulated invasion. *Integr Biol (Camb).* 2013;5(11):1334–1343.

96. Paszek MJ, DuFort CC, Rossier O, et al. The cancer glycocalyx mechanically primes integrin-mediated growth and survival. *Nature.* 2014;511(7509):319–325.

97. Ingber DE. Cellular basis of mechanotransduction. *Biol Bull.* 1998;194(3):323–325. discussion 325–7.

98. Tojkander S, Gateva G, Lappalainen P. Actin stress fibers–assembly, dynamics and biological roles. *J Cell Sci.* 2012;125(Pt 8):1855–1864.

99. Zemel A, Rehfeldt F, Brown AE, Discher DE, Safran SA. Optimal matrix rigidity for stress fiber polarization in stem cells. *Nat Phys.* 2010;6(6):468–473.

100. Oddou C, Wendling S, Petite H, Meunier A. Cell mechanotransduction and interactions with biological tissues. *Biorheology.* 2000;37(1–2):17–25.

101. Wang J, Su M, Fan J, Seth A, McCulloch CA. Transcriptional regulation of a contractile gene by mechanical forces applied through integrins in osteoblasts. *J Biol Chem.* 2002;277(25):22889–22895.

102. Ingber DE. Tensegrity: the architectural basis of cellular mechanotransduction. *Annu Rev Physiol.* 1997;59:575–599.

103. Schmidt CE, Horwitz AF, Lauffenburger DA, Sheetz MP. Integrin-cytoskeletal interactions in migrating fibroblasts are dynamic, asymmetric, and regulated. *J Cell Biol.* 1993;123(4):977–991.

104. Urbich C, Dernbach E, Reissner A, et al. Shear stress-induced endothelial cell migration involves integrin signaling via the fibronectin receptor subunits alpha(5) and beta(1). *Arterioscler Thromb Vasc Biol.* 2002;22(1):69–75.

105. Gupta M, Sarangi BR, Deschamps J, et al. Adaptive rheology and ordering of cell cytoskeleton govern matrix rigidity sensing. *Nat Commun.* 2015;6:7525.

106. Schwarz US, Gardel ML. United we stand: integrating the actin cytoskeleton and cell-matrix adhesions in cellular mechanotransduction. *J Cell Sci.* 2012;125(Pt 13):3051–3060.

107. Trichet L, Le Digabel J, Hawkins RJ, et al. Evidence of a large-scale mechanosensing mechanism for cellular

adaptation to substrate stiffness. *Proc Natl Acad Sci U S A.* 2012;109(18):6933–6938.

108. Prager-Khoutorsky M, Lichtenstein A, Krishnan R, et al. Fibroblast polarization is a matrix-rigidity-dependent process controlled by focal adhesion mechanosensing. *Nat Cell Biol.* 2011;13(12):1457–1465.

109. Mejillano MR, Kojima S, Applewhite DA, et al. Lamellipodial versus filopodial mode of the actin nanomachinery: pivotal role of the filament barbed end. *Cell.* 2004;118(3):363–373.

110. Chaudhuri PK, Pan CQ, Low BC, Lim CT. Differential depth sensing reduces cancer cell proliferation via Rho-Rac-regulated invadopodia. *ACS Nano.* 2017;11(7): 7336–7348.

111. Geiger B, Spatz JP, Bershadsky AD. Environmental sensing through focal adhesions. *Nat Rev Mol Cell Biol.* 2009;10(1):21–33.

112. Swift J, Ivanovska IL, Buxboim A, et al. Nuclear lamin-A scales with tissue stiffness and enhances matrix-directed differentiation. *Science.* 2013;341(6149):1240104.

113. Dahl KN, Kahn SM, Wilson KL, Discher DE. The nuclear envelope lamina network has elasticity and a compressibility limit suggestive of a molecular shock absorber. *J Cell Sci.* 2004;117(Pt 20):4779–4786.

114. Guilak F, Tedrow JR, Burgkart R. Viscoelastic properties of the cell nucleus. *Biochem Biophys Res Commun.* 2000; 269(3):781–786.

115. Jaalouk DE, Lammerding J. Mechanotransduction gone awry. *Nat Rev Mol Cell Biol.* 2009;10(1):63–73.

116. Zhang Q, Ragnauth C, Greener MJ, Shanahan CM, Roberts RG. The nesprins are giant actin-binding proteins, orthologous to Drosophila melanogaster muscle protein MSP-300. *Genomics.* 2002;80(5):473–481.

117. Stroud MJ, Banerjee I, Veevers J, Chen J. Linker of nucleoskeleton and cytoskeleton complex proteins in cardiac structure, function, and disease. *Circ Res.* 2014;114(3): 538–548.

118. Uzer G, Thompson WR, Sen B, et al. Cell mechanosensitivity to extremely low-magnitude signals is enabled by a LINCed nucleus. *Stem Cells.* 2015;33(6):2063–2076.

119. Heo SJ, Thorpe SD, Driscoll TP, et al. Biophysical regulation of chromatin architecture instills a mechanical memory in mesenchymal stem cells. *Sci Rep.* 2015;5:16895.

120. Guilluy C, Osborne LD, Van Landeghem L, et al. Isolated nuclei adapt to force and reveal a mechanotransduction pathway in the nucleus. *Nat Cell Biol.* 2014;16(4): 376–381.

121. Ingber DE. Mechanobiology and diseases of mechanotransduction. *Ann Med.* 2003;35(8):564–577.

122. Cox LG, van Rietbergen B, van Donkelaar CC, Ito K. Analysis of bone architecture sensitivity for changes in mechanical loading, cellular activity, mechanotransduction, and tissue properties. *Biomech Model Mechanobiol.* 2011;10(5):701–712.

123. Denais CM, Gilbert RM, Isermann P, et al. Nuclear envelope rupture and repair during cancer cell migration. *Science.* 2016;352(6283):353–358.

124. Pfeifer CR, Alvey CM, Irianto J, Discher DE. Genome variation across cancers scales with tissue stiffness — an invasion-mutation mechanism and implications for immune cell infiltration. *Curr Opin Struct Biol.* 2017;2: 103–114.

CHAPTER 2.1

# Cells as Functional Load Sensors and Drivers of Adaptation

MATTHEW GOELZER, MS • WILLIAM R. THOMPSON, DPT, PHD • GUNES UZER, PHD

## 1 CELLS AS LOAD-BEARING STRUCTURES

As the smallest load-bearing unit, cells both sense and adapt to dynamic mechanical environments to drive tissue-level adaptations. The type and intensity of forces to which cells are subjected to depend on the tissue type and physiologic function. For example, myocytes, found in cardiac and skeletal muscles, are exposed to tensile and shear forces generated by sarcomeres that slide against each other within myofibrils.[1] Endothelial cells, which line the inner layer of blood vessels, are subjected to pulsatile fluid shear stress as blood is pumped.[2] Osteocytes are also subjected to oscillatory fluid shear as a result of deformation of bone during locomotion; however, these specialized cells are embedded in the calcified bone matrix, making the mechanical environment of the fluid movement distinct from that of blood vessels.[3] The ability to adapt to the mechanical demands conveyed by these unique environments is best embodied in stem cells that can differentiate into a multitude of cell types based on physical environmental cues.[4] Stem cells are multipotent cells that can self-renew and differentiate into terminally differentiated cell types of the tissues that they reside in. As stem cells replace and regenerate tissues by replenishing resident cell populations, they not only hold great therapeutic promise for tissue engineering and regenerative approaches but also provide a robust model for studying how mechanical forces are sensed and adapt at the cellular level.

Cellular mechanosensation is accomplished through a variety of structures and proteins that reside within the plasma membrane, the cytoskeleton, and the nucleus. Depending on the sensory element, mechanical signals are either converted into biochemical signaling cascades or directly transduced to inner cellular structures. This conversion of extracellular forces into intracellular signals is called mechanotransduction. However, the process of mechanotransduction cannot be reduced to passive sensory functioning of these structures. As a functioning unit, cells adapt to alterations in mechanical stress, resulting in remodeling of internal structures. In turn, these structural adaptations provide another level of control by regulating the effectiveness of subsequent mechanotransduction events. In this way, maladaptation to mechanical stress is a common causative factor in many debilitating diseases.

This chapter discusses the function of different sensing elements in cells. The major mechanotransduction pathways and signaling elements activated by these structures will also be discussed. When appropriate, we will discuss how these cellular structures adapt to mechanical force in both health and disease. Our examples will focus on the influence of mechanical force on bone marrow mesenchymal stem cell (MSC) differentiation, as it relates to the osteogenic and adipogenic lineages. While stem cells in other tissue compartments are subjected to different mechanical environments, information presented in this chapter can be easily generalized to these stem cells and to other cell types.

## 2 CELL MEMBRANE

The plasma membrane is a semipermeable lipid bilayer that serves numerous functions including protection from external insults, maintenance of the electrochemical gradient, and communication between cell types. Mechanical signals are conveyed through mechanosensory structures embedded within the plasma membrane. Transmission of mechanical information from the external environment, through the cell membrane,

Mechanobiology. https://doi.org/10.1016/B978-0-12-817931-4.00005-4

and into the cytoplasm is known as mechanotransduction. This section discusses the functions of prominent mechanosensory elements found in the plasma membrane including, ion channels, integrin-bound focal adhesion (FA) attachments, and primary cilia.

## 2.1 Ion Channels

The high concentration of positively charged ions, such as calcium ($Ca^{2+}$), outside the cell creates a large electrochemical gradient. Ion channels take advantage of this electrochemical gradient to enable rapid transduction of mechanical signals into biochemical responses. Ion channels span the plasma membrane and are selective to specific ions such as $Ca^{2+}$ or potassium. $Ca^{2+}$, in particular, serves as a potent second messenger, activating downstream signaling in many cell types, including MSCs.[5] In fact, the first cellular response of bone cells to mechanical force is a rapid intracellular $Ca^{2+}$ influx.[6,7] Once inside the cell, $Ca^{2+}$ regulates a variety of processes including migration,[8] activation of transcription factors,[9] differentiation,[10] and mechanosensation.[11] $Ca^{2+}$ enters the cell through integral membrane pores. The various forms of $Ca^{2+}$ channels enable tight regulation of $Ca^{2+}$ influx in response to different signals including ligand binding, voltage changes, and mechanical force. Such tight regulation enables distinct control of cellular processes. The various types of $Ca^{2+}$ channels include piezo channels, transient receptor potential (TRP) channels, purinergic receptors, and voltage-sensitive calcium channels (VSCCs). Each of these channel types enable mechanosensation through different mechanisms.

There are currently two models to account for the ability of ion channels to respond to mechanical force. The first is the lipid bilayer model, where tension across the cell membrane generates a conformational change of the channel complex resulting in channel opening.[12,13] The second is the tethered model, which relies on direct connections between the channel and either intra- or extracellular connections.[14] Channels activated via the lipid bilayer model are also referred to as "inherently mechanosensitive," as they are directly altered by mechanical force. The most prominent example of channels responsive to changes in membrane deformations are PIEZO channels.

The mechanosensitive PIEZO1 and PIEZO2 cation channels are found in a variety of tissues. PIEZO1 is predominantly expressed in nonsensory tissues including the urinary bladder, kidney, lung, and chondrocytes.[15] PIEZO2 channels are primarily found in sensory, or electrically excitable, tissues such as dorsal root ganglia sensory neurons and Merkel cells.[16] PIEZO channels help regulate the volume of red blood cells, enable the sensation of light touch, and were found to serve important functions in MSCs.[17] Exposure of human MSCs to hydrostatic pressure, which induces osteogenic differentiation, resulted in increased expression of PIEZO1, but not PIEZO2, channels. Furthermore, knockdown of PIEZO1 impaired osteogenesis with a subsequent increase in adipogenesis.[17] As activation of PIEZO1 channels increased bone morphogenetic protein (Bmp) 2 expression, the ability of PIEZO1 channels to regulate MSC lineage may depend on Bmp2 signaling.[17]

In contrast to the bilayer model, the tethering model of ion channel mechanotransduction requires connections between the channel and either intracellular or extracellular tethers, which may include actin cytoskeletal elements or extracellular matrix (ECM) molecules, respectively.[14] One example of a channel that likely functions through the tethering model is the transient receptor potential vanilloid 4 (TRPV4) ion channel. TRP channels are a family of ion channels consisting of about 30 different types.[18] Each type of TRP channel conveys specific function, many of which regulate various sensations including taste, hot/cold, and stretch or vibration. In humans, TRPV4 is important for proper skeletal development and joint function.[19–21] TRPV4 functions in chondrocytes by regulating the response to dynamic mechanical loading.[22] Although TRPV4 has been implicated in multiple mechanosensation functions, this channel does not respond to indentation[23] and responds only poorly to direct membrane stretch.[24] However, TRPV4 is activated when mechanical force is transmitted through cell-substrate interactions,[23] suggesting that additional interacting components are necessary for force transmission, which supports the tethering model. Additionally, $Ca^{2+}$ signaling through TRPV4 is necessary for the secretion and subsequent alignment of collagen fibers from MSCs.[25] Importantly, $Ca^{2+}$ influx in MSCs, through TRPV4, generated tension across vinculin, a critical component of actin-bound FAs.[25] TRP channels also contain ankyrin repeat domains,[26] allowing for binding between TRP channels and the actin cytoskeleton. Binding of TRP channels between both ankyrins and vinculin further supports the ability of TRP channels to be activated through the tethered model of stretch-activated channels.

While PIEZO and TRP channels are thought of as the classically defined mechanosensitive ion channels, VSCCs are also influenced by mechanical signals. There are several classes of VSCCs including L-type, N-type, P/Q-type, R-type, and T-type channels.[27] Each class of channels has specific properties, for instance, depolarization of L-type channels is "long lasting," whereas that of T-type channels is "transient." Twelve different genes encode for the various types of pore-forming, $\alpha_1$ VSCC subunits.[27] Treatment with inhibitors of L-type VSCCs, such as verapamil and nifedipine, impairs skeletal structure[28] and inhibits osteogenesis, resulting in vertebral defects, decreased mineral apposition rates, and impaired bone formation in animal models.[29–31] Reports also indicate that VSCC blockers, taken during pregnancy, are associated with increased incidence of fetal limb defects.[32,33] In addition to their role in skeletal development, VSCCs are necessary for anabolic responses to skeletal loading. In one study, treatment with verapamil and nifedipine suppressed the load-induced increase in bone formation rate (BFR/BS [bone formation rate per unit of bone surface]) observed in vehicle-treated mice by 56%–61%,[34] highlighting the function of these channels in regulating anabolic responses to loading.

VSCCs are expressed in MSCs,[35] osteoblasts,[36] and osteocytes.[37] Knockdown[28] or inhibition[34] of VSCCs impairs MSC osteogenesis, whereas generation of constitutively active VSCCs in mesenchymal progenitors resulted in increased bone thickness by both activating osteogenesis and inhibiting osteoclast activity.[38] While these data show that VSCCs are crucial for osteogenic differentiation and bone formation, the ability of VSCCs to modulate mechanical effects in MSCs are less well defined. However, in osteoblasts, L-type VSCCs are activated in response to force and subsequently activate purinergic receptors to mediate anabolic responses.[39] As osteoblasts differentiate into osteocytes, expression of L-type channels is downregulated and T-type channels predominate.[37] Recent work shows that T-type VSCCs are critical for the mechanosensitive response of osteocytes.[40]

As VSCCs are defined by their ability to open following voltage changes across the cell membrane, one might question how a "voltage-sensitive channel" responds to mechanical force. One possibility is that gating of inherently mechanosensitive channels, such as TRP channels, results in a local depolarization at the cell membrane sufficient to activate VSCCs via voltage-gating. In this model, VSCCs themselves are not directly responsive to mechanical stimuli but rely on the mechanosensitive nature of other channels, which triggers the voltage sensors within VSCCs. Another possibility is that in addition to depolarization-mediated gating, VSCCs may respond to direct mechanical alterations, possibly through extra- and intracellular connections via auxiliary subunits.

VSCCs are a complex of multiple subunits. In addition to the $\alpha_1$ subunit, which forms the pore through which calcium influx occurs, VSCCs have auxiliary subunits.[41] These protein subunits are encoded by distinct genes and associate with the $\alpha_1$ pore to modulate channel activity. The auxiliary subunits include the intramembranous $\gamma$ subunit, the intracellular $\beta$ subunit, and the $\alpha_2\delta_1$ subunit.[41] Of particular interest, the $\alpha_2$ portion of $\alpha_2\delta_1$ is entirely extracellular, bound to the membrane-anchored $\delta$ peptide by a disulfide bond, which associates with the $\alpha_1$ pore.[42] In neurons, $\alpha_2\delta_1$ regulates the gating kinetics of the channel complex, and the open probability of the channel is diminished in the absence of $\alpha_2\delta_1$.[43] As the $\alpha_2$ portion of $\alpha_2\delta_1$ binds to ECM molecules in other cell types,[44] the $\alpha_2\delta_1$ subunit may serve as a critical tethering node that is capable of modulating VSCC activity in response to mechanical forces transmitted through the ECM. This seems to be true in bone cells, as $\alpha_2\delta_1$ is necessary for extracellular signal-regulated kinase (ERK) 1/2 phosphorylation in response to membrane stretch.[45] In addition to $\alpha_2\delta_1$, the intracellular $\beta$ subunit may enable tethering. In osteoblasts, the $\beta$ subunit binds to a large scaffolding protein called Ahnak,[46] which functionally tethers the channel complex to the actin cytoskeleton.[47] Through this association, mechanically induced cytoskeleton contraction may activate VSCCs and contribute directly to the mechanosensitive response.

## 2.2 Primary Cilia

Nearly all mammalian cells have a single, nonmotile primary cilia, distinct from the large number of motile cilia found in specialized cells, such as those within the tracheal lining.[48] Primary cilia emerge from the distal centriole of the centrosome and contain a microtubule-based cytoskeleton called the axoneme. The presence of motor proteins and adaptors within these structures make them prime candidates to function as transducers of mechanical signals. Indeed, in response to bending deformations, primary cilia initiate influx of extracellular $Ca^{2+}$ through channels, including TRPV4, enabling the activation of downstream signaling cascades.[49] Interestingly, lengthening of the primary cilia resulted in increased sensitivity to mechanical input,[50] further suggesting that deformation of these structures conveys essential mechanical information.

Impaired function of the primary cilia leads to developmental disorders,[51] including skeletal defects.[52] Several bone cell types, including MSCs, contain primary cilia.[53] In MSCs, the primary cilia regulate both osteogenic and adipogenic differentiation, as knockdown of polaris, an intraflagellar transport protein required for cilium formation and function, altered the production of Runx2 and Pparγ, which are the key regulators of osteogenic and adipogenic differentiation, respectively.[53] Furthermore, selective deletion of polaris in mesenchymal progenitors led to stunted limb growth due to disruptions in endochondral and intramembranous ossification,[54] further highlighting the importance of primary cilia in guiding lineage decisions of mesenchymal progenitors.

The function of primary cilia as mechanically sensitive structures was first recognized in the kidney, where bending of the cilia served an essential function in the sensation of fluid flow.[55] The role of primary cilia as fluid flow sensors was also demonstrated in osteocytes.[56] Deletion of Kif3α from bone marrow cells resulted in impaired anabolic responses of bones to mechanical loading, possibly due to impaired recruitment of progenitor cells to sites of bone formation. Furthermore, recruitment of MSCs, via transforming growth factor β, requires primary cilia, and this response is mediated through SMAD-3 signaling.[57] These data suggest that primary cilia contribute to the ability of mechanical signals to regulate osteogenic differentiation of progenitor cells.[58] The ability of primary cilia to transduce mechanical force into a biochemical signal in MSCs may be mediated through connections with TRPV4, as TRPV4 localizes to primary cilia and impairing cilium function resulted in a diminished osteogenic response via TRPV4 following fluid shear.[49]

## 2.3 Focal Adhesions

In contrast to the gating of ion channels or physical deformation of ciliated structures, FAs enable direct connections between the ECM and signaling effectors. FAs are made up of numerous proteins, forming mechanical signaling "platforms."[59] Central to the formation of these signaling platforms are integrins, which both bind to the ECM and enable initiation of intracellular signaling cascades.

Integrins are heterodimeric structures made up of α and β subunits, both of which contain small cytoplasmic domains.[60] Integrin activation induces a conformational change in the β subunit, such as that occurs following stretching of the matrix substrate. Integrin activation enables direct transmission of mechanical signals through the cell membrane, which results in recruitment of signaling adapters at the cytoplasmic regions.[60] These linker proteins contain both structural and biochemical functions, which further propagate mechanical signals through direct structural interactions with the cytoskeleton and by activating downstream signaling cascades.[61]

These adapter proteins include talin, paxillin, vinculin, p130Cas, and focal adhesion kinase (FAK). Talin and paxillin bind to FAK via the FA targeting sequence.[62] Association of talin with the cytoplasmic tail of β integrins brings FAK to the FA complex.[63] Additional adapter protein binding may be initiated by conformational changes in talin in response to a mechanical force. These conformational changes are thought to unmask binding sites for adapter proteins.[64] Paxillin binds FAK and Src kinase,[65] while p130Cas enhances the association of Src kinases with FAs following a mechanical strain.[66] Src kinase and FAK not only initiate downstream signaling events but also are essential for the composition of the FA complex. In particular, accumulation of FAK at FA sites reinforces the structure of the FA through integrin clustering.[62] Importantly, the number of integrins is directly proportional to the strength of adhesion binding to the ECM as well as their persistence[67]; as FAs act as signaling relays for extracellular signals, the number of active FAs strongly influences the intracellular mechanotransduction cascades that control MSC differentiation.

Lineage allocation of mesenchymal progenitors is strongly influenced by the stiffness of the underlying substrate,[68] which is directly related to FA attachment. Transmission of mechanical signals through integrin-based FAs, and the subsequent activation of intracellular signaling, directs the lineage fate of marrow-derived MSCs, partly through an increased number of FAs following a mechanical input.[69] Mechanical force activates pathways that restrict adipogenesis while promoting osteogenic differentiation.[70] ERK1/2 is one such signal that promotes osteogenesis through the activation of Runx2.[71] Another distinctly regulated pathway is initiated by the recruitment of FAK and Fyn (an Src-like kinase) to FA sites.[72] The mechanical strain applied to MSCs enhanced the association of Fyn with vinculin at FAs, resulting in the coactivation of FAK and Fyn through the phosphorylation of tyrosines 397 and 418, respectively. FAK and Fyn were both necessary for mTORC2-dependent phosphorylation of Akt at serine 473. While both pharmacologic and small interfering RNA (siRNA)-mediated knockdown confirmed that mTORC2 is necessary for Fyn/FAK activation of Akt, the regulatory sites of mTORC2 responsible for mechanical activation of Akt remain unclear. Importantly, mechanical force also recruits mTORC2 and Akt to FA sites, where it is presumably positioned to amplify additional mechanical input.[73]

Activation of mTORC2 in MSCs, by mechanical force, culminates in the repression of adipogenesis through two distinct pathways, one through the regulation of adipogenic genes and the other through the enhancement of cytoskeletal structure, leading to increased cellular stiffness. In the former case, transcription of adipogenic genes is altered through nuclear translocation of GSK3β. mTORC2-mediated activation of pAkt phosphorylates GSK3β at serine 9, leading to the inactivation and subsequent release of β-catenin from the proteasomal degradation complex.[74] Release of β-catenin from proteasomal targeting enables nuclear translocation, allowing for the activation of LEF (lymphoid enhancer factor) and TCF (T-cell factor) targeted transcription factors, or through the direct repression of Pparγ transcriptional targets, ultimately resulting in decreased adipogenesis.[75]

In addition to regulating adipogenic gene transcription via GSK3β, mechanical activation of mTORC2 influences actin cytoskeletal structure through activation of RhoA.[72] Inhibition of both mTORC2 and Fyn resulted in abrogation of strain-induced actin stress fiber formation in MSCs, as well as increased adipogenic commitment.[72,73] Formation of actin stress fibers in response to mechanical force requires RhoA,[69] a guanosine triphosphatase (GTPase) that organizes actin monomers into filamentous structures.[76] The on/off balance of RhoA is controlled by guanine nucleotide exchange factors (GEFs) and GTPase-activating proteins (GAPs).[77] GEFs enable the exchange of GDP for GTP, resulting in the activation of RhoA GTPase activity. Conversely, GAPs facilitate release of GTP from the active RhoA. This balance controls the formation of structural components of the actin cytoskeleton, which are particularly important in cells that rely on actin structures for motility, where the leading and lagging edges need to be dynamically remodeled. Juxtaposed to the on/off dynamics required for cell motility, mechanical forces imposed on the skeleton during exercise are delivered in a dynamic fashion. As such, GEFs and GAPs provide a means by which the cellular mechanostat can regulate the necessary stiffness of the cell to respond to additional mechanical stimuli, or to prepare for entry into the appropriate lineage programming. In MSCs, mechanical strain activates RhoA through a GEF called leukemia-associated Rho guanine nucleotide exchange factor (Larg).[78] Knockdown of Larg impaired the mechanical activation of RhoA, while also preventing the ability of mechanical strain to suppress adipogenesis. Furthermore, the Rho GAP, ARHGAP18, was found to be essential for lineage allocation of MSCs. Although ARHGAP18 did not influence the ability of RhoA to be activated in response to force, knockdown of ARHGAP18 reduced the tonic suppression of RhoA activity, resulting in increased actin cytoskeletal structure with enhanced osteogenesis and decreased adipogenesis.[78]

The increased accumulation of FA platforms following the application of mechanical force not only brings signaling effectors to the sites of force transmission but also provides a signaling hub capable of amplifying signal transmission. This effect can be seen in the substantial increase in the activation of Akt and β-catenin signaling following the insertion of a rest period between bouts of mechanical force.[79] As such, in response to the initial mechanical stimulus, signals are recruited to FA platforms,[73] where the enhanced cytoskeletal structure provides the necessary scaffolding to retain such signals where they are amplified in response to the second bout of mechanical input.[69,72] The increased efficacy of a rest period in between two bouts of loading is also evident in tissue responses in vivo where a twice-daily mechanical input, separated by 5 hours, attenuated adipose accumulation and improved glucose metabolism in response to high-fat diet compared with a once-daily mechanical regimen of equal overall duration.[80]

## 3 CYTOSKELETON

The cytoskeleton provides a highly adaptable dynamic scaffolding by which cells move and interact with their environment. Remodeling of cytoskeletal elements, in response to mechanical force, not only provides a mechanism by which cell structure adapts to its physical environment but also repositions the mechanosensitive machinery in cells to amplify the activation of downstream mechanotransduction cascades upon subsequent loading events. In this section, we will discuss the force-bearing cytoskeletal networks including actin, microtubules, and intermediate filaments. We will focus on how force alters the structure and signaling in these networks.

Polymerization of new actin filaments is largely modulated by the actin-related protein (Arp) 2/3 complex, which acts as a nucleation core.[81] Arp2/3-complex-mediated actin branching events are also assisted by nucleation promoting factors from the Wiskott-Aldrich syndrome (WAS) family of proteins, including WASP, N-WASP, and SCAR, which enable rapid polymerization.[81–83] Arp2/3 complex and WAS family of proteins also play a role in binding newly formed actin filaments to the existing actin network.[84,85] Formins, on the other hand, regulate the end-to-end polymerization of F-actin. Formins are classified by the presence of formin homology 1 (FH1) and 2 (FH2) domains.[86] Myosin-related cytoskeletal contractility in stem cells is regulated by small Rho GTPases such as RhoA, Ras, and CDC42A.[87] RhoA activity increases the cell tension through its effector protein

Rock, activating myosin light-chain kinase, which in turn activates the dimerized motor protein myosin II.[88] RhoA can be activated by a variety of mechanical cues such as fluid flow, ECM, ECM compliance, and vibrations.[69,89,90] The end result of RhoA activation is stress fiber formation in cells. Formation of these RhoA-mediated stress fibers is context dependent. Under laminar fluid shear, for example, stress fiber formation is parallel to the flow direction.[89,91] Under cyclic strain, cells reorient their stress fibers perpendicular to the principal stretch direction.[92,93] Different than both fluid shear and strain, low-intensity vibrations result in robust F-actin bundling at the perinuclear region of cells.[94] This perinuclear F-actin formation is concomitant to the activation of F-actin modulators such as RhoA stimulator ARHFGEF11 (Rho guanine nucleotide exchange factor 11, +6-fold) and Arp2/3 complex regulatory protein WAS (+43-fold).[95] Therefore the F-actin cytoskeleton not only is load bearing but also can adapt to mechanical force in a context-dependent manner.

Aside from the "slow" adaptations of cytoskeletal networks through Rho GTPases, the cytoskeleton also functions to transmit mechanical signals within the cell, and this transmittance appears to be dependent on the cellular tension. In airway smooth muscle cells, for example, integrin-bound surface deformation of 0.4 μm (peak) deformed the nucleus, and stress propagation within the cell was controlled by intracellular stress levels.[96] This effect of intracellular stress levels on force propagation can be explained by the concept of tensegrity. Tensegrity dictates that cells are under a constant force balance and thus even small local deformations to the cytoskeleton are compensated by the whole structure.[97] Consistent with the hypothesis, fluid-flow-induced deformations in a single cell were weakly correlated with the direction of the flow but instead propagated through cell structure in a multidirectional way.[98] The tensegrity model also predicts instantaneous and nonlinear cell stiffening under magnetically induced membrane deformations.[99,100] These key studies open up a new perspective in which mechanical signals are the first signals that propagate through the whole cell body even earlier than biochemical signals. Therefore effective transmittance of these forces may rely on a prestressed cytoskeleton.

F-actin fibers are under constant stress; for example, initiation of F-actin polymerization and signaling requires forces in the piconewton range ($\sim 10-50$ pN).[101] When F-actin fibers bundle, they generate more and more force. In smooth muscle cells, dissection of a single apical actin stress fiber generates a force of 65 nN on the nucleus.[102] Similarly, switching between weak and strong actin

coupling with the nuclear envelope (NE) can generate up to 40 nN force differentials. These forces generated by the F-actin cytoskeleton result in micro-/nanoscale damage of individual fibers. One important molecule that plays a role in the repair of these small breaks is zyxin.[103] Zyxin is a prominent member of FA, but in contrast to other FA proteins that get recruited when force is present, zyxin leaves the FA site upon mechanical force and localizes to actin fibers that are damaged due to cellular contractions or extracellular deformation. This zyxin localization to stress fibers then recruits other F-actin initiators such as α-actinin to fortify and repair F-actin fibers.[104–107]

Intermediate filaments act as stabilizers of the cytoskeleton, play a role in resisting shear stress,[108] and provide stiffness to the cell nucleus.[109] We will discuss intermediate filaments in later sections within the context of nuclear structure. Microtubules are the stiffest of the three building blocks of the cytoskeleton and can form long tracks that span the length of the cell. During mitosis, these long and stiff networks assemble into radial arrays that function as central hubs and facilitate intracellular transport. It is not entirely clear if microtubules are directly involved in mechanosensory functions. However, microtubules are known to buckle under compressive loads in cells[110] and these compressive loads can increase the overall curvature and create fractures in microtubules. The relationship between microtubule fracture and intracellular forces suggests a regulatory mechanism.[111] Additionally, microtubules act to stabilize the cytoplasm and nucleus.[112] When endothelial cells were stretched by pulling actin filaments, removal of functional microtubules resulted in the release of normal restriction of nuclear movement,[109] suggesting that microtubules may indirectly be involved in mechanosensory functions by protecting the nucleus from extreme deformations. Experimental evidence regarding the regulatory function of microtubule integrity on mechanosensing is not clear. For example, when microtubules were disrupted by colchicine, cell sensitivity to fluid flow was enhanced, by further increasing prostaglandin (PG) $E_2$ levels released by osteoblast cells.[113] A different study, however, showed muting of the $PGE_2$ response.[114] These studies show that microtubules play a role in mechanotransduction, but the mechanism by which the regulation is controlled is less clear.

## 4 NUCLEUS

The nucleus is the largest and most dense organelle found in our cells. In the past decade, the understanding of nuclear functionality has shifted from a passive organelle to an active participant in the mechanisms of mechanosensing and mechanosignaling. In this section, we will divide the nucleus into three main compartments: the NE, the nucleoskeleton, and the chromatin. First, the NE will be discussed, as the NE is a unit that maintains a dynamic connectivity between the cytoskeleton and the nucleoskeleton. Second, the nucleoskeleton will be discussed from the perspective of nuclear actins, microtubules, and intermediate filaments (lamin A/C and lamin B). Finally, the chromatin will be introduced as a structural element, and we will discuss how chromatin organization may play a role in mechanosensing. Throughout these sections, the focus will be on both structural sensing mechanisms and identified signaling proteins that facilitate nuclear mechanotransduction. Additionally, important diseases that affect nuclear organization and mechanosignaling will be highlighted, as modern medicine and science has been linking mechanosignaling and mechanotransduction to known human pathologic conditions.

## 4.1 Nuclear Envelope

The NE is a double-membraned barrier that is composed of the outer nuclear membrane (ONM) and the inner nuclear membrane (INM), which are separated by a 20- to 100-nm perinuclear space. Despite continuity between the NE and the endoplasmic reticulum (ER), the ONM and INM house diverse protein groups that are not common to the ER.[115] As the NE physically separates cytoplasmic and intranuclear compartments, a vital function of the NE is to interact with both the cytoskeleton and chromatin to enable mechanical and biochemical communication between these two compartments.[116]

At the ONM the nucleus interacts with cytoskeletal networks via the linker of nucleoskeleton and cytoskeleton (LINC) complex.[117] The LINC complex provides physical coupling of the nucleus to the cytoskeleton by maintaining interactions with actin, microtubule, and intermediate filament networks.[118] Giant isoforms of nesprin-1 and nesprin-2 reside at the cytoplasmic face of the ONM and provide the cytoskeletal tethering points for the LINC complex. N termini of giant nesprins share a calponin homology (CH) domain that binds to actin with high affinity. The CH domains found in nesprin-1 and nesprin-2 giant isoforms are identical to those found in α-actinin.[119,120] Presenting CH domain of nesprins to nonpolymerized actin filaments promotes actin polymerization in vitro,[120] suggesting that CH domains may act as initiation points for actin polymerization at the perinuclear region of the cytoplasm. Nesprins have been found to interact with microtubules through intermediate proteins such

as dynein and kinesin.[121,122] The N and C termini of giant nesprin isoforms are connected together with larger spectrin repeat (SR) domains. The giant isoform of nesprin-1 has 72 SR domains, whereas the giant isoform of nesprin-2 has 56 SR domains,[123] making the size of the proteins ~1000 and ~800 kDa, respectively. These SR regions in giant nesprins are large, but the role these regions play in mechanosensing is not well understood. It has been suggested that these regions may play a role in mechanical load transmission.[124] Interestingly, these SR domains are highly conserved across different species[123] and thus present putative binding sites for protein-protein interactions[124] that may be important for mechanosensing. For example, FH1-/FH2-domain-containing protein 1 (FHOD1) binds to the SR region of the giant isoform of nesprin-2. FHOD1 binding has been shown to increase the coupling strength between LINC complex and F-actin.[125] Other cytoskeleton-related binding partners of the LINC complex include the actin bundling protein fascin.[126] Indeed, findings utilizing fluorescence resonance energy transfer (FRET)-based force sensors have shown that giant nesprin-2 was subject to myosin-dependent tension.[127] These force sensors are based on the so-called nesprin-mini, a designed nesprin isoform that lacks an SR region. Overexpression of nesprin-mini increases the number of F-actin connections to the nucleus and decreases nuclear size.[128] There are a number of naturally occurring isoforms of nesprin-1 and nesprin-2 that lack CH and large SR regions.[129] These smaller α and β isoforms are found in both the cytoplasm and the nucleus.[130] The LINC complex also interacts with intermediate filaments through plectin; however, this interaction is facilitated by nesprin-3, a structurally distinct isoform of nesprin.[118]

The C termini of nesprin proteins traverse the ONM and end in a conserved KASH (Klarsicht, ANC-1, Syne homology) domain.[131] The KASH domain interacts with sun (Sad1p, UNC-84) proteins. In mammalian cells, two isoforms are highly expressed: sun-1[132] and sun-2.[133] The N terminals of sun-1 and sun-2 proteins start at the perinuclear space, traverse the INM, and connect to nuclear lamina at the C terminal. The physical lengths of sun-1 and sun-2 are predicted to be similar to the distance between ONM and INM[134,135] and possibly play a role in maintaining the size of the perinuclear space. For example, shorter sun isoforms shorten the distance between INM and ONM, but in contrast, depleting sun proteins increases the spacing between INM and ONM.[136] Proteins sun-1 and sun-2 are structurally similar and capable of forming strong trimeric structures that connect to the KASH domain of LINC complexes.[137] Furthermore, depleting either sun-1 or sun-2 is not sufficient to disrupt LINC complex connections.[117] However, findings indicate that there are distinct functional differences between sun-1 and sun-2. For example, sun-2 anchorage to the INM requires lamin A/C, whereas sun-1 localization to the INM is less dependent on lamin A/C.[138] Also, only sun-1 associates with the nuclear pore complexes (NPCs).[139] Functionally, sun-2 was found to be involved in actomyosin-dependent nuclear connectivity, whereas sun-1 function was preferentially important for meiosis[140] and dynein-mediated connectivity with microtubules.[141]

As discussed earlier, LINC complexes play a critical role in connecting the nucleus to the cytoskeleton. However, in the recent years, LINC complex connections emerged as important mechanotransduction elements for signaling molecules. The nucleus relies on both direct mechanical input and its molecular transducers to sense external stimuli and regulate gene expression. In MSCs that reside in bone tissue, β-catenin and Yes-associated protein/PDZ-binding motif (YAP/TAZ) have been identified as two important molecular "transducers" of mechanical information and they rely on LINC-mediated connectivity to be functional. Residing in the cytoplasm, both β-catenin and YAP/TAZ are activated (dephosphorylated) by mechanical strain[142,143] and enter into the nucleus to function as transcription factors. Depleting nesprin-1 impairs strain-induced nuclear entry of YAP1 in MSCs.[144] It has been reported that the access of transcriptional coactivator YAP1 to the nucleus is facilitated through the stretching of NPCs via F-actin contractility to facilitate its nuclear accumulation.[145] The LINC complex has an important role in β-catenin nuclear access as well. β-Catenin does not possess a classic nuclear localization signal; instead, it transits through the nuclear leaflets via direct contact with the NPC.[146,147] β-Catenin is localized on the LINC element nesprin-2 that appears to provide a "launching pad" for a subsequent nuclear entry.[148] It has been reported that untethering of nesprin-2 from the NE displaces β-catenin from the NE, impeding its nuclear entry rate, reducing its nuclear levels, and lowering the transcription of β-catenin gene target Axin-2.[149]

LINC-mediated physical coupling to the nucleus through the cytoskeleton is mechanoadaptive and plays a role in cell mechanosensitivity. Sun-1- or nesprin-4-deficient mice fail to maintain the nuclear positioning in mechanosensory epithelial cells, resulting in gradual hearing loss.[150] This suggests that the LINC complex's connections to the nucleus has a part to play in the sensation of mechanical vibrations, such as those sensed as sound. When MSCs are subjected to high frequency (30−200 Hz), low-intensity (0.1−1 g) vibrations that mimic muscle contractions,[151,152] FAK-mediated RhoA activation and subsequent cytoskeletal remodeling is activated in MSCs.[94] Depleting the LINC complex functionality via sun-1- and sun-2-specific siRNA sequences, or overexpressing a dominant negative form of nesprin KASH domain, impairs vibration-induced FAK and Akt phosphorylation.[94] Importantly, stretching the substrate to which the cells are attached to also activates FAK and Akt signaling.[153] However, depleting LINC function[94] or removing nuclei[154] was not sufficient to inhibit strain-activated FAK and Akt signaling. Strain-induced signaling was also intact in LINC-deficient fibroblasts,[155] but not in cardiomyocytes.[156] These findings suggest that the LINC complex may play a role in how cells sense certain mechanical information. Therefore mechanical influences may cause gain or loss of function in the LINC complex to alter cell mechanosensitivity. In support of these observations, conditions that simulate unloading events, such as microgravity, result in the reduced expression of LINC complex elements.[157] In contrast, LINC complex elements are more highly expressed in response to low-intensity vibrations.[158] The ability of mechanical signals to regulate the LINC complex implies that positive regulation of LINC complex elements through the application of mechanical therapies, such as low-intensity vibrations, may be leveraged to improve cell sensitivity to external mechanical and biochemical demands. Not surprisingly, the application of low-intensity vibration has been shown to amplify the response to other mechanical or biochemical factors in mice.[4]

An important protein that plays a prominent role in NE mechanotransduction is emerin.[159] Functionally, emerin is an actin capping protein that binds to the pointed ends of actin fibers and accelerates actin polymerization.[160] When force is applied to epidermal stem cells, emerin enrichment at the ONM is accompanied by the recruitment of nonmuscle myosin IIA to promote local actin polymerization at the perinuclear region.[161] Emerin also plays a role in the nuclear export of β-catenin, as emerin depletion results in nuclear β-catenin accumulation.[148,162−164] Furthermore, direct application of force to isolated nuclei results in Src-dependent emerin phosphorylation leading to increased nuclear stiffness in a lamin A/C dependent manner.[165] Clinically, mutations of emerin are associated with Emery-Dreifuss muscular dystrophy

(EDMD). EDMD is characterized by early contracture of the elbows and Achilles tendons, slow progressive muscle weakening and wasting, and cardiomyopathy with conduction block. There are two main genetic forms of EDMD: X-linked EDMD (X-EDMD) and autosomal dominant EDMD (AD-EDMD). The X-EDMD version is caused by a mutation in emerin, whereas the AD-EDMD inherited version is caused by mutations in *LMNA*.

Another important regulator of mechanosensing is torsin A. Torsin A is a NE protein that belongs to AAA+ family (ATPases associated with various cellular activities) that utilizes ATP to unfold other proteins.[166,167] Torsin A interacts with sun-1,[168] nesprin-3α,[169] lamina-associated polypeptide (LAP) 1,[170] and emerin.[171] While the exact function of torsin A in nuclear mechanotransduction is unclear, its functional role in the formation of perinuclear actin cables during rearward nuclear movement[172] suggests that it is important in regulating the cytoskeletal dynamics at the NE.

## 4.2 Nucleoskeleton

The term nucleoskeleton is used to describe the pellet of proteins that are precipitated during salt buffer extractions of cell nuclei.[173] This insoluble fraction of nuclear structure houses the nuclear lamina, chromatin, and cytoskeletal scaffolding. These proteins play a vital role in providing structural support to the nucleus, regulating gene expression, and regulating chromatin structure. Additionally, nucleoskeleton proteins have been linked to mechanical signal sensing and transduction. Nucleoskeleton proteins also affect human health, as multiple human diseases have been linked to the misregulation and synthesis of nucleoskeleton proteins, causing a wide range of diseases ranging from deformed muscle tissues to inducing rapid premature aging. This section investigates the role of nucleoskeleton proteins in mechanical sensing and signal transduction, human health, and the regulation of chromatin.

### 4.2.1 Nuclear actin and microtubules

Although the role of actin within the nucleus is largely uncharted, the nucleus houses a large number of regulators of actin polymerization and depolymerization.[174] These components include mDia (mammalian Diaphanous-related) formins that catalyze end-on-end actin polymerization and Arp2/3 complex elements necessary for actin branching.[175] Actin can be found in filamentous forms[176,177] and as actin-cofilin rods[178] within the nucleus. Importantly, nuclear F-actin structures form in a LINC-complex-dependent manner during cell spreading,[179] suggesting actin polymerization as the

potential target for mechanosignaling. In MSCs that are treated with cytochalasin D, cytoplasmic actin depolymerizes and rapidly enters into the nucleus.[180] The influx of G-actin results in secondary intranuclear scaffolding that results in osteogenic differentiation in MSCs.[181] The effect of cytochalasin D can be abolished when actin branching is inhibited via arp2/3 ck666.[181] A consistent target for cytochalasin D targeted transcriptions is vestigial-like 4 (VGLL4),[182] a regulator of the Hippo pathway that controls the interactions of YAP/TAZ interaction with the transcription factor TEAD (transcriptional enhanced associate domain). While cytochalasin D represents a chemically induced effect of actin influx into the nucleus, persistent actin polymerization in the nucleus has been shown to reduce messenger RNA transcription,[183] suggesting a common, actin-related pathway. Not surprisingly, nuclear actin has also directly been implicated in gene transcription through its association with RNA polymerases. Immunoprecipitation assays have shown that nuclear actin directly interacts with RNA polymerase II.[184] As such, the intranuclear state of actin controls both the availability of genes to their transcription factors and the transcription.

Nuclear cytoskeletal elements, in conjunction with LINC complex elements, play a role in directing DNA repair. For example, inhibiting actin polymerization alters telomere dynamics.[185] During postmitotic genome reorganization, the LINC protein sun-1 and GTPase rap1 (Ras-proximate-1 or Ras-related protein 1) mediate telomere tethering onto the NE.[186] As telomeres function to protect the ends of chromatin from DNA repair machinery,[187] mechanically induced changes to the nuclear actin structure may play a role in cellular health. Microtubules also contribute to the DNA repair machinery from both inside and outside the nucleus. It has been reported that microtubules, in conjunction with the LINC complex and double-strand break repair protein 53BP, generate pockets within the nuclear interior to direct double-stranded DNA break mobility.[188] While the extent of nuclear microtubule machinery and its responsiveness to mechanical cues is unknown, the existence of DNA-damage-inducible intranuclear microtubule filaments was recently discovered,[189] suggesting that traditional cytoskeletal proteins may play an important role in sensing and responding to mechanics within the cell nucleus.

### 4.2.2 Lamin A/C

Lamin A and lamin C are part of the lamin protein family within which they can be further classified as A-type lamins. All lamin proteins are considered intermediate filaments. A-type lamins are expressed by the gene *LMNA*.[190] Alternative splicing produces either lamin A or lamin C, most often termed lamin A/C proteins. Both A-type lamins and B-type lamins, which will be discussed in the next section, share similar protein domains with other intermediate filaments. They are composed of an N-terminal head domain, coiled-coil central rod domain, and a C-terminal domain. Both lamin types have unique features such as a nuclear localization signal, an immunoglobulin (Ig)-fold domain, and a chromatin binding site.[191] During maturation, lamin A undergoes posttranslational modification in the CAAX motif located in the C terminal. The CAAX motif allows for modifications at the cysteine residues where farnesylation and proteolytic processing occur. Other modifications include carboxymethylation of new C-terminal residues and isoprenylation. Finally, a second proteolytic event occurs that is mediated by zinc metalloproteinase ZMPTSE24, which removes 15 amino acids from the C-terminal end. This secondary proteolytic cleavage removes any previously modified cysteines to form the mature lamin A.[191] Interestingly, lamin C escapes these modifications because it lacks a C terminal due to the alternative splicing of the *LMNA* gene. These posttranslational modifications during maturation of lamin A play an important role in human pathologic conditions related to lamin proteins, termed laminopathies, as most laminopathies are related to an altered posttranslational modification or a loss of function of ZMPTSE24.[192]

All A-type lamins are not constitutively expressed but are expressed during and after differentiation into specific tissues. For example, lamin A is not detected in the brain cells of mice until several days after birth[193] and embryonic cells display very low levels of lamin A/C.[194] One of the prominent roles of lamin A/C is maintaining the mechanical competency of the nucleus based on extracellular mechanosensing. Depletion of lamin A and lamin C results in a softer nuclei,[195] and lamin A/C levels in resident tissue types positively correlate with the tissue stiffness.[196] One of the mechanisms that regulate lamin A/C levels based on physical force appears to be its phosphorylation. Fluid shear stress, for example, has been shown to expose the $Cys^{522}$ residue of the lamin A/C Ig tail domain, and Ser/Thr phosphorylation of these residues appear to be inversely correlated with lamin A/C stability.[196] In this way, increasing matrix stiffness reduces the Ser/Thr phosphorylation and results in increased protein levels of lamin A/C.[197] This correlation between matrix stiffness and lamin A/C levels may be seen as a protective

mechanism to shield cell nuclei from excessive environmental force and may serve to blunt the effectiveness of mechanical force on the chromatin in order to maintain normal transcriptional activity. Cells with lamin A/C depletion or mutation are more sensitive to mechanical damage,[196,198] and lamin A/C was also shown to recruit perinuclear actin cables upon cyclic mechanical stretching to shield the nucleus from mechanical damage.[199] However, alternative protective mechanisms also exist to shield the genome in embryonic stem cells with very low levels of lamin A/C.[194] As embryonic stem cells have an open chromatin state with little or no heterochromatin,[200] during the transition phase from undifferentiated to predifferentiation states, the nucleus becomes auxetic, an intrinsic material behavior described by the ability to shrink transversely when compressed in the longitudinal direction (i.e., Poisson's ratio is negative).[201] This auxetic property of embryonic nuclei increases the compression modulus considerably,[202] thus protecting the chromatin from mechanical perturbation.

Interestingly, cancer cells also display reduced levels of lamin A/C and LINC complex elements.[203] As cancer cells are characterized by uncontrolled gene expression, loss of lamin A/C in these cells supports the important role of lamin A/C in regulating chromatin dynamics as well as the subsequent control of cell fate.[187]

Laminopathies describe a breadth of human pathologic conditions. Currently, there are 15 diseases directly linked to mutations in *LMNA* and the altered synthesis of lamin A/C.[192] Changes in lamin A/C produce four main clinical phenotypes: striated muscle diseases, lipodystrophic syndromes, peripheral neuropathy, and accelerated aging diseases. One of the most well-known laminopathies is the Hutchinson-Gilford progeria syndrome, also known as progeria, which induces premature aging. It is hallmarked by postnatal growth retardation, premature atherosclerosis, generalized osteodysplasia with osteolysis, micrognathia, absence of subcutaneous fat, alopecia, and pathologic fractures. The cause of progeria is a dominant silent mutation in exon 11.[204] This silent mutation leads to a splicing defect and the generation of a smaller prelamin A variant, creating a new splice variant called progerin. Progerin has permanent farnesylation and carboxymethylation that causes a loss of the cleavage site, not allowing the secondary proteolytic event to occur.[205] This induces progerin accumulation in the nuclear periphery. Interestingly, progerin accumulation is also accompanied by upregulation of the LINC complex element sun-1. For example, farnesylated-lamin-producing mice[206] with the LMNA$^{L530P/L530P}$ (LmnaΔ9) mutation[207] show increased presence of sun-1 in their NE and ER membranes, and sun-1 depletion was sufficient to partially rescue the bone phenotype.[208] Although the reason why sun-1 accumulation leads to a decreased bone phenotype is unclear, both fibroblasts from patients with progeria and progerin-expressing fibroblasts show abnormal sun-1 accumulation. This accumulation of sun-1 results in preferential cell polarity defects due to increased sun-1 interaction with microtubules.[141] Importantly, similar effects are seen in fibroblasts from aged patients, which show low levels of progerin production and display irregular nuclear geometry.[209] Critically, skeletal tissues display increased bone marrow fat and reduced bone quality during aging.[210] One possible mechanism that explains why both progeric and aged cells cannot respond to mechanical signaling may be the relation between β-catenin mobility and the LINC complex. Both LmnaΔ9 mice and LmnaΔ50 cells display reduced effectiveness of Wnt activity and decreased β-catenin levels.[211] Our work in vitro has further shown that disabling the LINC complex function by depleting sun-1 and sun-2 results in decreased β-catenin function and increased adipogenic bias in MSCs.[94] These findings suggest that age-related alterations in the NE and nucleoskeleton may alter how cells respond to their mechanical environments.

### 4.2.3 Lamin B

Other lamins are classified as B-type lamins. B-type lamins are made of two different proteins, lamin B1 and lamin B2, which are also intermediate filaments. Unlike lamin A/C, lamin B1 and lamin B2 are expressed from two different genes *LMNB1* and *LMNB2*, which are located on chromosome 5 and 19, respectively.[193] As noted previously, B-type lamins also have the same domains as A-type lamins. Additionally, B-type lamins experience similar posttranslational modifications such as farnesylation at cysteine residues, proteolytic processing, and carboxymethylation during maturation. However, B-type lamins do not have any secondary proteolytic processing in which 15 amino acids are cleaved to remove the farnesylated and carboxymethylated sites. Instead, mature B-type lamins permanently maintain farnesylated and carboxymethylated tails. Known laminopathies are also linked to B-type lamin mutations.[212]

One B-type lamin disease is adult-onset leukodystrophy. This disease is linked to autosomal dominant mutations in the gene *LMNB1*. Adult-onset leukodystrophy is hallmarked by demyelination of the central nervous system. This disease is similar to multiple sclerosis,

where there is a widespread demyelination of the central nervous system. The cause of adult-onset leukodystrophy is linked to duplication of the genomic region containing *LMNB1*, leading to an increase in lamin B1. Although it is not fully known how the overexpression of lamin B1 leads to adult-onset leukodystrophy, it is hypothesized that when lamin B1 is overexpressed, it causes an abnormal nuclear shape that is more susceptible to physical disturbances.[213] Not to be left out, lamin B2 is indicated in acquired partial lipodystrophy.[212] Acquired partial lipodystrophy is the loss of subcutaneous adipose tissue in areas around the neck, face, arms, and legs. Studies have discovered that patients with heterozygous mutations in *LMNB2* are susceptible to acquiring partial lipodystrophy.[213]

Unlike lamin A/C, lamin B proteins associate with neither the NPCs nor lamin A/C. Instead, lamin B proteins bind to the lamin B receptor (LBR) located in the nuclear membrane.[215] B-type lamins' role in mechanical signal sensing and transduction is contrastingly small compared with that of lamin A/C. In cells with defects in lamin B1, few changes in the cell's mechanically sensitive genes or changes in the nuclear mechanics are detected when a mechanical force is applied.[195] In relation to chromatin regulation, when chromatin was removed from cells, lamin A/C was uncovered but lamin B was not.[216]

### 4.3 Chromatin

Chromatin is the level of DNA packaging where the DNA is wrapped around histone proteins that are then wrapped more tightly to form into nucleosomes. Euchromatin and heterochromatin describe the origination of the nucleosomes. Euchromatin is lightly packed nucleosomes, commonly referred to as "beads on a string," that allows transcription to occur on exposed genes. Heterochromatin is tightly packed nucleosomes forming a 30-nm fiber that does not allow transcription of enclosed genes to occur. Chromatin binding domains have been found on both A-type and B-type lamins. Chromatin close to lamins in the nucleoskeleton forms heterochromatin.[217] This is a result of the lamins, along with a class of proteins called LEM (LAP2, emerin, MAN1) proteins that anchor the chromatin to the INM. The LBR and A-type lamins are very important to chromatin localization, as their loss causes heterochromatin to localize in the interior of the nucleus.[217] This occurs because the LBR and LAPs tether the chromatin to lamins.[218] Lamins can also directly bind to the DNA through matrix-associated regions (MARs) in the DNA. These MARs are AT-rich, are located at the beginning and the end of protein-coding genes, and organize the chromatin into distinct functional chromosomal territories.[219]

The role of chromatin as a load-bearing structure is a relatively new concept. Earlier studies with micropipette aspiration and atomic force microscopic assays showed that chromatin significantly contributes to the apparent nuclear modulus.[220] As cells differentiate from the embryonic state, they accumulate heterochromatin in order to silence unwanted gene regions,[221] consequently inhibiting heterochromatin regulators, such as ezh2, to uncover histone-embedded gene regions, which is a robust way to affect cell differentiation.[222] When stem cells differentiate, their nuclei become stiffer.[223,224] Importantly, chromatin is directly subjected to mechanical force from the extracellular environment. For example, deformation applied to the cell membrane via magnetic beads was transmitted to the nuclear surface through LINC complexes and directly regulated the displacements of coilin and SMN (survival motor neuron) proteins in nuclear Cajal bodies.[225] Mechanical forces may also cause detachment of the chromatin from lamins, allowing genes to be accessible to transcription initiation factors.[215] In this way, not only pharmacologic regulation of histone modifications regulates nuclear stiffness[226] but also mechanical forces applied to the cell membrane directly alter heterochromatin dynamics[161] and chromatin organization.[227] While the mechanisms by which heterochromatin is altered by mechanical challenges is an underexplored question, chromatin should be considered as a load-sensitive compartment with self-regulatory functions.

## 5 SUMMARY

Cells possess numerous load-bearing structures that actively sense and respond to mechanical challenges. The adaptations of and interactions between these load-bearing elements are integral to the cellular function. In the following chapter, some new techniques for measuring these forces within the cell will be presented. The field of mechanobiology is just beginning to accumulate enough knowledge to utilize these mechanisms in regenerative medicine and tissue engineering applications. Ultimately, it may be possible to combat conditions, such as aging or space travel, that suffer from altered mechanosensitivity of tissues by leveraging the mechanosensitive pathways.

## FUNDING SUPPORT

NIH 5P2CHD086843-03, P20GM109095 and P20GM103408 (GU). DOD BC150678P1, NIH R15AR069943-01, and NIH 1R01AR074473-01 (WRT).

## REFERENCES

1. Lemke SB, Schnorrer F. Mechanical forces during muscle development. *Mech Dev*. 2017;144:92−101. https://doi.org/10.1016/j.mod.2016.11.003.
2. Fung YC. *Biomechanics: Mechanical Properties of Living Tissues*. 2nd ed.. Springer-Verlag; 1993.
3. Riddle RC, Donahue HJ. From streaming potentials to shear stress: 25 years of bone cell mechanotransduction. *J Orthop Res*. 2009;27:143−149. https://doi.org/10.1002/jor.20723.
4. Pagnotti GM, et al. Combating osteoporosis and obesity with exercise: leveraging cell mechanosensitivity. *Nat Rev Endocrinol*. 2019. https://doi.org/10.1038/s41574-019-0170-1.
5. Pchelintseva E, Djamgoz MBA. Mesenchymal stem cell differentiation: control by calcium-activated potassium channels. *J Cell Physiol*. 2018;233:3755−3768. https://doi.org/10.1002/jcp.26120.
6. Ajubi NE, Klein-Nulend J, Alblas MJ, Burger EH, Nijweide PJ. Signal transduction pathways involved in fluid flow-induced PGE(2) production by cultured osteocytes. *Am J Physiol*. 1999;276:E171−E178.
7. Hung CT, Allen FD, Pollack SR, Brighton CT. Intracellular $Ca^{2+}$ stores and extracellular $Ca^{2+}$ are required in the real-time $Ca^{2+}$ response of bone cells experiencing fluid flow. *J Biomech*. 1996;29:1411−1417.
8. Jiang LH, Mousawi F, Yang X, Roger S. ATP-induced $Ca(2+)$-signalling mechanisms in the regulation of mesenchymal stem cell migration. *Cell Mol Life Sci*. 2017;74:3697−3710. https://doi.org/10.1007/s00018-017-2545-6.
9. Kawano S, et al. ATP autocrine/paracrine signaling induces calcium oscillations and NFAT activation in human mesenchymal stem cells. *Cell Calcium*. 2006;39:313−324. https://doi.org/10.1016/j.ceca.2005.11.008.
10. Sun S, Liu Y, Lipsky S, Cho M. Physical manipulation of calcium oscillations facilitates osteodifferentiation of human mesenchymal stem cells. *FASEB J*. 2007;21:1472−1480. https://doi.org/10.1096/fj.06-7153com.
11. Kim TJ, et al. Substrate rigidity regulates Ca2+ oscillation via RhoA pathway in stem cells. *J Cell Physiol*. 2009;218:285−293. https://doi.org/10.1002/jcp.21598.
12. Kloda A, Martinac B. Molecular identification of a mechanosensitive channel in archaea. *Biophys J*. 2001;80:229−240. https://doi.org/10.1016/S0006-3495(01)76009-2.
13. Martinac B, Adler J, Kung C. Mechanosensitive ion channels of *E. coli* activated by amphipaths. *Nature*. 1990;348:261−263. https://doi.org/10.1038/348261a0.
14. Gillespie PG, Walker RG. Molecular basis of mechanosensory transduction. *Nature*. 2001;413:194−202. https://doi.org/10.1038/35093011.

15. Geng J, Zhao Q, Zhang T, Xiao B. In touch with the mechanosensitive piezo channels: structure, ion permeation, and mechanotransduction. *Curr Top Membr*. 2017;79:159–195. https://doi.org/10.1016/bs.ctm.2016.11.006.

16. Wu J, Lewis AH, Grandl J. Touch, tension, and transduction - the function and regulation of piezo ion channels. *Trends Biochem Sci*. 2017;42:57–71. https://doi.org/10.1016/j.tibs.2016.09.004.

17. Sugimoto A, et al. Piezo type mechanosensitive ion channel component 1 functions as a regulator of the cell fate determination of mesenchymal stem cells. *Sci Rep*. 2017;7:17696. https://doi.org/10.1038/s41598-017-18089-0.

18. Startek JB, Boonen B, Talavera K, Meseguer V. TRP channels as sensors of chemically-induced changes in cell membrane mechanical properties. *Int J Mol Sci*. 2019;20. https://doi.org/10.3390/ijms20020371.

19. Kang SS, Shin SH, Auh CK, Chun J. Human skeletal dysplasia caused by a constitutive activated transient receptor potential vanilloid 4 (TRPV4) cation channel mutation. *Exp Mol Med*. 2012;44:707–722. https://doi.org/10.3858/emm.2012.44.12.080.

20. Lamande SR, et al. Mutations in TRPV4 cause an inherited arthropathy of hands and feet. *Nat Genet*. 2011;43:1142–1146. https://doi.org/10.1038/ng.945.

21. Nilius B, Voets T. The puzzle of TRPV4 channelopathies. *EMBO Rep*. 2013;14:152–163. https://doi.org/10.1038/embor.2012.219.

22. O'Conor CJ, Leddy HA, Benefield HC, Liedtke WB, Guilak F. TRPV4-mediated mechanotransduction regulates the metabolic response of chondrocytes to dynamic loading. *Proc Natl Acad Sci U S A*. 2014;111:1316–1321. https://doi.org/10.1073/pnas.1319569111.

23. Servin-Vences MR, Moroni M, Lewin GR, Poole K. Direct measurement of TRPV4 and PIEZO1 activity reveals multiple mechanotransduction pathways in chondrocytes. *Elife*. 2017;6. https://doi.org/10.7554/eLife.21074.

24. Strotmann R, Harteneck C, Nunnenmacher K, Schultz G, Plant TD. OTRPC4, a nonselective cation channel that confers sensitivity to extracellular osmolarity. *Nat Cell Biol*. 2000;2:695–702. https://doi.org/10.1038/35036318.

25. Gilchrist CL, et al. TRPV4-mediated calcium signaling in mesenchymal stem cells regulates aligned collagen matrix formation and vinculin tension. *Proc Natl Acad Sci U S A*. 2019;116:1992–1997. https://doi.org/10.1073/pnas.1811095116.

26. Takahashi N, et al. TRPV4 channel activity is modulated by direct interaction of the ankyrin domain to PI(4,5)P(2). *Nat Commun*. 2014;5:4994. https://doi.org/10.1038/ncomms5994.

27. Catterall WA, Seagar MJ, Takahashi M, Nunoki K. Molecular properties of voltage-sensitive calcium channels. *Adv Exp Med Biol*. 1989;255:101–109.

28. Li J, et al. Skeletal phenotype of mice with a null mutation in Cav 1.3 L-type calcium channel. *J Musculoskelet Neuronal Interact*. 2010;10:180–187.

29. Duriez J, Flautre B, Blary MC, Hardouin P. Effects of the calcium channel blocker nifedipine on epiphyseal growth plate and bone turnover: a study in rabbit. *Calcif Tissue Int*. 1993;52:120–124.

30. Ridings JE, Palmer AK, Davidson EJ, Baldwin JA. Prenatal toxicity studies in rats and rabbits with the calcium channel blocker diproteverine. *Reprod Toxicol*. 1996;10:43–49.

31. Li J, Duncan RL, Burr DB, Gattone VH, Turner CH. Parathyroid hormone enhances mechanically induced bone formation, possibly involving L-type voltage-sensitive calcium channels. *Endocrinology*. 2003;144:1226–1233.

32. Scott Jr WJ, Resnick E, Hummler H, Clozel JP, Burgin H. Cardiovascular alterations in rat fetuses exposed to calcium channel blockers. *Reprod Toxicol*. 1997;11:207–214.

33. Ariyuki F. Effects of diltiazem hydrochloride on embryonic development: species differences in the susceptibility and stage specificity in mice, rats, and rabbits. *Okajimas Folia Anat Jpn*. 1975;52:103–117.

34. Li J, Duncan RL, Burr DB, Turner CH. L-type calcium channels mediate mechanically induced bone formation in vivo. *J Bone Miner Res*. 2002;17:1795–1800. https://doi.org/10.1359/jbmr.2002.17.10.1795.

35. Wen L, et al. L-type calcium channels play a crucial role in the proliferation and osteogenic differentiation of bone marrow mesenchymal stem cells. *Biochem Biophys Res Commun*. 2012;424:439–445. https://doi.org/10.1016/j.bbrc.2012.06.128.

36. Ryder KD, Duncan RL. Parathyroid hormone enhances fluid shear-induced Ca2+ (i) signaling in osteoblastic cells through activation of mechanosensitive and voltage-sensitive Ca2+ channels. *J Bone Miner Res*. 2001;16:240–248. https://doi.org/10.1359/jbmr.2001.16.2.240.

37. Shao Y, Alicknavitch M, Farach-Carson MC. Expression of voltage sensitive calcium channel (VSCC) L-type Cav1.2 (alpha1C) and T-type Cav3.2 (alpha1H) subunits during mouse bone development. *Dev Dyn*. 2005;234:54–62. https://doi.org/10.1002/dvdy.20517.

38. Cao C, et al. Increased Ca2+ signaling through CaV1.2 promotes bone formation and prevents estrogen deficiency-induced bone loss. *JCI Insight*. 2017;2. https://doi.org/10.1172/jci.insight.95512.

39. Liu DW, et al. Activation of extracellular-signal regulated kinase (ERK1/2) by fluid shear is Ca2+- and ATP-dependent in MC3T3-E1 osteoblasts. *Bone*. 2008;42:644–652. https://doi.org/10.1016/j.bone.2007.09.058.

40. Brown GN, Leong PL, Guo XE. T-Type voltage-sensitive calcium channels mediate mechanically-induced intracellular calcium oscillations in osteocytes by regulating endoplasmic reticulum calcium dynamics. *Bone*. 2016;88:56–63. https://doi.org/10.1016/j.bone.2016.04.018.

41. Dolphin AC. Voltage-gated calcium channels and their auxiliary subunits: physiology and pathophysiology and pharmacology. *J Physiol*. 2016;594:5369–5390. https://doi.org/10.1113/JP272262.

42. Marais E, Klugbauer N, Hofmann F. Calcium channel alpha(2)delta subunits-structure and Gabapentin binding. *Mol Pharmacol*. 2001;59:1243–1248.

43. Dolphin AC. *Jasper's Basic Mechanisms of the Epilepsies* (eds. Noebels JL., et al.); 2012.

44. Garcia K, Nabhani T, Garcia J. The calcium channel alpha2/delta1 subunit is involved in extracellular signalling. *J Physiol*. 2008;586:727–738. https://doi.org/10.1113/jphysiol.2007.147959.

45. Thompson WR, et al. Association of the α2δ1 subunit with Cav3.2 enhances membrane expression and regulates mechanically induced ATP release in MLO-Y4 osteocytes. *J Bone Miner Res*. 2011;26:2125–2139. https://doi.org/10.1002/jbmr.437.

46. Shao Y, et al. Dynamic interactions between L-type voltage-sensitive calcium channel Cav1.2 subunits and ahnak in osteoblastic cells. *Am J Physiol Cell Physiol*. 2009;296:C1067–C1078. https://doi.org/10.1152/ajpcell.00427.2008.

47. Hohaus A, et al. The carboxyl-terminal region of ahnak provides a link between cardiac L-type Ca2+ channels and the actin-based cytoskeleton. *FASEB J*. 2002;16:1205–1216. https://doi.org/10.1096/fj.01-0855com.

48. Anvarian Z, Mykytyn K, Mukhopadhyay S, Pedersen LB, Christensen ST. Cellular signalling by primary cilia in development, organ function and disease. *Nat Rev Nephrol*. 2019;15:199–219. https://doi.org/10.1038/s41581-019-0116-9.

49. Corrigan MA, et al. TRPV4-mediates oscillatory fluid shear mechanotransduction in mesenchymal stem cells in part via the primary cilium. *Sci Rep*. 2018;8:3824. https://doi.org/10.1038/s41598-018-22174-3.

50. Spasic M, Jacobs CR. Lengthening primary cilia enhances cellular mechanosensitivity. *Eur Cell Mater*. 2017;33:158–168. https://doi.org/10.22203/eCM.v033a12.

51. Christensen ST, Ott CM. Cell signaling. A ciliary signaling switch. *Science*. 2007;317:330–331. https://doi.org/10.1126/science.1146180.

52. Kaku M, Komatsu Y. Functional diversity of ciliary proteins in bone development and disease. *Curr Osteoporos Rep*. 2017;15:96–102. https://doi.org/10.1007/s11914-017-0351-6.

53. Tummala P, Arnsdorf EJ, Jacobs CR. The role of primary cilia in mesenchymal stem cell differentiation: a pivotal switch in guiding lineage commitment. *Cell Mol Bioeng*. 2010;3:207–212. https://doi.org/10.1007/s12195-010-0127-x.

54. Moore ER, Yang Y, Jacobs CR. Primary cilia are necessary for Prx1-expressing cells to contribute to postnatal skeletogenesis. *J Cell Sci*. 2018;131. https://doi.org/10.1242/jcs.217828.

55. Praetorius HA, Frokiaer J, Nielsen S, Spring KR. Bending the primary cilium opens Ca2+-sensitive intermediate-conductance K+ channels in MDCK cells. *J Membr Biol*. 2003;191:193–200. https://doi.org/10.1007/s00232-002-1055-z.

56. Temiyasathit S, Jacobs CR. Osteocyte primary cilium and its role in bone mechanotransduction. *Ann N Y Acad Sci*. 2010;1192:422–428. https://doi.org/10.1111/j.1749-6632.2009.05243.x.

57. Labour MN, Riffault M, Christensen ST, Hoey DA. TGFbeta1 - induced recruitment of human bone mesenchymal stem cells is mediated by the primary cilium in a SMAD3-dependent manner. *Sci Rep*. 2016;6:35542. https://doi.org/10.1038/srep35542.

58. Chen JC, Hoey DA, Chua M, Bellon R, Jacobs CR. Mechanical signals promote osteogenic fate through a primary cilia-mediated mechanism. *FASEB J*. 2016;30:1504–1511. https://doi.org/10.1096/fj.15-276402.

59. Burridge K, Guilluy C. Focal adhesions, stress fibers and mechanical tension. *Exp Cell Res*. 2016;343:14–20. https://doi.org/10.1016/j.yexcr.2015.10.029.

60. Katsumi A, Orr AW, Tzima E, Schwartz MA. Integrins in mechanotransduction. *J Biol Chem*. 2004;279:12001–12004. https://doi.org/10.1074/jbc.R300038200.

61. Thompson WR, Rubin CT, Rubin J. Mechanical regulation of signaling pathways in bone. *Gene*. 2012;503:179–193. https://doi.org/10.1016/j.gene.2012.04.076.

62. Schaller MD. The focal adhesion kinase. *J Endocrinol*. 1996;150:1–7. https://doi.org/10.1677/joe.0.1500001.

63. Tamkun JW, et al. Structure of integrin, a glycoprotein involved in the transmembrane linkage between fibronectin and actin. *Cell*. 1986;46:271–282.

64. del Rio A, et al. Stretching single talin rod molecules activates vinculin binding. *Science*. 2009;323:638–641. https://doi.org/10.1126/science.1162912.

65. Subauste MC, et al. Vinculin modulation of paxillin–FAK interactions regulates ERK to control survival and motility. *J Cell Biol*. 2004;165:371–381. https://doi.org/10.1083/jcb.200308011.

66. Sawada Y, et al. Force sensing by mechanical extension of the Src family kinase substrate p130Cas. *Cell*. 2006;127:1015–1026. https://doi.org/10.1016/j.cell.2006.09.044.

67. Gallant ND, Michael KE, Garcia AJ. Cell adhesion strengthening: contributions of adhesive area, integrin binding, and focal adhesion assembly. *Mol Biol Cell*. 2005;16:4329–4340. https://doi.org/10.1091/mbc.e05-02-0170.

68. Engler AJ, Sen S, Sweeney HL, Discher DE. Matrix elasticity directs stem cell lineage specification. *Cell*. 2006;126:677–689. https://doi.org/10.1016/j.cell.2006.06.044.

69. Sen B, et al. Mechanically induced focal adhesion assembly amplifies anti-adipogenic pathways in mesenchymal stem cells. *Stem Cells*. 2011;29:1829–1836. https://doi.org/10.1002/stem.732.

70. Uzer G, Fuchs RK, Rubin J, Thompson WR. Concise review: plasma and nuclear membranes convey mechanical information to regulate mesenchymal stem cell lineage. *Stem Cells*. 2016;34:1455–1463. https://doi.org/10.1002/stem.2342.

71. Zhang P, Wu Y, Jiang Z, Jiang L, Fang B. Osteogenic response of mesenchymal stem cells to continuous mechanical strain is dependent on ERK1/2-Runx2 signaling. *Int J Mol Med*. 2012;29:1083–1089. https://doi.org/10.3892/ijmm.2012.934.

72. Thompson WR, et al. Mechanically activated Fyn utilizes mTORC2 to regulate RhoA and adipogenesis in mesenchymal stem cells. *Stem Cells*. 2013;31:2528–2537. https://doi.org/10.1002/stem.1476.

73. Sen B, et al. mTORC2 regulates mechanically induced cytoskeletal reorganization and lineage selection in

marrow-derived mesenchymal stem cells. *J Bone Miner Res*. 2014;29:78−89. https://doi.org/10.1002/jbmr.2031.

74. Case N, et al. Mechanical regulation of glycogen synthase kinase 3beta (GSK3beta) in mesenchymal stem cells is dependent on Akt protein serine 473 phosphorylation via mTORC2 protein. *J Biol Chem*. 2011;286: 39450−39456. https://doi.org/10.1074/jbc.M111.265330.

75. Case N, et al. Mechanical activation of β-catenin regulates phenotype in adult murine marrow-derived mesenchymal stem cells. *J Orthop Res*. 2010;28:1531−1538. https://doi.org/10.1002/jor.21156.

76. Burridge K, Wennerberg K. Rho and Rac take center stage. *Cell*. 2004;116:167−179.

77. Bos JL, Rehmann H, Wittinghofer A. GEFs and GAPs: critical elements in the control of small G proteins. *Cell*. 2007; 129:865−877. https://doi.org/10.1016/j.cell.2007.05.018.

78. Thompson WR, et al. LARG GEF and ARHGAP18 orchestrate RhoA activity to control mesenchymal stem cell lineage. *Bone*. 2018;107:172−180. https://doi.org/ 10.1016/j.bone.2017.12.001.

79. Sen B, et al. Mechanical signal influence on mesenchymal stem cell fate is enhanced by incorporation of refractory periods into the loading regimen. *J Biomech*. 2011;44: 593−599. https://doi.org/10.1016/j.jbiomech.2010.11.022.

80. Patel VS, et al. Incorporating refractory period in mechanical stimulation mitigates obesity-induced adipose tissue dysfunction in adult mice. *Obesity*. 2017;25:1745−1753. https://doi.org/10.1002/oby.21958.

81. Machesky LM, Insall RH. Scar1 and the related Wiskott−Aldrich syndrome protein, WASP, regulate the actin cytoskeleton through the Arp2/3 complex. *Curr Biol*. 1998;8: 1347−1356.

82. Machesky LM, et al. Scar, a WASp-related protein, activates nucleation of actin filaments by the Arp2/3 complex. *Proc Natl Acad Sci U S A*. 1999;96: 3739−3744. https://doi.org/10.1073/pnas.96.7.3739.

83. Rohatgi R, Nollau P, Ho H-YH, Kirschner MW, Mayer BJ. Nck and phosphatidylinositol 4,5-bisphosphate synergistically activate actin polymerization through the N-WASP-Arp2/3 pathway. *J Biol Chem*. 2001;276: 26448−26452. https://doi.org/10.1074/jbc.M103856200.

84. Marchand J-B, Kaiser DA, Pollard TD, Higgs HN. Interaction of WASP/Scar proteins with actin and vertebrate Arp2/3 complex. *Nat Cell Biol*. 2001;3:76−82. http:// www.nature.com/ncb/journal/v3/n1/suppinfo/ncb0101_ 76_S1.html.

85. Mullins RD, Heuser JA, Pollard TD. The interaction of Arp2/3 complex with actin: nucleation, high affinity pointed end capping, and formation of branching networks of filaments. *Proc Natl Acad Sci*. 1998;95: 6181−6186.

86. Blanchoin L, Boujemaa-Paterski R, Sykes C, Plastino J. Actin dynamics, architecture, and mechanics in cell motility. *Physiol Rev*. 2014;94:235−263. https://doi.org/ 10.1152/physrev.00018.2013.

87. Jaffe AB, Hall A. Rho GTPases: biochemistry and biology. *Annu Rev Cell Dev Biol*. 2005;21:247−269. https:// doi.org/10.1146/annurev.cellbio.21.020604.150721.

88. Riddick N, Ohtani K, Surks HK. Targeting by myosin phosphatase-RhoA interacting protein mediates RhoA/ ROCK regulation of myosin phosphatase. *J Cell Biochem*. 2008;103:1158−1170. https://doi.org/10.1002/jcb.21488.

89. Arnsdorf EJ, Tummala P, Kwon RY, Jacobs CR. Mechanically induced osteogenic differentiation - the role of RhoA, ROCKII and cytoskeletal dynamics. *J Cell Sci*. 2009;122:546−553. https://doi.org/10.1242/jcs.036293.

90. Khatiwala CB, Kim PD, Peyton SR, Putnam AJ. ECM compliance regulates osteogenesis by influencing MAPK signaling downstream of RhoA and ROCK. *J Bone Miner Res*. 2009;24:886−898. https://doi.org/10.1359/jbmr.081240.

91. Horikawa A, Okada K, Sato K, Sato M. Morphological changes in osteoblastic cells (MC3T3-E1) due to fluid shear stress: cellular damage by prolonged application of fluid shear stress. *Tohoku J Exp Med*. 2000;191: 127−137. https://doi.org/10.1620/tjem.191.127.

92. Greiner AM, Chen H, Spatz JP, Kemkemer R. Cyclic tensile strain controls cell shape and directs actin stress fiber formation and focal adhesion alignment in spreading cells. *PLoS One*. 2013;8:e77328. https://doi.org/ 10.1371/journal.pone.0077328.

93. Chancellor TJ, Lee J, Thodeti CK, Lele T. Actomyosin tension exerted on the nucleus through nesprin-1 connections influences endothelial cell adhesion, migration, and cyclic strain-induced reorientation. *Biophys J*. 2010; 99:115−123.

94. Uzer G, et al. Cell mechanosensitivity to extremely low-magnitude signals is enabled by a LINCed nucleus. *Stem Cells*. 2015;33:2063−2076. https://doi.org/ 10.1002/stem.2004.

95. Uzer G, Pongkitwitoon S, Ete Chan M, Judex S. Vibration induced osteogenic commitment of mesenchymal stem cells is enhanced by cytoskeletal remodeling but not fluid shear. *J Biomech*. 2013;46:2296−2302. https://doi.org/ 10.1016/j.jbiomech.2013.06.008.

96. Hu SH, Chen JX, Butler JP, Wang. N. Prestress mediates force propagation into the nucleus. *Biochem Biophys Res Commun*. 2005;329:423−428. https://doi.org/10.1016/ j.bbrc.2005.02.026.

97. Ingber DE, Wang N, Stamenovic D. Tensegrity, cellular biophysics, and the mechanics of living systems. *Rep Prog Phys*. 2014;77:046603. https://doi.org/10.1088/ 0034-4885/77/4/046603.

98. Cooper LF, Harris CT, Bruder SP, Kowalski R, Kadiyala S. Incipient analysis of mesenchymal stem-cell-derived osteogenesis. *J Dent Res*. 2001;80:314−320. https:// doi.org/10.1177/00220345010800010401.

99. Wang N, Butler JP, Ingber DE. Mechanotransduction across the cell-surface and through the cytoskeleton. *Science*. 1993;260:1124−1127. https://doi.org/10.1126/ science.7684161.

100. McGarry JG, Prendergast PJ. A three-dimensional finite element model of an adherent eukaryotic cell. *Eur Cells Mater*. 2004;7:27−33. discussion 33−24.

101. Wang X, Ha T. Defining single molecular forces required to activate integrin and notch signaling. *Science*. 2013; 340:991−994. https://doi.org/10.1126/science.1231041.

102. Nagayama K, Yamazaki S, Yahiro Y, Matsumoto T. Estimation of the mechanical connection between apical stress fibers and the nucleus in vascular smooth muscle cells cultured on a substrate. *J Biomech.* 2014;47(6):1422–1429. https://doi.org/10.1016/j.jbiomech.2014.01.042.

103. Smith MA, et al. A zyxin-mediated mechanism for actin stress fiber maintenance and repair. *Dev Cell.* 2010;19: 365–376. https://doi.org/10.1016/j.devcel.2010.08.008.

104. Reinhard M, et al. An α-actinin binding site of zyxin is essential for subcellular zyxin localization and α-actinin recruitment. *J Biol Chem.* 1999;274:13410–13418. https://doi.org/10.1074/jbc.274.19.13410.

105. Yoshigi M, Hoffman LM, Jensen CC, Yost HJ, Beckerle MC. Mechanical force mobilizes zyxin from focal adhesions to actin filaments and regulates cytoskeletal reinforcement. *J Cell Biol.* 2005;171:209–215. https://doi.org/10.1083/jcb.200505018.

106. Hirata H, Tatsumi H, Sokabe M. Mechanical forces facilitate actin polymerization at focal adhesions in a zyxin-dependent manner. *J Cell Sci.* 2008;121:2795–2804. https://doi.org/10.1242/jcs.030320.

107. Smith M, Blankman E, Deakin NO, Hoffman LM, Jensen CC. LIM domains target actin regulators paxillin and zyxin to sites of stress fiber strain. *PLoS One.* 2013; 8:e69378. https://doi.org/10.1371/journal.pone.0069378.

108. Flitney EW, Kuczmarski ER, Adam SA, Goldman RD. Insights into the mechanical properties of epithelial cells: the effects of shear stress on the assembly and remodeling of keratin intermediate filaments. *FASEB J.* 2009;23: 2110–2119. https://doi.org/10.1096/fj.08-124453.

109. Maniotis AJ, Chen CS, Ingber DE. Demonstration of mechanical connections between integrins cytoskeletal filaments, and nucleoplasm that stabilize nuclear structure. *Proc Natl Acad Sci U S A.* 1997;94:849–854.

110. Brangwynne CP, et al. Microtubules can bear enhanced compressive loads in living cells because of lateral reinforcement. *J Cell Biol.* 2006;173:733–741. https://doi.org/10.1083/jcb.200601060.

111. Odde DJ, Ma L, Briggs AH, DeMarco A, Kirschner MW. Microtubule bending and breaking in living fibroblast cells. *J Cell Sci.* 1999;112:3283–3288.

112. Ingber DE. Tensegrity: the architectural basis of cellular mechanotransduction. *Annu Rev Physiol.* 1997;59: 575–599.

113. McGarry JG, Klein-Nulend J, Prendergast PJ. The effect of cytoskeletal disruption on pulsatile fluid flow-induced nitric oxide and prostaglandin E2 release in osteocytes and osteoblasts. *Biochem Biophys Res Commun.* 2005;330: 341–348. https://doi.org/10.1016/j.bbrc.2005.02.175.

114. Norvell SM, Ponik SM, Bowen DK, Gerard R, Pavalko FM. Fluid shear stress induction of COX-2 protein and prostaglandin release in cultured MC3T3-E1 osteoblasts does not require intact microfilaments or microtubules. *J Appl Physiol.* 2004;96:957–966. https://doi.org/10.1152/japplphysiol.00869.2003.

115. De Magistris P, Antonin W. The dynamic nature of the nuclear envelope. *Curr Biol.* 2018;28:R487–R497. https://doi.org/10.1016/j.cub.2018.01.073.

116. Rubin J, Styner M, Uzer G. Physical signals may affect mesenchymal stem cell differentiation via epigenetic controls. *Exerc Sport Sci Rev.* 2018;46:42–47. https://doi.org/10.1249/jes.0000000000000129.

117. Crisp M, et al. Coupling of the nucleus and cytoplasm: role of the LINC complex. *J Cell Biol.* 2006;172:41–53. https://doi.org/10.1083/jcb.200509124.

118. Ketema M, Sonnenberg A. Nesprin-3: a versatile connector between the nucleus and the cytoskeleton. *Biochem Soc Trans.* 2011;39:1719–1724. https://doi.org/10.1042/bst20110669.

119. Padmakumar VC, et al. Enaptin, a giant actin-binding protein, is an element of the nuclear membrane and the actin cytoskeleton. *Exp Cell Res.* 2004;295:330–339. https://doi.org/10.1016/j.yexcr.2004.01.014.

120. Zhen Y-Y, Libotte T, Munck M, Noegel AA, Korenbaum E. NUANCE, a giant protein connecting the nucleus and actin cytoskeleton. *J Cell Sci.* 2002;115:3207–3222.

121. Wilson MH, Holzbaur EL. Nesprins anchor kinesin-1 motors to the nucleus to drive nuclear distribution in muscle cells. *Development.* 2015;142:218–228. https://doi.org/10.1242/dev.114769.

122. Roux KJ, et al. Nesprin 4 is an outer nuclear membrane protein that can induce kinesin-mediated cell polarization. *Proc Natl Acad Sci.* 2009;106:2194–2199. https://doi.org/10.1073/pnas.0808602106.

123. Simpson JG, Roberts RG. Patterns of evolutionary conservation in the nesprin genes highlight probable functionally important protein domains and isoforms. *Biochem Soc Trans.* 2008;36:1359–1367. https://doi.org/10.1042/bst0361359.

124. Autore F, et al. Large-scale modelling of the divergent spectrin repeats in nesprins: giant modular proteins. *PLoS One.* 2013;8:e63633. https://doi.org/10.1371/journal.pone.0063633.

125. Kutscheidt S, et al. FHOD1 interaction with nesprin-2G mediates TAN line formation and nuclear movement. *Nat Cell Biol.* 2014;16:708–715. https://doi.org/10.1038/ncb2981.

126. Antoku S, Gundersen GG. Analysis of nesprin-2 interaction with its binding partners and actin. *Methods Mol Biol.* 2018;1840:35–43. https://doi.org/10.1007/978-1-4939-8691-0_4.

127. Arsenovic PT, et al. Nesprin-2G, a component of the nuclear LINC complex, is subject to myosin-dependent tension. *Biophys J.* 2016;110:34–43. https://doi.org/10.1016/j.bpj.2015.11.014.

128. Lu W, et al. Nesprin interchain associations control nuclear size. *Cell Mol Life Sci.* 2012;69:3493–3509. https://doi.org/10.1007/s00018-012-1034-1.

129. Duong NT, et al. Nesprins: tissue-specific expression of epsilon and other short isoforms. *PLoS One.* 2014;9: e94380. https://doi.org/10.1371/journal.pone.0094380.

130. Rajgor D, Mellad JA, Autore F, Zhang Q, Shanahan CM. Multiple novel nesprin-1 and nesprin-2 variants act as versatile tissue-specific intracellular scaffolds. *PLoS One.* 2012;7:e40098. https://doi.org/10.1371/journal.pone.0040098.

131. Starr DA, Han M. Role of ANC-1 in tethering nuclei to the actin cytoskeleton. *Science*. 2002;298:406−409. https://doi.org/10.1126/science.1075119.

132. Dreger M, Bengtsson L, Schöneberg T, Otto H, Hucho F. Nuclear envelope proteomics: novel integral membrane proteins of the inner nuclear membrane. *Proc Natl Acad Sci*. 2001;98:11943−11948. https://doi.org/10.1073/pnas.211201898.

133. Hodzic DM, Yeater DB, Bengtsson L, Otto H, Stahl PD. Sun2 is a novel mammalian inner nuclear membrane protein. *J Biol Chem*. 2004;279:25805−25812. https://doi.org/10.1074/jbc.M313157200.

134. Sosa BA, Kutay U, Schwartz TU. Structural insights into LINC complexes. *Curr Opin Struct Biol*. 2013;23:285−291. https://doi.org/10.1016/j.sbi.2013.03.005.

135. Cain NE, Starr DA. SUN proteins and nuclear envelope spacing. *Nucleus*. 2015;6:2−7. https://doi.org/10.4161/19491034.2014.990857.

136. Cain NE, Tapley EC, McDonald KL, Cain BM, Starr DA. The SUN protein UNC-84 is required only in force-bearing cells to maintain nuclear envelope architecture. *J Cell Biol*. 2014;206:163−172. https://doi.org/10.1083/jcb.201405081.

137. Sosa BA, Rothballer A, Kutay U, Schwartz TU. LINC complexes form by binding of three KASH peptides to the interfaces of trimeric SUN proteins. *Cell*. 2012;149:1035−1047. https://doi.org/10.1016/j.cell.2012.03.046.

138. Haque F, et al. Mammalian SUN protein interaction networks at the inner nuclear membrane and their role in laminopathy disease processes. *J Biol Chem*. 2010;285:3487−3498. https://doi.org/10.1074/jbc.M109.071910.

139. Liu Q, et al. Functional association of Sun1 with nuclear pore complexes. *J Cell Biol*. 2007;178:785−798. https://doi.org/10.1083/jcb.200704108.

140. Link J, et al. Analysis of meiosis in SUN1 deficient mice reveals a distinct role of SUN2 in mammalian meiotic LINC complex formation and function. *PLoS Genet*. 2014;10:e1004099. https://doi.org/10.1371/journal.pgen.1004099.

141. Chang W, et al. Imbalanced nucleocytoskeletal connections create common polarity defects in progeria and physiological aging. *Proc Natl Acad Sci U S A*. 2019;116:3578−3583. https://doi.org/10.1073/pnas.1809683116.

142. Sen B, et al. Mechanical strain inhibits adipogenesis in mesenchymal stem cells by stimulating a durable beta-catenin signal. *Endocrinology*. 2008;149:6065−6075. https://doi.org/10.1210/en.2008-0687.

143. Codelia VA, Sun G, Irvine KD. Regulation of YAP by mechanical strain through Jnk and Hippo signaling. *Curr Biol*. 2014;24:2012−2017. https://doi.org/10.1016/j.cub.2014.07.034.

144. Driscoll, T.P., Cosgrove, B.D., Heo, S.-J., Shurden, Z.E. & Mauck, R.L. Cytoskeletal to nuclear strain transfer regulates YAP signaling in mesenchymal stem cells. Biophys J 108, 2783-2793, doi:10.1016/j.bpj.2015.05.010.

145. Elosegui-Artola A, et al. Force triggers YAP nuclear entry by regulating transport across nuclear pores. *Cell*. 2017;

171:1397−1410.e1314. https://doi.org/10.1016/j.cell.2017.10.008.

146. Koike M, et al. β-Catenin shows an overlapping sequence requirement but distinct molecular interactions for its bidirectional passage through nuclear pores. *J Biol Chem*. 2004;279:34038−34047. https://doi.org/10.1074/jbc.M405821200.

147. Tolwinski NS, Wieschaus E. A nuclear function for armadillo/beta-catenin. *PLoS Biol*. 2004;2:E95. https://doi.org/10.1371/journal.pbio.0020095.

148. Neumann S, et al. Nesprin-2 interacts with α-catenin and regulates Wnt signaling at the nuclear envelope. *J Biol Chem*. 2010;285:34932−34938. https://doi.org/10.1074/jbc.M110.119651.

149. Uzer G, et al. Sun-mediated mechanical LINC between nucleus and cytoskeleton regulates βcatenin nuclear access. *J Biomech*. 2018;74:32−40. https://doi.org/10.1016/j.jbiomech.2018.04.013.

150. Horn HF, et al. The LINC complex is essential for hearing. *J Clin Invest*. 2013;123:740−750. https://doi.org/10.1172/JCI66911.

151. Rubin C, Turner AS, Bain S, Mallinckrodt C, McLeod K. Anabolism. Low mechanical signals strengthen long bones. *Nature*. 2001;412:603−604. https://doi.org/10.1038/35088122.

152. Gilsanz V, et al. Low-level, high-frequency mechanical signals enhance musculoskeletal development of young women with low BMD. *J Bone Miner Res*. 2006;21:1464−1474. https://doi.org/10.1359/jbmr.060612.

153. Sen B, et al. mTORC2 regulates mechanically induced cytoskeletal reorganization and lineage selection in marrow-derived mesenchymal stem cells. *J Bone Miner Res*. 2014;29:78−89. https://doi.org/10.1002/jbmr.2031.

154. Graham DM, et al. Enucleated cells reveal differential roles of the nucleus in cell migration, polarity, and mechanotransduction. *J Cell Biol*. 2018;217:895−914. https://doi.org/10.1083/jcb.201706097.

155. Lombardi ML, et al. The interaction between nesprins and sun proteins at the nuclear envelope is critical for force transmission between the nucleus and cytoskeleton. *J Biol Chem*. 2011;286:26743−26753. https://doi.org/10.1074/jbc.M111.233700.

156. Banerjee I, et al. Targeted ablation of nesprin 1 and nesprin 2 from murine myocardium results in cardiomyopathy, altered nuclear morphology and inhibition of the biomechanical gene response. *PLoS Genet*. 2014;10:e1004114. https://doi.org/10.1371/journal.pgen.1004114.

157. Touchstone H, et al. Recovery of stem cell proliferation by low intensity vibration under simulated microgravity requires intact LINC complex. *NPJ Microgravity*. 2019;5:11. https://doi.org/10.1038/s41526-019-0072-5. eCollection.

158. Pongkitwitoon S, Uzer G, Rubin J, Judex S. Cytoskeletal configuration modulates mechanically induced changes in mesenchymal stem cell osteogenesis, morphology, and stiffness. *Sci Rep*. 2016;6:34791. https://doi.org/10.1038/srep34791.

159. Manilal S, Man Nt, Sewry CA, Morris GE. The emery-dreifuss muscular dystrophy protein, emerin, is a nuclear membrane protein. *Hum Mol Genet.* 1996;5:801–808. https://doi.org/10.1093/hmg/5.6.801.

160. Holaska JM, Kowalski AK, Wilson KL. Emerin caps the pointed end of actin filaments: evidence for an actin cortical network at the nuclear inner membrane. *PLoS Biol.* 2004;2: E231. https://doi.org/10.1371/journal.pbio.0020231.

161. Le HQ, et al. Mechanical regulation of transcription controls Polycomb-mediated gene silencing during lineage commitment. *Nat Cell Biol.* 2016;18:864–875. https://doi.org/10.1038/ncb3387.

162. Stubenvoll A, Rice M, Wietelmann A, Wheeler M, Braun T. Attenuation of Wnt/beta-catenin activity reverses enhanced generation of cardiomyocytes and cardiac defects caused by the loss of emerin. *Hum Mol Genet.* 2015;24:802–813. https://doi.org/10.1093/hmg/ddu498.

163. Tilgner K, Wojciechowicz K, Jahoda C, Hutchison C, Markiewicz E. Dynamic complexes of A-type lamins and emerin influence adipogenic capacity of the cell via nucleocytoplasmic distribution of β-catenin. *J Cell Sci.* 2009;122:401–413. https://doi.org/10.1242/jcs.026179.

164. Markiewicz E, et al. The inner nuclear membrane protein Emerin regulates [beta]-catenin activity by restricting its accumulation in the nucleus. *EMBO J.* 2006;25: 3275–3285. http://www.nature.com/emboj/journal/v25/n14/suppinfo/7601230a_S1.html.

165. Guilluy C, et al. Isolated nuclei adapt to force and reveal a mechanotransduction pathway in the nucleus. *Nat Cell Biol.* 2014;16:376–381. https://doi.org/10.1038/ncb2927.

166. Padmakumar VC, et al. The inner nuclear membrane protein Sun1 mediates the anchorage of Nesprin-2 to the nuclear envelope. *J Cell Sci.* 2005;118:3419–3430. https://doi.org/10.1242/jcs.02471.

167. Tanabe LM, Kim CE, Alagem N, Dauer WT. Primary dystonia: molecules and mechanisms. *Nat Rev Neurol.* 2009; 5:598–609.

168. Jungwirth M, Kumar D, Jeong D, Goodchild R. The nuclear envelope localization of DYT1 dystonia torsinA-DeltaE requires the SUN1 LINC complex component. *BMC Cell Biol.* 2011;12:24.

169. Nery FC, et al. TorsinA binds the KASH domain of nesprins and participates in linkage between nuclear envelope and cytoskeleton. *J Cell Sci.* 2008;121: 3476–3486. https://doi.org/10.1242/jcs.029454.

170. Sosa BA, et al. How lamina-associated polypeptide 1 (LAP1) activates Torsin. *Elife.* 2014;3:e03239. https://doi.org/10.7554/eLife.03239.

171. Shin JY, et al. Lamina-associated polypeptide-1 interacts with the muscular dystrophy protein emerin and is essential for skeletal muscle maintenance. *Dev Cell.* 2013;26: 591–603. https://doi.org/10.1016/j.devcel.2013.08.012.

172. Saunders CA, et al. TorsinA controls TAN line assembly and the retrograde flow of dorsal perinuclear actin cables during rearward nuclear movement. *J Cell Biol.* 2017;216: 657–674. https://doi.org/10.1083/jcb.201507113.

173. Cook PR. The nucleoskeleton: artefact, passive framework or active site? *J Cell Sci.* 1988;90(Pt 1):1–6.

174. Grosse R, Vartiainen MK. To be or not to be assembled: progressing into nuclear actin filaments. *Nat Rev Mol Cell Biol.* 2013;14:693–697. https://doi.org/10.1038/nrm3681.

175. Hotulainen P, Lappalainen P. Stress fibers are generated by two distinct actin assembly mechanisms in motile cells. *J Cell Biol.* 2006;173:383–394. https://doi.org/10.1083/jcb.200511093.

176. Baarlink C, Wang H, Grosse R. Nuclear actin network assembly by formins regulates the SRF coactivator MAL. *Science.* 2013;340:864–867. https://doi.org/10.1126/science.1235038.

177. Belin BJ, Cimini BA, Blackburn EH, Mullins RD. Visualization of actin filaments and monomers in somatic cell nuclei. *Mol Biol Cell.* 2013;24:982–994. https://doi.org/10.1091/mbc.E12-09-0685.

178. Munsie LN, Desmond CR, Truant R. Cofilin nuclear-cytoplasmic shuttling affects cofilin-actin rod formation during stress. *J Cell Sci.* 2012;125:3977–3988. https://doi.org/10.1242/jcs.097667.

179. Plessner M, Melak M, Chinchilla P, Baarlink C, Grosse R. Nuclear F-actin formation and reorganization upon cell spreading. *J Biol Chem.* 2015;290:11209–11216. https://doi.org/10.1074/jbc.M114.627166.

180. Sen B, et al. Intranuclear actin regulates osteogenesis. *Stem Cells.* 2015. https://doi.org/10.1002/stem.2090.

181. Sen B, et al. Intranuclear actin structure modulates mesenchymal stem cell differentiation. *Stem Cells.* 2017; 35:1624–1635. https://doi.org/10.1002/stem.2617.

182. Samsonraj RM, et al. Validation of osteogenic properties of cytochalasin D by high-resolution RNA-sequencing in mesenchymal stem cells derived from bone marrow and adipose tissues. *Stem Cells Dev.* 2018. https://doi.org/10.1089/scd.2018.0037.

183. Serebryannyy LA, et al. Persistent nuclear actin filaments inhibit transcription by RNA polymerase II. *J Cell Sci.* 2016;129:3412–3425. https://doi.org/10.1242/jcs.195867.

184. Zhu X, Zeng X, Huang B, Hao S. Actin is closely associated with RNA polymerase II and involved in activation of gene transcription. *Biochem Biophys Res Commun.* 2004; 321:623–630.

185. Spichal M, et al. Evidence for a dual role of actin in regulating chromosome organization and dynamics in yeast. *J Cell Sci.* 2016;129:681–692. https://doi.org/10.1242/jcs.175745.

186. Crabbe L, Cesare AJ, Kasuboski JM, Fitzpatrick JA, Karlseder J. Human telomeres are tethered to the nuclear envelope during postmitotic nuclear assembly. *Cell Rep.* 2012;2: 1521–1529. https://doi.org/10.1016/j.celrep.2012.11.019.

187. Czapiewski R, Robson MI, Schirmer EC. Anchoring a Leviathan: how the nuclear membrane tethers the genome. *Front Genet.* 2016;7:82. https://doi.org/10.3389/fgene.2016.00082.

188. Lottersberger F, Karssemeijer RA, Dimitrova N, de Lange T. 53BP1 and the LINC complex promote

microtubule-dependent DSB mobility and DNA repair. *Cell.* 2015;163:880−893. https://doi.org/10.1016/j.cell.2015.09.057.

189. Oshidari R, et al. Nuclear microtubule filaments mediate non-linear directional motion of chromatin and promote DNA repair. *Nat Commun.* 2018;9:2567. https://doi.org/10.1038/s41467-018-05009-7.

190. Broers JL, Ramaekers FC, Bonne G, Yaou RB, Hutchison CJ. Nuclear lamins: laminopathies and their role in premature ageing. *Physiol Rev.* 2006;86:967−1008. https://doi.org/10.1152/physrev.00047.2005.

191. Rusinol AE, Sinensky MS. Farnesylated lamins, progeroid syndromes and farnesyl transferase inhibitors. *J Cell Sci.* 2006;119:3265−3272. https://doi.org/10.1242/jcs.03156.

192. Schreiber KH, Kennedy BK. When lamins go bad: nuclear structure and disease. *Cell.* 2013;152:1365−1375. https://doi.org/10.1016/j.cell.2013.02.015.

193. Rober RA, Weber K, Osborn M. Differential timing of nuclear lamin A/C expression in the various organs of the mouse embryo and the young animal: a developmental study. *Development.* 1989;105:365−378.

194. Eckersley-Maslin MA, Bergmann JH, Lazar Z, Spector DL. Lamin A/C is expressed in pluripotent mouse embryonic stem cells. *Nucleus.* 2013;4:53−60. https://doi.org/10.4161/nucl.23384.

195. Lammerding J, et al. Lamins A and C but not lamin B1 regulate nuclear mechanics. *J Biol Chem.* 2006;281:25768−25780. https://doi.org/10.1074/jbc.M513511200.

196. Swift J, et al. Nuclear lamin-A scales with tissue stiffness and enhances matrix-directed differentiation. *Science.* 2013;341. https://doi.org/10.1126/science.1240104.

197. Buxboim A, et al. Matrix elasticity regulates lamin-A,C phosphorylation and turnover with feedback to actomyosin. *Curr Biol.* 2014;24:1909−1917. https://doi.org/10.1016/j.cub.2014.07.001.

198. Verstraeten VLRM, Ji JY, Cummings KS, Lee RT, Lammerding J. Increased mechanosensitivity and nuclear stiffness in Hutchinson−Gilford progeria cells: effects of farnesyltransferase inhibitors. *Aging Cell.* 2008;7:383−393. https://doi.org/10.1111/j.1474-9726.2008.00382.x.

199. Kim J-K, et al. Nuclear lamin A/C harnesses the perinuclear apical actin cables to protect nuclear morphology. *Nat Commun.* 2017;8:2123. https://doi.org/10.1038/s41467-017-02217-5.

200. Gaspar-Maia A, Alajem A, Meshorer E, Ramalho-Santos M. Open chromatin in pluripotency and reprogramming. *Nat Rev Mol Cell Biol.* 2011;12:36−47.

201. Chiang FP, Uzer G. Mapping full field deformation of auxetic foams using digital speckle photography. *Phys Status Solidi B.* 2008;245:2391−2394. https://doi.org/10.1002/pssb.200880254.

202. Pagliara S, et al. Auxetic nuclei in embryonic stem cells exiting pluripotency. *Nat Mater.* 2014;13:638−644. https://doi.org/10.1038/nmat3943.

203. Matsumoto A, et al. Global loss of a nuclear lamina component, lamin A/C, and LINC complex components SUN1, SUN2, and nesprin-2 in breast cancer. *Cancer Med.* 2015. https://doi.org/10.1002/cam4.495.

204. De Sandre-Giovannoli A, et al. Lamin a truncation in Hutchinson-Gilford progeria. *Science.* 2003;300:2055. https://doi.org/10.1126/science.1084125.

205. Liu Q, et al. Dynamics of lamin-A processing following precursor accumulation. *PLoS One.* 2010;5:e10874. https://doi.org/10.1371/journal.pone.0010874.

206. Hale CM, et al. Dysfunctional connections between the nucleus and the actin and microtubule networks in laminopathic models. *Biophys J.* 2008;95:5462−5475. https://doi.org/10.1529/biophysj.108.139428.

207. Mounkes LC, Kozlov S, Hernandez L, Sullivan T, Stewart CL. A progeroid syndrome in mice is caused by defects in A-type lamins. *Nature.* 2003;423:298−301. http://www.nature.com/nature/journal/v423/n6937/suppinfo/nature01631_S1.html.

208. Chen C-Y, et al. Accumulation of the inner nuclear envelope protein Sun1 is pathogenic in progeric and dystrophic laminopathies. *Cell.* 2012;149:565−577. https://doi.org/10.1016/j.cell.2012.01.059.

209. Scaffidi P, Misteli T. Lamin A-dependent nuclear defects in human aging. *Science.* 2006;312:1059−1063. https://doi.org/10.1126/science.1127168.

210. Mulvihill BM, Prendergast PJ. An algorithm for bone mechanoresponsiveness: implementation to study the effect of patient-specific cell mechanosensitivity on trabecular bone loss. *Comput Methods Biomech Biomed Eng.* 2008;11:443−451. https://doi.org/10.1080/10255840802136150.

211. Hernandez L, et al. Functional coupling between the extracellular matrix and nuclear lamina by Wnt signaling in progeria. *Dev Cell.* 2010;19:413−425. https://doi.org/10.1016/j.devcel.2010.08.013.

212. Padiath QS, et al. Lamin B1 duplications cause autosomal dominant leukodystrophy. *Nat Genet.* 2006;38:1114−1123. https://doi.org/10.1038/ng1872.

213. Hegele RA, et al. Sequencing of the reannotated LMNB2 gene reveals novel mutations in patients with acquired partial lipodystrophy. *Am J Hum Genet.* 2006;79:383−389. https://doi.org/10.1086/505885.

214. Deleted in review

215. de Leeuw R, Gruenbaum Y, Medalia O. Nuclear lamins: thin filaments with major functions. *Trends Cell Biol.* 2018;28:34−45. https://doi.org/10.1016/j.tcb.2017.08.004.

216. Hozak P, Sasseville AM, Raymond Y, Cook PR. Lamin proteins form an internal nucleoskeleton as well as a peripheral lamina in human cells. *J Cell Sci.* 1995;108(Pt 2):635−644.

217. Mattout-Drubezki A, Gruenbaum Y. Dynamic interactions of nuclear lamina proteins with chromatin and transcriptional machinery. *Cell Mol Life Sci.* 2003;60:2053−2063. https://doi.org/10.1007/s00018-003-3038-3.

218. Solovei I, et al. LBR and lamin A/C sequentially tether peripheral heterochromatin and inversely regulate differentiation. *Cell.* 2013;152:584−598. https://doi.org/10.1016/j.cell.2013.01.009.

219. Bode J, Goetze S, Heng H, Krawetz SA, Benham C. From DNA structure to gene expression: mediators of nuclear compartmentalization and dynamics. *Chromosome Res.* 2003;11:435−445.

220. Dahl KN, Engler AJ, Pajerowski JD, Discher DE. Power-law rheology of isolated nuclei with deformation mapping of nuclear substructures. *Biophys J.* 2005;89:2855−2864. https://doi.org/10.1529/biophysj.105.062554.

221. Ugarte F, et al. Progressive chromatin condensation and H3K9 methylation regulate the differentiation of embryonic and hematopoietic stem cells. *Stem Cell Rep.* 2015;5: 728−740. https://doi.org/10.1016/j.stemcr.2015.09.009.

222. Dudakovic A, et al. Epigenetic control of skeletal development by the histone methyltransferase Ezh2. *J Biol Chem.* 2015;290:27604−27617.    https://doi.org/10.1074/ jbc.M115.672345.

223. Pajerowski JD, Dahl KN, Zhong FL, Sammak PJ, Discher DE. Physical plasticity of the nucleus in stem cell differentiation. *Proc Natl Acad Sci U S A.* 2007;104: 15619−15624.    https://doi.org/10.1073/ pnas.0702576104.

224. Heo SJ, et al. Biophysical regulation of chromatin architecture instills a mechanical memory in mesenchymal stem cells. *Sci Rep.* 2015;5:16895. https://doi.org/ 10.1038/srep16895.

225. Poh Y-C, et al. Dynamic force-induced direct dissociation of protein complexes in a nuclear body in living cells. *Nat Commun.* 2012;3:866. http://www.nature.com/ncomms/ journal/v3/n5/suppinfo/ncomms1873_S1.html.

226. Stephens AD, et al. Chromatin histone modifications and rigidity affect nuclear morphology independent of lamins. *Mol Biol Cell.* 2018;29:220−233. https:// doi.org/10.1091/mbc.E17-06-0410.

227. Makhija E, Jokhun DS, Shivashankar GV. Nuclear deformability and telomere dynamics are regulated by cell geometric constraints. *Proc Natl Acad Sci U S A.* 2016;113:E32−E40. https://doi.org/10.1073/ pnas.1513189113.

# Primary Cilia Mechanobiology

DANIEL P. AHERN, MCH, MRCSI, PHD[A] • MEGAN R. MC FIE, MSC, PHD[A] •
CLARE L. THOMPSON, PHD • MICHAEL P. DUFFY, MSC, PHD •
JOSEPH S. BUTLER, PHD, FRCS • DAVID A. HOEY, PHD

## 1 INTRODUCTION

Physical forces and changes in cell and tissue mechanics can have profound effects on development, physiology, and disease. The study of such phenomena that lie at the interface of traditional disciplines such as biology, physics, and engineering has resulted in a relatively new discipline termed "mechanobiology."[1] This new field has grown over the past 20 years[2] revealing the importance of physical stimuli in regulating physiology in all systems and tissues. In particular, the importance of mechanics is very much evident in the urinary, musculoskeletal, and circulatory systems where physical stimuli applied to resident cells are required to maintain normal kidney function, maintain bone and cartilage mass and architecture, and maintain a healthy vasculature, respectively. Given the importance of mechanics in regulating tissue physiology, it is not surprising that alterations in mechanics or mechanosensing can also lead to disease. Disuse of the skeleton leads to bone loss and increased risk of fracture, while abnormal flow profiles in the vasculature can lead to atherosclerosis. It is therefore clear that there is a need to fully understand the role of mechanics in regulating the biology of our body.

A key question that has driven the mechanobiology field over the past number of years is how do tissues and resident cells sense these mechanical forces and translate these biophysical cues into a homeostatic or regenerative biochemical response, a process known as mechanotransduction. Mechanotransduction is primarily mediated by components of cellular structures/organelles subjected to mechanical stimuli, such as integrins and associated proteins at focal adhesion sites, stretch-activated channels, and the cytoskeleton. These mechanosensitive sites can facilitate the transfer of load from the extracellular milieu to the intracellular environment, triggering a biochemical signaling cascade.[1] Another

cellular organelle that has recently gained much interest as a site of mechanotransduction is the primary cilium. The cilium is a solitary cellular extension that protrudes from the cell body into the extracellular environment and thus is ideally positioned to act as a cellular sensory structure.[3] The significance of the primary cilium in mechanobiology and biology as a whole became clear when defects in the cilium were linked with abnormal urine flow sensing within the kidney, leading to polycystic kidney disease.[4] This finding fueled decades of research that has now revealed this organelle to be an important cellular sensor and coordinator of multiple signaling pathways critical for development, physiology, and mechanobiology across multiple tissues.[5]

In this chapter, we introduce the structure and function of the primary cilium (Section 2); explore the role of the cilium in the mechanobiology of the urinary system, circulatory system, and musculoskeletal system (Section 3); discuss the mechanics of the cilium and how this may influence ciliary function (Section 4); and finish with a discussion of the current unknowns and potential therapeutic options associated with the primary cilium (Section 5).

## 2 THE PRIMARY CILIUM
### 2.1 Structure

The primary cilium is a membrane-bound cellular organelle that represents a distinct compartment commonly identified extending out into the extracellular environment (Fig. 2.2.1). It is present ubiquitously across almost all cell types of the human body and has been highly preserved throughout evolution—thought to be present in the last eukaryotic common ancestor.[6,7] Unlike motile cilia, the primary cilium is an immotile structure that lacks accessory components such as dynein arms and radial spokes that facilitate force generation. Moreover, motile cilia typically occur in large

---

[a]Authors contributed equally to this chapter.

Mechanobiology. https://doi.org/10.1016/B978-0-12-817931-4.00006-6

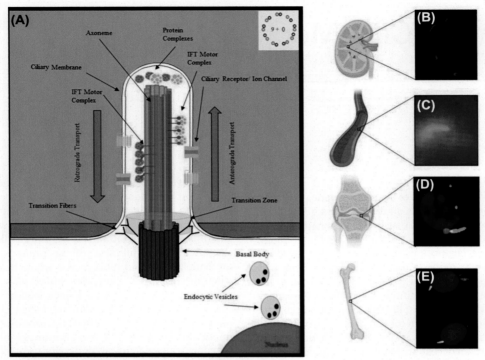

FIG. 2.2.1 **(A)** The primary cilium and its organization. **(B)** Kidney MDCK cells, primary cilium identified with acetylated α-tubulin (*red*) under confocal microscopy. **(C)** Endothelial cells transfected with a tubulin-eGFP (enhanced green fluorescent protein) fusion construct to identify the primary cilium (*green*). **(D)** Chondrocyte primary cilium from 2-month-old mice, stained with anti-Arl13b antibody (*green*); basal bodies were stained with anti-γ-tubulin antibody (*red*) and nuclei were stained with DAPI (4′,6-diamidino-2-phenylindole) (*blue*). **(E)** Primary cilia of rat calvarial osteoblast cells. *IFT*, intraflagellar transport; *MDCK*, Madin-Darby Canine Kidney. ((**B**) Adapted from Low SH, et al. Polycystin-1, STAT6, and P100 function in a pathway that transduces ciliary mechanosensation and is activated in polycystic kidney disease. *Dev Cell*. 2006;10(1):57–69. **(C)** Adapted from Egorova AD, et al. Lack of primary cilia primes shear-induced endothelial-to-mesenchymal transition. *Circ Res*. 2011;108(9):1093–1101. **(D)** Adapted from Chang CF, Ramaswamy G, Serra R. Depletion of primary cilia in articular chondrocytes results in reduced Gli3 repressor to activator ratio, increased Hedgehog signaling, and symptoms of early osteoarthritis. *Osteoarthr Cartil*. 2012;20(2):152–161. **(E)** Adapted from Xie YF, et al. Pulsed electromagnetic fields stimulate osteogenic differentiation and maturation of osteoblasts by upregulating the expression of BMPRII localized at the base of primary cilium. *Bone*. 2016;93:22–32.)

numbers on individual cells, with the notable exception of sperm flagellum, whereas the primary cilia occur as a solitary organelle.[8] The cilium is separated from the extracellular space by a membrane lipid bilayer that is continuous with the plasma membrane, but it contains a unique localization of membrane receptors and ion channels. Enclosing the ciliary compartment from the cytosol is the ciliary necklace or transition zone.[9] The structure of the cilium is maintained by a core cylindric skeleton, which is composed of nine microtubule doublets termed the ciliary axoneme. The cylindric axoneme is denoted as 9 + 0 in structure, which refers to the peripherally located microtubule doublets. This stands in contrast to the 9 + 2 structure of motile cilia, which contain a pair of central microtubules. At the base of the cilium lies the basal body, which acts as the anchor for the ciliary axoneme within the cell body. The basal body is composed of nine microtubule triplets, two of which contribute to the formation of the axoneme during cilium formation. The basal body and the associated transition zone proteins control regulation of protein entry and exit into the cilia.[8,10,11] Although slight differences exist between cilia of various cell types, presumably for specialized function, proteins common to all cilia include tubulins, microtubule motors, intraflagellar transport (IFT) complexes,

Bardet-Biedl syndrome (BBS)-related protein complex, and vesicular trafficking-related small GTPases.[12]

The specialized compartment of the primary cilium has a wide variety of functions because of the heterogeneous mixture of receptors, ion channels, and proteins uniquely localized to or concentrated at this organelle. This ciliary spatial organization mediates chemosensation and mechanosensation of the extracellular milieu, leading to transduction and regulation of key signaling pathways important in organ development, physiology, and disease. The presence of primary cilia on a cell is intimately related to the cell cycle. The ciliary microtubules are derived from the same template that coordinates the centrioles of the mitotic spindle.[13] Cilia are found in nearly all cells in the $G_0$ state and cells in interphase.[14] They are assembled during $G_1$ and remain until the formation of the mitotic spindle commences. At this point, the primary cilia are disassembled until the $G_1$ of the next cell cycle.[13,15] IFT, the process that facilitates protein movement along the cilium and is responsible for cilium length control, was discovered in the green alga, *Chlamydomonas*.[16,17] IFT refers to the transport of protein complexes between the base of the cilium through the transition zone to the distal tip. This transport may be anterograde, from the base to the tip, or retrograde, from the tip to the base. This mechanism is essential for the development and maintenance of the cilia owing to the lack of protein synthesis machinery within the cilium. The bidirectional transport is controlled by individual polypeptide subcomplexes, namely, Complex A and B. The polypeptide Complex B is responsible for anterograde transport and relies on proteins belonging to the kinesin 2 family (KIF3A and KIF3B). Complex A controls retrograde transport and is composed of proteins belonging to the dynein 2 protein family (DYNC2 proteins).[18,19]

Controlling cilium length is a dynamic process, with continuous turnover of tubulin at the ciliary tip. The initial cilium growth is rapid, until a steady-state length is achieved, whereby the rate of assembly and disassembly is equal.[17,20] IFT is therefore essential for the maintenance of cilium length. Huang et al. demonstrated this by arresting ciliary anterograde transport, using a flagellar assembly (FLA) 10 mutant gene in *Chlamydomonas*, with subsequent impairment of flagellar regeneration.[21] Typical cilium length ranges between 1–10μm and 0.2 μm in diameter; however, the cilium length is specific to individual cell types and environmental condition (Table 2.2.1).[22] Both genetic and sensory cues affect the cilium length, with subsequent impact on function.

## 2.2 Function

Owing to the distinct cellular compartment and unique lipid and receptor composition of the primary cilium, this organelle has evolved to have multiple diverse functions including chemosensation, mechanosensation, and signaling pathway regulation, all of which are dynamic and dependent on the differentiation state and the microenvironment of the cell.[23] Primary cilia coordinate several signaling pathways, including the hedgehog, transforming growth factor (TGF) β, G-protein-coupled receptor (GPCR), and Wnt pathways,[23–26] that are critical for development, physiology, and regeneration. Moreover, all these pathways are known to be mechanically regulated. One of the earliest known sensory roles for the cilium was in photoreception and olfaction.[8] For light detection, photoreceptors use a modified cilium connected to an outer segment composed of discs derived from the plasma membrane. The maintenance of this outer segment relies on IFT, as demonstrated in *kif3a* and Oak Ridge Polycystic Kidney (ORPK) mutant mice, in which absence of IFT resulted in blindness.[27] Furthermore, IFT is an established method of protein transport from the outer segment to the inner cell, via the cilium. In olfaction, the dendritic endings of sensory neurons may have 15–30 nonmotile cilia, which provide sensory transduction through an adenylyl cyclase type III (ACIII)-mediated cyclic AMP (cAMP) process, with signal amplification via $Ca^{2+}$-activated chloride channels.[28] The role of the cilia as a mechanosensor was first demonstrated in kidney epithelium, whereby deflection of the cilia in response to fluid movement triggered an intracellular calcium signal.[29] The role of the cilia as a mechanosensor has subsequently been demonstrated in multiple organ types and developmental processes, which will be discussed in more detail in Section 3.

## 2.3 Ciliopathies

Given the diverse prevalence and function of primary cilia in human physiology, it is not surprising that defects in this organelle lead to an array of pathologic conditions commonly termed ciliopathies. An important discovery linking ciliary dysfunction to human pathology was in the ORPK mouse/*ift88*[orpk], which presents with autosomal dominant polycystic kidney disease (ADPKD).[30] IFT88 is a component of anterograde IFT and is required for ciliogenesis,[4] and thus ORPK mice have abnormally short or missing cilia,[27] linking primary cilia and ADPKD. Human diseases associated with ciliary dysfunction have been extensively reviewed in Refs. 31 and 32 and will not be discussed here. However, further details related to ciliopathies associated with abnormal cilium mechanobiology will be discussed in Section 3.

**TABLE 2.2.1**
**Length and Incidence of Primary Cilium in Various Tissues, Both In Vivo and In Vitro**

| System | Tissue | Cell | Species | Observation | Cilium Length (μm) | Incidence | Citation |
|---|---|---|---|---|---|---|---|
| Urinary | Kidney | Epithelium, wild-type | Mice | In vitro | 1.8–5.2 | – | Pazour et al.[30] |
|  | Kidney | MDCK | Mice | In vitro | 5.0–8.0 | – | Praetorius and Spring[34] |
|  | Kidney | Epithelium, wild-type | Mice | In vitro | 11.8–12.7 | – | Nauli et al.[38] |
| Circulatory | Aortic arch | Endothelium | Mice | In vivo | 1.0–2.0 | 25% | Van der Heiden et al.[55] |
| Musculoskeletal | Sternum | Chondrocyte | Chick embryo | In vivo | 1–4.0 | 100% | Poole et al.[70] |
|  | Articular cartilage | Chondrocyte | Cow | In vivo | 1–1.5 | 40–65% | McGlashan et al.[72] |
|  | Articular cartilage | Chondrocyte | Canine | In vivo | 1.1–2.6 | – | Wilsman et al.[67] |
|  | Cortical bone | Osteocyte | Rat | In vivo | 0.8 | 94% | Uzbekove et al.[95] |
|  | Cervical vertebrae | Osteocyte | Sheep | In vivo | 1.46–1.86 | 1.5–4.6% | Coughlin et al.[96] |
|  | Vertebrae | Osteoblast | Human | In vitro | 1.25–2.86 | 59–75% | Oliazadeh et al.[103] |

# 3  PRIMARY CILIARY MECHANOBIOLOGY AND MECHANOTRANSDUCTION

The primary cilium first emerged as a site of mechanotransduction in the urinary system, where the kidney epithelial cell primary cilium extends out into the lumen where it deflects under urinary flow. This bending of the primary cilium was shown to trigger a flux in intracellular calcium levels, demonstrating a transduction of biophysical cellular deformation into a biochemical response.[33,34] Interestingly, this mechanotransduction response was lost with removal of the primary cilium, highlighting the critical role of the cilium in kidney mechanobiology. From this pioneering early work, the cilium has been shown to be required for mechanotransduction in multiple organisms and across many tissues responding to a diverse range of mechanical stimuli such as fluid shear, pressure, compression, and vibration.[35,36] Although the cilium has been shown to be involved in mechanotransduction in nearly all systems within the body, this chapter will aim to review our understanding of cilium mechanobiology within the urinary system, where it was first established; the circulatory system; and the musculoskeletal system (Fig. 2.2.2).

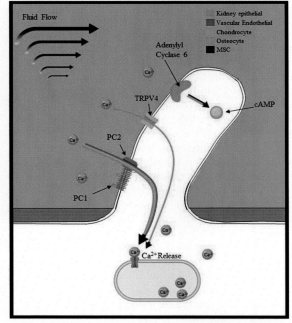

FIG. 2.2.2 The proposed ciliary mechanosignaling mechanisms in different cell types. *cAMP*, cyclic AMP; *MSC*, mesenchymal stem cell; *PC*, polycystin; *TRPV4*, transient receptor potential vanilloid 4.

## 3.1 Urinary System

### 3.1.1 Kidney

The kidney is a highly complex organ that participates in the control of whole-body homeostasis, osmolality, acid-base balance, and electrolyte concentrations and in the removal of various toxins. The kidney receives approximately 20% of the cardiac output, which is processed through filtration, reabsorption, secretion, and excretion to control the aforementioned parameters and urine is produced as a waste product. The microscopic functional unit of the kidney is the nephron, of which there are approximately 1 million in the average human adult kidney. The lumen of the nephron is lined with epithelial cells. These epithelial cells are exposed to a wide range of fluid shear stresses that are reliably transduced into intracellular responses. For example, the epithelial cells of the cortical collecting ducts have the ability to detect shear stresses at negligibly low velocities.[29,37] Praetorius and Spring's[29] work first demonstrated that this is achieved via the primary cilium, where bending of the ciliary axoneme under flow triggered an intracellular calcium flux using Madin-Darby Canine Kidney (MDCK) epithelial cells, a response that was lost when the cilium was removed. From this work, it was clear that these cells possess a primary cilium that is a highly specialized mechanosensory cell structure. Before this discovery, little thought had been given to the role of these cells in mechanotransduction because of the low fluid velocities present within the lumen. This work led to the considerable work focused on primary cilia within kidney mechanobiology.

The primary cilium is nearly ubiquitous in the kidney, with this organelle present on the apical surface of every epithelial cell extending into the tubular lumen. Several in vivo studies have established the role of the primary cilia in renal organogenesis and homeostasis. An important discovery linking the primary cilium to renal organogenesis and tissue homeostasis was that the mutant gene, which led to the ORPK mouse/$ift88^{orpk}$, was orthologous to the *Chlamydomonas* IFT88, an intraflagellar motor protein.[4] Pazour et al. demonstrated a role for IFT88 in flagellar/ciliary assembly using tg737/$ift88$ mutant mice. These mice died shortly after birth from ADPKD and were shown to have abnormally short or missing cilia.[27,30] This therefore implies that defects in primary cilium assembly can lead to polycystic kidney disease. ADPKD is due to an inherited mutation of polycystin (PC) 1, a large, integral membrane protein, and/or PC2, a $Ca^{2+}$-permeable cation channel, encoded by the *PKD1* and *PKD2* genes, respectively.

Nauli et al.[38] demonstrated that PC1 and PC2 are localized to the primary cilium, linking these two discoveries. This was achieved by comparing wild-type (WT) mice to mice with a targeted deletion of exon 34 ($Pkd1^{tm1Jzh}$ or $Pkd1^{del34/del34}$ mice). Here, PC1 was shown to colocalize with the specific ciliary axoneme marker acetylated α-tubulin in the kidneys of WT mice but not in mutant mice. The zebra fish is another in vivo animal model commonly used to study renal pathogenesis.[39] Low et al.[40] used this model to further decipher the detailed function of PC1. They proposed that PC1 undergoes proteolytic cleavage whereby the cytoplasmic tail undergoes nuclear translocation. This PC1 tail interacts with STAT6, a transcription factor, and the coactivator P100, which stimulates STAT6-dependent gene expression. This pathway was found to be inappropriately elevated in ADPKD, whereby elevated levels of nuclear STAT6, P100, and the PC1 tail were seen in cyst-lining epithelial cells. Furthermore, the overexpression of the human PC1 tail was sufficient to induce the formation of kidney cysts.

Experimental evidence that renal primary cilia had the potential to act as fluid flow sensors was first provided by Roth et al. in 1988.[41] Using a flexible substratum technique, Roth et al. visualized kidney epithelial cells side-on. It was observed that the renal cilia were deflected when exposed to flow and recoiled following the cessation of flow, acting not unlike a cantilevered beam.[42] It was not until Praetorius and Spring's work, which demonstrated that this primary cilium deflection in renal epithelial cells triggered intracellular biochemical signaling, that the cilium became known as a mechanosensory organelle. This was achieved using in vitro experiments in which MDCK primary cilia underwent bending with either exposure to fluid flow or micropipette suction leading to an initiation of an intracellular $Ca^{2+}$ flux.[29] Furthermore, the intracellular $Ca^{2+}$ rise was eliminated with $Ca^{2+}$-free media and $Gd^{3+}$ (a mechanosensitive ion channel blocker), thus indicating that this process is dictated by an extracellular $Ca^{2+}$-dependent influx via stretch-activated ion channels triggering an intracellular $Ca^{2+}$ release. Nauli et al.[38] demonstrated that renal epithelial cells with a mutation in *Pkd1* failed to elicit a $Ca^{2+}$ influx in response to fluid flow. Furthermore, extracellular $Ca^{2+}$ influx through PC2 was demonstrated with in vitro flow experiments using $Ca^{2+}$-depleted media and antibodies to inactivate PC2. Therefore Nauli et al. concluded that flow-induced bending of the cilia results in a conformational change to PC1, which leads to subsequent activation of the $Ca^{2+}$ channel PC2, thus facilitating mechanotransduction. They also suggested that the pathogenesis of ADPKD may be a result of abnormal mechanosensation due to the failure of this process.

In addition to ADPKD, the renal primary cilia are also implicated in the pathogenesis of a number of other renal cystic diseases. Nephronophthisis-related ciliopathies (NPHP—RC) are recessive cystic kidney diseases and the most common cause of end-stage renal disease in the first three decades of life.[43] They often have associated retinal and cerebellar involvement. In contrast to ADPKD, the macroscopic appearance of the kidney is often normal or reduced in size. The cysts in this disease occur at the corticomedullary junction of the kidney. A total of 10 causative genes (*NPH1—9* and *AHI1*) have thus far been identified, each contributing to individual variations of the disease. These nephrocystin proteins have been shown to be located at the primary cilium-centrosome complex and thus play important functions in ciliary sensation, cell polarity, and cell division. Otto et al.[44] showed using homozygosity mapping of 120 affected families that many more genes are likely to cause NPHP-RC than have thus far been identified. As our understanding of the various ciliopathies, such as ADPKD, has increased rapidly since the turn of the century, recent work has now turned to potential therapeutic strategies. The mammalian target of rapamycin (mTOR) pathway has been shown to be upregulated in ADPKD.[45] Under normal physiologic conditions, cilia regulate the mTOR pathway via PC1-mediated suppression. Rapamycin, an mTOR inhibitor, has been shown to reduce cyst growth and preserve renal function in in vivo models.[45,46] Furthermore, rapamycin increases both ciliary length and sensitivity in in vitro studies, further highlighting its potential as a therapeutic option for ciliopathies.[47]

## 3.2 Circulatory System
### 3.2.1 Endothelium
A natural progression for cilium mechanobiology research, following on from the extensive work demonstrating its role in fluid shear mechanosensation in the renal system, is within the circulatory system, specifically the vascular endothelium. Modifications of blood vessel diameter serve an important function as physiologic regulation of blood flow and pressure. This is achieved by contraction and relaxation of vascular smooth muscles located in the walls of bloods vessels. Control is regulated at both a systemic and local level. The systemic or central regulation is achieved through neuronal control by the central and peripheral nervous systems.[48,49] Local control is termed autoregulation and allows for an immediate control of vascular tone and blood flow in a specific region of tissue. Advantages of this system include immediate control, in addition to having minimal impact on neighboring systems or the global physiology.[50] Endothelial cells, which line the lumen of blood vessels, have the ability to sense blood flow, which can generate shear stresses as high as 5 Pa in the main arteries.[51] The physiologic role of these blood shear stresses on endothelial cells and the underlying smooth muscle function has long been established, as altered blood flow profiles or abnormal endothelial cell mechanosensation can lead to diseases such as atherosclerosis and hypertension.[52,53]

Bystrevskaya et al.[54] first identified vascular primary cilia in adult and embryonic human aortic endothelial cells in 1988. Unlike the renal epithelial cilia, ciliated endothelial cells are far more limited in prevalence (<25%).[55] In endothelial cells, primary cilia are found in areas of low or disturbed flow and are typically absent in areas of high flow or shear stress. Van der Heiden et al.[56] demonstrated in chick embryonic endocardium, by using the high shear stress marker Kruppel-like factor 2 (KLF2), that the presence of cilia was inversely proportional to shear stress. Atherosclerosis typically occurs at areas of low and oscillating shear stress. Van der Heiden et al.[55] further demonstrated increased primary cilium number at atherosclerotic predilection sites, such as the aortic arch and carotid arteries, in both WT C57BL/6 adult and apolipoprotein-E-deficient mice. Interestingly, increased cilia were observed on and around atherosclerotic lesions in the apolipoprotein-E-deficient mice, thus suggesting a role for endothelial activation and dysfunction in this disease.[55] However, Dinsmore and Reiter,[57] using a conditional IFT88 knockout model, found increased atherosclerosis in $Apoe^{-/-}$ mice, thus highlighting a protective effect of cilia. They further demonstrated that these mice without primary cilia (IFT88 knockout) had increased inflammatory gene expression and decreased endothelial nitric oxide synthase activity, further indicating that endothelial cilia inhibit proatherosclerotic signaling. Iomini et al.[58] demonstrated that high fluid shear stress induces cultured human endothelial cilia to disassemble, which may be a normal finding in larger arteries or pathologic in smaller vessels rendering them unable to sense minute alterations in blood pressure.

Primary cilia function to convert local blood flow information into functional responses, such as amplification of gene expression and control of nitric oxide (NO) production. NO results in smooth muscle relaxation, therefore resulting in vasodilation and a reduction in total peripheral resistance. PC1 and PC2 are present on endothelial cilia, similar to those seen in the kidney, and play a central role in the conversion of fluid shear stress to NO production and release into the surrounding smooth muscle cells. The ability to sense fluid flow and produce a NO response was lost in both knockdown and knockout mouse models of PKD2, in addition to vascular endothelia from patients with ADPKD with absent PC2.[59] This, therefore, suggests that abnormal endothelial mechanosensation of blood flow may lead to the pathogenesis of vascular defects associated with ADPKD. Moreover, Egorova et al.[60] demonstrated the pivotal role of the primary cilia in rendering endothelial cells susceptible to shear-induced activation of TGFβ/ALK5, thus providing a link between the primary cilia and flow-related endothelial activity.

One of the vascular pathologic conditions associated with ADPKD include aneurysms. An aneurysm is a pathologic, localized swelling of the wall of an artery. AbouAlaiwi et al. demonstrated both low *survivin* and chromosomal instability in the cysts and aneurysms from patients with ADPKD, *Pkd* mouse, and zebra fish model of ADPKD. Moreover, they demonstrated *survivin* downregulation as a result of abnormal mechanosensory cilia via the PKC-Akt-NF-κB-survivin pathway, which induces aberrant ploidy formation. Low *survivin* expression also results in abnormal cytokinesis and disorientated cell division. This asymmetric cell division is hypothesized to contribute to the expansion of tissue architecture and arterial diameter in aneurysmal formation.[61] Furthermore, another vascular pathologic condition associated with ADPKD is hypertension. A key hormone regulating hypertension is dopamine, and consequently, dopamine receptor type 5 (DR5) has received considerable attention, as it localizes to primary cilia of endothelial cells both in vitro and in vivo. Abdul-Majeed and Nauli[62] demonstrated that DR5 activation affects both ciliary length and function, where activation of DR5 increases ciliary length and silencing of DR5 significantly shortens ciliary length and inhibits mechanotransduction. Elegant studies have utilized nanoparticles that target the cilium via DR5 to deliver fenoldopam directly to the vascular endothelia, an approach that was more effective than using dopamine alone.[63] Moreover, the nanoparticles enabled remote control of the movement and function of a cilium with an external magnetic field, as a potential mechanism to correct for altered blood flow. These therapies combined significantly improved cardiac function in ciliopathic hypertensive mouse models.[63] Therefore similar to other systems, such as the urinary system, cilium length is intimately related to cilium function; however, further work is required to fully delineate the mechanisms at play in the endothelial system. Importantly, as our understanding of the underlying mechanisms at play increases, so too does the potential therapeutic targets, as evidenced by the aforementioned work using fenoldopam and dopamine.

## 3.3 Musculoskeletal System
### 3.3.1 Cartilage
Articular cartilage is fundamentally a mechanoresponsive tissue that is exposed to complex, dynamic mechanical loading during everyday activity. The physical response of the tissue to this loading is critical to its biomechanical function, while the mechanobiological response of the cells ensures maintenance of this tissue. Disruption of either the mechanical environment (e.g., abnormally high or low levels of loading) or the cellular mechanosignaling behavior leads to cartilage degradation and loss of functionality, associated with musculoskeletal joint disease. Cartilage is an aneural, avascular connective tissue composed predominantly of extracellular matrix (ECM). Articular cartilage, covering the surface of bones in diarthrodial joints such as the knee, provides a smooth, low-friction surface and functions to distribute the compressive forces experienced by normal articulating joints, thereby protecting the underlying bone from high stress. Articular cartilage has a limited capacity for repair, with the chronic degenerative disease, osteoarthritis, developing as a result of tissue damage. During normal physiologic movement, articular cartilage is subjected to dynamic mechanical loading, which promotes cartilage matrix production, joint health, and joint maintenance. The articular cartilage ECM composition is regulated by the singular cartilage cell type, the chondrocyte, which makes up only ~10% of the tissue's total volume. In particular, chondrocytes are responsible for both the synthesis and degradation of the ECM. These processes are influenced by biochemical and biomechanical stimuli by interacting with the ECM via integrins and other receptors and ion channels expressed at the cell surface.[64]

During joint loading, the extracellular environment of chondrocytes is altered such that the cells are subjected to shear, tensile, and compressive strain; fluid flow; and changes in pH, hydrostatic pressure, and osmolarity. Chondrocytes are specially adapted to sense and respond to physicochemical stimuli and regulate their metabolic profile according to the frequency and amplitude of the load applied.[65–67] The mechanisms by which chondrocytes convert mechanical signals into intracellular responses that modulate cell behavior are not well understood. Given the variety of mechanical stimuli the chondrocyte experiences,[65] it is unlikely that any one mechanism will prove to be wholly responsible for the multitude of responses that have been observed upon loading. Numerous pathways involving the cellular cytoskeleton, integrins, and mechanosensitive ions have been identified in the process of chondrocyte mechanotransduction. Indeed many excellent reviews of chondrocyte mechanosignaling have been published.[65] More recently, primary cilia have been suggested as providing a nexus for mechanosignaling in a variety of different cell types including chondrocytes.

The majority of articular chondrocytes elaborate a primary cilium.[67-74] The prevalence, length, orientation, and positioning of chondrocyte primary cilia in situ varies with cartilage depth. In bovine patellae, less than 40% of cells in the superficial zone exhibit a primary cilium with a mean length of 1.1 μm.[72] This increases to 65% in the deep zone where the mean ciliary length is 1.5 μm.[72] Primary cilia in the middle and deep zones are frequently found on the medial or lateral surface of the cell, with respect to the articular surface.[72,75] While in the superficial zone, where cells experience the greatest levels of compressive strain, ciliary length is at its lowest and the cilia are almost exclusively found pointing away from the articular surface.[72,75,76] Chondrocyte primary cilia are significantly longer when cells are isolated from their ECM and cultured in a two-dimensional monolayer, where cilia have a mean length of 2–3 μm and prevalence is up to 90%. These findings suggest a relationship between the mechanical environment and cilium expression.

A variety of in vivo and in vitro models have been used to investigate the role of the primary cilium in chondrocyte mechanotransduction. In vivo genetic mutation models have suggested a role for the primary cilium in the development and maintenance of articular cartilage. In particular, the ever important Tg737[ORPK] mouse, introduced earlier, has increased articular cartilage cellularity and the tissue exhibits altered expression and distribution of collagen II.[71] Furthermore, mice with a cartilage-specific deletion of primary cilia (Col2a-Cre; Ift88[loxP/loxP]) have been reported to have thicker, more cellular articular cartilage with increased collagen II and aggrecan in the pericellular region.[77] Other studies have investigated cartilage in BBS, a human ciliopathy presenting with a musculoskeletal polydactyly phenotype. In mice harboring mutations in several of the BBS proteins, the proteoglycan content of the cartilage matrix was reported to be lower and the tissue exhibited a reduced thickness relative to WT mice.[78]

Owing to the complexity of the mechanical environment within the articular cartilage in situ, numerous in vitro studies have used isolated chondrocyte models to investigate mechanotransduction and the role of the primary cilium in this process. In these experimental systems the influence of different mechanical stimuli can be examined in isolation from the physicochemical changes associated with compression of the charged ECM. These models include either isolated chondrocytes in a monolayer subjected to tensile strain[79-81] or fluid flow[82,83] or chondrocytes seeded within 3D constructs (e.g., agarose) and subjected to compressive strain[73,84,85] or hydrostatic pressure.[86,87] Wann et al.[88] provided the first direct evidence supporting a role for the primary cilium in chondrocyte mechanotransduction using a mutant cell line generated from the Tg737[ORPK] mouse. In WT cells, cyclic mechanical compression triggered ATP release and activated a $Ca^{2+}$ signaling cascade, culminating in the upregulation of aggrecan gene expression and proteoglycan synthesis. In the mutant (ORPK) chondrocytes, no significant changes in aggrecan expression were observed and proteoglycan production was not altered in response to mechanical loading. Although the mechanosensitive release of ATP was observed in both WT and ORPK chondrocytes, the $Ca^{2+}$ signaling downstream of this event was disrupted in ORPK hypomorphic IFT88 cells. This study indicated that the chondrocyte primary cilium and/or IFT88 is an essential component of the downstream mechanotransduction response but intriguingly is not involved in initial mechanosensing. Cleavage of the PC1 C-terminal tail was increased in IFT88[ORPK] chondrocytes lacking cilia, suggesting PC1 may have a novel function in this mechanosignaling response.[88] Using the chondrogenic cell line ATDC5, a model for growth plate chondrocytes, Rais et al. demonstrated the importance of KIF3A, a fundamental ciliary motor protein, in chondrocyte mechanotransduction. KIF3A expression was knocked down in ATDC5 cells, mechanical stretch was applied, and a panel of mechanosensitive genes was examined. While some mechanosensitive genes were either completely or partially independent of KIF3A, others such as AGC1 and collagen type X (major ECM components) had their responsivity abolished with KIF3A knockdown. Additionally, regulation of the primary-cilium-related genes PKD1, PKD2, IFT88, and IFT172, in response to mechanical load, is also KIF3A-dependent, such that when KIF3A expression is disrupted, these genes are also downregulated.[89] Although the mechanistic pathway has begun to be elucidated, the exact mechanism and components behind the role of the primary cilium in chondrocyte mechanotransduction remain unclear.

### 3.3.2 Bone

Like all tissues within the musculoskeletal system, bone is a highly mechanoresponsive and dynamic tissue that adapts to changes in mechanical loading. Macroscale loading of the skeleton generates a complex mechanical environment at the cellular level, consisting of microstrain deformations, which in turn generate pressures in the kilopascal range and fluid shear predicted to be in the pascal range. Osteocytes are fully differentiated, highly mechanosensitive cells embedded in the mineral matrix and are the most abundant cell type within bones. As such the osteocyte is believed to be the master orchestrator of bone metabolism,[90,91] coordinating bone resorption via osteoclasts, which originate from hematopoietic origin,[92] and bone formation via osteoblasts, which originate from the mesenchymal origin.[90,93] Disruption of this osteocyte driven-tissue mechanoadaptation, as is seen with age, can lead to bone loss diseases such as osteoporosis and an increased risk of fracture.

The presence of a primary cilium on osteocytes was first described over 40 years ago using electron microscopy.[94] Despite a relatively low incidence—1.7% in adolescent, 3.6% in adult, and 3.5% in geriatric murine specimens—the study demonstrated that the osteoblastic lineage could form a primary cilium in vivo.[94] Two recent and conflicting microscopy studies found an osteocyte primary cilium incidence of 94% in adult Wistar rat cortical bone, but only 4% in sheep cervical vertebrae.[95,96] In the vertebral samples, the incidence was also determined for lining cells, 4.6%, and for marrow cells, 1.5%.[96] Supplementing the few in vivo studies, in vitro studies with primary and immortalized cell lines found a primary cilium incidence of 60% −85% in bone-marrow-derived mesenchymal stem cells (MSCs),[26,97−102] 50%−65% in osteoblasts,[103,104] and 55%−62% in osteocytes.[105−107] Like osteoblastic cells, osteoclasts are mechanoresponsive,[108] but the response is unlikely mediated by the primary cilium because osteoclasts are not known to form one.[109] However, primary cilia have recently been found on over 90% of human peripheral blood mononuclear cells and bone marrow cells.[110] The role of the primary cilium in these osteoclast precursors is unknown.

The primary cilium in the osteoblastic cell lineage is critical to bone mechanotransduction and load-induced bone formation. In one study, bone marrow cells with a *Kif3a* (a key anterograde ciliary transport motor protein

necessary for cilium formation) knockout were transplanted into bone-marrow-irradiated mice and ulnar loading was applied. In response to loading, mice with ciliary defects had reduced recruitment and differentiation of progenitor cells and decreased mineralization of bone.[111] Osteoblast and osteocyte primary cilia were similarly disrupted using a *Colα1(I)* Cre recombinase promoter to knock out *Kif3a*. Although no skeletal abnormalities were noted in adult mice, ulnar loading resulted in a 32% reduction in the relative mineral apposition rate, a measure of the amount of osteoblast activity.[112] Interpretation of *Kif3a* knockout models is nuanced, as Kif3a has been found to have an extraciliary role leading to unrestrained Wnt signaling,[113] which is a key pathway regulating bone homeostasis. IFT88 knockouts are another method to disrupt primary cilium formation.[114] Loss of *IFT88*, a component of anterograde IFT, inhibits proper cilium extension. Periosteal cells expressing the cell marker *Prx1* were identified as a mechanosensing cell in the adult murine skeleton, and loss of *IFT88* in this cell population abrogates fluid flow mechanosensing.[115] Instead of inhibiting primary cilium formation, other models have knocked out key proteins implicated in ciliary mechanosensing. Adenylyl cyclase (AC) 6, a cAMP catalyzing enzyme, localizes to the osteocyte and stem cell primary cilia.[98,116] Mice lacking AC6 have reduced bone formation in response to loading. This is attributed to defects in bone cell mechanosensing because these mice exhibited normal osteoblastic bone formation in response to parathyroid hormone treatment.[117] PC1 (Pkd1) is a mechanosensitive transmembrane protein that is highly enriched in the primary cilium. Conditional deletion of *Pkd1* in mature osteoblasts and osteocytes results in reduced bone mineral density and diminished load-induced bone formation.[118]

By isolating cells from the multitude of factors in vivo, in vitro studies have clearly demonstrated that the primary cilium is a bone cell mechanosensing organelle. These studies use a combination of primary and immortalized cell lines from human and murine origin. In MSCs, osteoblasts, and osteocytes, inhibiting primary cilium formation with small interfering RNA-mediated knockdown of *IFT88* or chloral hydrate treatment causes a reduction in fluid-flow-induced osteogenic gene expression (e.g., cyclooxygenase 2, bone morphogenic protein 2, and osteopontin) and cytokine signaling (e.g., osteoprotegerin, RANKL, and prostaglandin $E_2$).[97,99,105,119] Although fluid flow is the dominant in vivo stimulus, pulsed electromagnetic fields are a potential osteogenic mechanical intervention that increase osteoblastic activity in a primary cilium-dependent manner.[120,121] Primary cilium length is positively correlated with mechanosensitivity and adapts based on loading history. Osteoblast cilia become shorter after fluid flow stimulation, possibly protecting the cell from overstimulation.[119] Pharmacologically lengthening the primary cilium in MSCs and osteocytes increases the osteogenic mechanoresponse.[107,122] Interestingly, stimulated microgravity causes ciliary shortening and reduced osteoblastic activity,[123] despite the fact that primary cilia elongate in areas of decreased stimulation in other contexts.[124,125] Other aspects of cellular mechanics adapt to loading in a primary cilium-dependent manner—loss of the primary cilium prevents a flow-induced increase in microtubule network density[126] and focal adhesion staining.[127] Additionally, osteogenic induction of MSCs by mechanically stimulated osteocytes is inhibited with a loss of osteocyte primary cilia.[128] It is not yet clear if there is a unique primary cilium-mediated signaling pathway in osteoblastic cells, but the predominant ciliary signaling model involves a mechanoresponsive influx of ciliary calcium through the ion channel transient receptor potential vanilloid 4 (TRPV4) that may in turn regulate cAMP signaling of ciliary localized AC3 and/or AC6,[97,98,106,116,129,130] although this cAMP signaling may also be potentially regulated via GPCRs.[131,132]

Primary ciliary defects are implicated in skeletal disease. Skeletal ciliopathies, or ciliary chondroplasias, are caused by a multitude of genes, and various mutations can lead to the same disease phenotype.[133] Skeletal phenotypic defects are primarily developmental and include short rib syndrome, polydactyly, thoracic dystrophy, craniofacial defects, spinal abnormalities, short stature, and ossification defects.[133,134] Skeletal ciliopathies involve all six components of the retrograde axoneme IFT-A transport complex,[135] several anterograde components of the IFT-B complex,[135–137] motor proteins,[138] and components of the ciliary base.[139] Owing to the significant developmental defects and involvement of other systems, it is difficult to assess effects of bone cell mechanosensing on disease progression. One study examined osteoblast function in pediatric patients with idiopathic scoliosis, which is associated with primary cilium mutated genes. Isolated osteoblasts had longer primary cilia, decreased proliferation, and decreased mechanosensitivity.[103] Much remains to be learned about the role of mechanotransduction in the cause and course of these diseases.

## 4 MECHANICS OF PRIMARY CILIUM BIOLOGY

The primary cilium is now an established sensory organelle capable of transducing a number of biophysical extracellular signals. Central to ciliary deformation and mechanotransduction is the mechanics of this antenna-like organelle, which is dictated by its internal structural components and length. The mechanics of the primary cilium has been reviewed previously,[35] detailing the intricate and complex dynamic structure of the cilium. Although studies have shown that ciliary stiffness and thus deflection can be altered by both applied mechanical and chemical means,[140] a major contributor to ciliary deformation and associated signaling is axoneme length. Increasingly, studies are reporting that the physicochemical environment to which cells are exposed can regulate their primary ciliary expression, with changes observed in prevalence, length, orientation, and position. Primary cilia are known to disassemble and shorten in response to mechanical loading, in the form of compression,[73] tensile strain,[141,142] fluid shear,[143] or hydrostatic pressure.[143] This behavior appears to be conserved in multiple cell types including chondrocytes,[73,144] tenocytes,[142] epithelial cells,[143] endothelial cells,[145] and MSCs.[119] However, some cell types appear to be more sensitive to mechanically induced ciliary disassembly. For example, isolated tenocytes show almost complete loss of cilia in response to 5% cyclic tensile strain,[142] whereas chondrocytes exhibit much more modest reduction in ciliary length in response to 20% strain.[144] These differences may be due to the in vivo mechanical environment to which different cell types are exposed during normal physiologic activity. Tenocytes are unlikely to routinely experience strains above 2%–3%, whereas chondrocytes routinely experience 10%–20% compressive strain during normal activity. In endothelial cells, alterations in fluid flow regulate cilium expression in vivo such that increased prevalence of elongated cilia is observed in areas of disturbed flow, such as bifurcation points.[55,57]

This mechanism of mechanically induced ciliary disassembly appears to involve activation of the tubulin deacetylase, HDAC6, which causes deacetylation of the axoneme and increased disassembly, leading to shorter cilia.[140] It is unclear whether this reflects a snapshot of a process leading to complete disassembly or a new stable shorter cilium set length. Further studies have identified actin tension as being a key regulator of ciliogenesis and set length.[57,146–148] Thus changes in cilium expression may be initiated by changes in actin organization, which are well known to occur in response to mechanical loading.[149,150] The upstream mechanism associated with this alteration in actin may include mechanosensitive activation of calcium signaling, leading to conformational changes in actin-associated proteins.[151] Other studies have described the importance of the Yap/Taz pathway in modulating actin organization and cilium expression.[152] Both direct mechanical forces and the physicochemical environment can regulate cilium expression. Studies have reported temporal modulation of ciliary length in response to osmotic challenge,[153] changes in substrate stiffness, or changes in surface topography.[154]

Alterations in cilium prevalence and/or ciliary length can have profound influence on ciliary signaling. It is well established that genetic disruption of both motile and primary cilium structure is associated with loss of function in the so-called ciliopathies (for review see Refs. 155 and 31). However, more recent evidence suggests that mechanical, physicochemical, or pharmaceutical changes in primary cilium length are also associated with alterations in cilium-mediated signaling. For example, a slight ciliary disassembly induced by mechanical loading of chondrocytes (0% −20% cyclic tensile strain), inhibits hedgehog signaling.[144] Inhibition of this mechanically induced ciliary disassembly via the HDAC6 inhibitor, tubacin, prevented the disruption in hedgehog signaling. However, despite the evidence that changes in ciliary expression and length are associated with alterations in ciliary function, the mechanisms for this environmental regulation of ciliary signaling remain unclear. In the case of mechanosignaling in response to fluid flow, as occurs in epithelial cells, osteocytes, and MSCs,[119,143] increases in the length of apical cilia are likely to result in greater levels of ciliary deflection.[107,122] This may increase activation of mechanosensitive ion channels, such as the PC1-PC2 calcium channel complex or TRPV4, located at the base of the cilium where bending strains will be highest.[42,59,156] Although in other cell types, such as chondrocytes, IFT-dependent mechanosignaling does not appear to involve ciliary deflection,[88] it is less clear how changes in ciliary length may impact signaling.

For other cilium- or IFT-dependent pathways such as hedgehog signaling, Wnt signaling, growth factor signaling, and inflammatory signaling, the mechanism through which changes in length regulate signaling is unknown. However, previous studies have identified changes in ciliary length associated with alterations in hedgehog signaling induced by pharmaceutical antagonists[157] and by high levels of mechanical strain.[144] Similarly, cilium elongation has been associated with the modulation of IFT-dependent proinflammatory signaling by various chemical and physicochemical stimuli.[74]

One possible mechanism suggests a relationship between the length of the ciliary axoneme and the rate of IFT protein delivered. According to the balance-point model and other related models of ciliary assembly, as the cilium lengthens, the rate of insertion of IFT cargoes is reduced.[158,159] When these IFT cargoes are carrying tubulin for assembly into the axoneme at the tip, this causes the rate of ciliogenesis to slow and eventually stop at a stable set length where the rate of anterograde delivery of tubulin is matched by the constant rate of disassembly and retrograde transport. It is therefore possible that the same mechanism controlling IFT cargo injection and tubulin delivery may also regulate the delivery on and off the axoneme and associated accumulation of signaling proteins necessary for ciliary function. Alternatively, changes in ciliary length may be indicative of alterations in IFT, which may be the real driver for the modulation of signaling activity. In this scenario the alterations in ciliary length are purely a readout of changes in ciliary function and are not the direct cause.

Whatever the mechanism, there is clearly increasing evidence of a correlation between ciliary length and ciliary function including mechanosignaling as well as regulation of other pathways. In addition to axoneme length, other changes in ciliary expression and morphology may also influence the associated signaling. For example, it is unclear what drives cilia to be expressed on the apical or basal cell surfaces[56] and yet a switch from one to the other could have profound consequences on ciliary deflection and mechanosignaling in response to fluid flow. Similarly, alterations in the mechanical properties of the axoneme, possibly through posttranslational modifications of the tubulin or structural changes in its doublet structure, may lead to changes in deflection and the ability to respond to fluid shear.[140,156] Other studies have measured the projection of the cilium from the ciliary pocket[160] (for review see Ref. 161). In cells where the axoneme is largely invaginated within the pocket, the responsiveness to external mechanical and chemical stimuli is likely to be attenuated.

## 5 DISCUSSION AND FUTURE DIRECTIONS

Although great strides have been taken in understanding primary ciliary mechanobiology since the early work of Praetorius and Spring, we are only beginning to decipher the diverse and far reaching significance of this cellular organelle within mechanobiology. Some of the open questions and areas for future research are discussed in the following.

Although a prominent role for the cilium in mechanotransduction has emerged in tissues such as the kidney, vasculature, liver, cartilage, and bone, the potential role for the cilium in the mechanobiology of numerous mechanoresponsive tissues remains to be explored. For example, primary cilia are present on tenocytes and are orientated with respect to the collagen direction and long axis of the tendon.[162] The tenocyte primary cilium also undergoes disassembly following the application of mechanical stimuli, suggesting that the cilia are mechanoresponsive in this tissue.[163] However, whether the cilium is required for mechanotransduction remains unknown. The cilia are also present on numerous other cell types within the musculoskeletal system, such as those found within all forms of cartilage, meniscus, perichondrium, muscle, and ligament, yet the potential mechanobiological role of the cilium in these tissues, and many more, has yet to be investigated.

Once a role for the cilium in the mechanobiology of a given tissue has been demonstrated, the next step will be to ascertain the mechanisms by which this organelle mediates mechanotransduction. This is of particular importance owing to the complex and dynamic makeup of these organelles, which can change depending on the cell type and environmental condition, resulting in mechanisms of cilium-mediated mechanotransduction that can be unique to a tissue or cell type. For example, the sensing of fluid flow via the primary cilium can occur via a calcium-dependent or calcium-independent mechanism, likely via cAMP, in tissues such as bone and the liver.[98,105,130,164,165] Identifying the ciliary components mediating these second messenger responses has been greatly aided by the development of advanced proteomics and imaging techniques. For example, utilizing a proximity labeling technique to biotinylate proteins that are trafficked into the cilium, primary cilium-specific proteomics can be performed, resulting in a complete map of proteins within this organelle that may be involved in mechanotransduction.[166] Combined with the development of cilium-localized biosensors that enable radiometric measurements of second messengers, such as calcium and cAMP, within the ciliary compartment during mechanical stimulation, we are now in a position to define the mechanisms of action of ciliary mechanotransduction. These biosensors have proved enormously powerful and have helped highlight the ciliary compartment as a calcium and cAMP microdomain within the cell.[63,129,130] It is through the identification and validation of the molecular mechanisms of cilium-mediated mechanotransduction that we will be able to better understand the role of cilium in the plethora of diseases associated with this organelle. It will also enable the targeting of these mechanisms as a therapeutic strategy to treat these pathologic conditions, an approach that is showing real promise through ciliotherapies.[63,122,167]

## ACKNOWLEDGMENTS

European Research Council (ERC) Starting Grant (#336882), Irish Research Council Government of Ireland Postgraduate Scholarship (GOIPG/2019/4469), Science Foundation Ireland ERC Support Grant SFI 13/ERC/L2864, the National Institute of Arthritis and Musculoskeletal and Skin Diseases at the National Institute of Health R01-AR062177, and a SEMS QMUL studentship.

## REFERENCES

1. Petridou NI, Spiro Z, Heisenberg CP. Multiscale force sensing in development. *Nat Cell Biol.* 2017;19(6):581−588.
2. Schwarz US. Mechanobiology by the numbers: a close relationship between biology and physics. *Nat Rev Mol Cell Biol.* 2017;18(12):711−712.
3. Satir P, Pedersen LB, Christensen ST. The primary cilium at a glance. *J Cell Sci.* 2010;123(4):499−503.
4. Lehman JM, Michaud EJ, Schoeb TR, Aydin-Son Y, Miller M, Yoder BK. The Oak Ridge Polycystic Kidney mouse: modeling ciliopathies of mice and men. *Dev Dyn.* 2008;237(8):1960−1971.
5. Bloodgood RA. From central to rudimentary to primary: the history of an underappreciated organelle whose time has come. The primary cilium. Primary cilia. *Methods Cell Biol.* 2009;94:3−52.
6. Fliegauf M, Benzing T, Omran H. When cilia go bad: cilia defects and ciliopathies. *Nat Rev Mol Cell Biol.* 2007;8(11):880.
7. Mitchell DR. The evolution of eukaryotic cilia and flagella as motile and sensory organelles. In: *Eukaryotic Membranes and Cytoskeleton.* Springer; 2007:130−140.
8. Berbari NF, O'Connor AK, Haycraft CJ, Yoder BK. The primary cilium as a complex signaling center. *Curr Biol.* 2009;19(13):R526−R535.
9. Gilula NB, Satir P. The ciliary necklace: a ciliary membrane specialization. *J Cell Biol.* 1972;53(2):494−509.
10. Marshall WF. Basal bodies: platforms for building cilia. *Curr Top Dev Biol.* 2008;85:1−22.
11. Pazour GJ, Bloodgood RA. Targeting proteins to the ciliary membrane. *Curr Top Dev Biol.* 2008;85:115−149.
12. Avasthi P, Maser RL, Tran PV. Primary cilia in cystic kidney disease. In: *Kidney Development and Disease.* Springer; 2017:281−321.

13. Rieder CL, Jensen CG, Jensen LC. The resorption of primary cilia during mitosis in a vertebrate (PtK1) cell line. *J Ultrastruct Res.* 1979;68(2):173–185.

14. Wheatley DN, Wang AM, Strugnell GE. Expression of primary cilia in mammalian cells. *Cell Biol Int.* 1996;20(1):73–81.

15. Pazour GJ, Witman GB. The vertebrate primary cilium is a sensory organelle. *Curr Opin Cell Biol.* 2003;15(1):105–110.

16. Kozminski KG, Beech PL, Rosenbaum JL. The Chlamydomonas kinesin-like protein FLA10 is involved in motility associated with the flagellar membrane. *J Cell Biol.* 1995;131(6):1517–1527.

17. Marshall WF, Rosenbaum JL. Intraflagellar transport balances continuous turnover of outer doublet microtubules: implications for flagellar length control. *J Cell Biol.* 2001;155(3):405–414.

18. Pedersen LB, Rosenbaum JL. Chapter two intraflagellar transport (IFT): role in ciliary assembly, resorption and signalling. *Curr Top Dev Biol.* 2008;85:23–61.

19. Scholey JM. Intraflagellar transport. *Annu Rev Cell Dev Biol.* 2003;19(1):423–443.

20. Rosenbaum JL, Child F. Flagellar regeneration in protozoan flagellates. *J Cell Biol.* 1967;34(1):345–364.

21. Huang B, Rifkin M, Luck D. Temperature-sensitive mutations affecting flagellar assembly and function in Chlamydomonas reinhardtii. *J Cell Biol.* 1977;72(1):67–85.

22. Poole CA, Zhang Z-J, Ross JM. The differential distribution of acetylated and detyrosinated alpha-tubulin in the microtubular cytoskeleton and primary cilia of hyaline cartilage chondrocytes. *J Anat.* 2001;199(4):393–405.

23. Anvarian Z, Mykytyn K, Mukhopadhyay S, Pedersen LB, Christensen ST. Cellular signalling by primary cilia in development, organ function and disease. *Nat Rev Nephrol.* 2019;15(4):199–219.

24. Liu H, Kiseleva AA, Golemis EA. Ciliary signalling in cancer. *Nat Rev Cancer.* 2018:1.

25. Bangs F, Anderson KV. Primary cilia and mammalian hedgehog signaling. *Cold Spring Harb Perspect Biol.* 2017;9(5):a028175.

26. Labour MN, Riffault M, Christensen ST, Hoey DA. TGFbeta1 – induced recruitment of human bone mesenchymal stem cells is mediated by the primary cilium in a SMAD3-dependent manner. *Sci Rep.* 2016;6:35542.

27. Pazour GJ, Baker SA, Deane JA, et al. The intraflagellar transport protein, IFT88, is essential for vertebrate photoreceptor assembly and maintenance. *J Cell Biol.* 2002;157(1):103–114.

28. McEwen DP, Jenkins PM, Martens JR. Olfactory cilia: our direct neuronal connection to the external world. *Curr Top Dev Biol.* 2008;85:333–370.

29. Praetorius H, Spring KR. Bending the MDCK cell primary cilium increases intracellular calcium. *J Membr Biol.* 2001;184(1):71–79.

30. Pazour GJ, Dickert BL, Vucica Y, et al. Chlamydomonas IFT88 and its mouse homologue, polycystic kidney disease gene tg737, are required for assembly of cilia and flagella. *J Cell Biol.* 2000;151(3):709–718.

31. Hildebrandt F, Benzing T, Katsanis N. Ciliopathies. *N Engl J Med.* 2011;364(16):1533–1543.

32. Waters AM, Beales PL. Ciliopathies: an expanding disease spectrum. *Pediatr Nephrol.* 2011;26(7):1039–1056.

33. Praetorius HA, Spring KR. The renal cell primary cilium functions as a flow sensor. *Curr Opin Nephrol Hypertens.* 2003;12(5):517–520.

34. Praetorius HA, Spring KR. Bending the MDCK cell primary cilium increases intracellular calcium. *J Membr Biol.* 2001;184(1):71–79.

35. Hoey DA, Downs ME, Jacobs CR. The mechanics of the primary cilium: an intricate structure with complex function. *J Biomech.* 2012;45(1):17–26.

36. Spasic M, Jacobs CR. Primary cilia: cell and molecular mechanosensors directing whole tissue function. *Semin Cell Dev Biol.* 2017;71:42–52.

37. Basmadjian D, Dykes D, Baines A. Flow through brushborders and similar protuberant wall structures. *J Membr Biol.* 1980;56(3):183–190.

38. Nauli SM, Alenghat FJ, Luo Y, et al. Polycystins 1 and 2 mediate mechanosensation in the primary cilium of kidney cells. *Nat Genet.* 2003;33(2):129.

39. Drummond IA. Kidney development and disease in the zebrafish. *J Am Soc Nephrol.* 2005;16(2):299–304.

40. Low SH, Vasanth S, Larson CH, et al. Polycystin-1, STAT6, and P100 function in a pathway that transduces ciliary mechanosensation and is activated in polycystic kidney disease. *Dev Cell.* 2006;10(1):57–69.

41. Roth KE, Rieder CL, Bowser SS. Flexible-substratum technique for viewing cells from the side: some in vivo properties of primary (9 + 0) cilia in cultured kidney epithelia. *J Cell Sci.* 1988;89(4):457–466.

42. Schwartz EA, Leonard ML, Bizios R, Bowser SS. Analysis and modeling of the primary cilium bending response to fluid shear. *Am J Physiol Renal Physiol.* 1997;272(1):F132–F138.

43. Hildebrandt F, Zhou W. Nephronophthisis-associated ciliopathies. *J Am Soc Nephrol.* 2007;18(6):1855–1871.

44. Otto EA, Hurd TW, Airik R, et al. Candidate exome capture identifies mutation of SDCCAG8 as the cause of a retinal-renal ciliopathy. *Nat Genet.* 2010;42(10):840.

45. Shillingford JM, Piontek KB, Germino GG, Weimbs T. Rapamycin ameliorates PKD resulting from conditional inactivation of Pkd1. *J Am Soc Nephrol.* 2010;21(3):489–497.

46. Tao Y, Kim J, Schrier RW, Edelstein CL. Rapamycin markedly slows disease progression in a rat model of polycystic kidney disease. *J Am Soc Nephrol.* 2005;16(1):46–51.

47. Sherpa RT, Atkinson KF, Ferreira VP, Nauli SM. Rapamycin increases length and mechanosensory function of primary cilia in renal epithelial and vascular endothelial cells. *Int Educ Res J.* 2016;2(12):91.

48. Taylor EW, Jordan D, Coote JH. Central control of the cardiovascular and respiratory systems and their interactions in vertebrates. *Physiol Rev.* 1999;79(3):855–916.

49. Korner PI, Head GA, Badoer E, Bobik A, Angus JA. Role of brain amine transmitters and some neuromodulators in blood pressure, heart rate, and baroreflex control. *J Cardiovasc Pharmacol.* 1987;10:S26–S32.

50. Greisen G. Autoregulation of cerebral blood flow in newborn babies. *Early Hum Dev.* 2005;81(5):423–428.

51. Papaioannou TG, Stefanadis C. Vascular wall shear stress: basic principles and methods. *Hellenic J Cardiol.* 2005; 46(1):9–15.

52. Rubanyi G. Ionic mechanisms involved in the flow-and pressure-sensor function of the endothelium. *Zeitschrift fur Kardiologie.* 1991;80:91–94.

53. Ingber D. Mechanobiology and diseases of mechanotransduction. *Ann Med.* 2003;35(8):564–577.

54. Bystrevskaya VB, Lichkun VV, Antonov AS, Perov NA. An ultrastructural study of centriolar complexes in adult and embryonic human aortic endothelial cells. *Tissue Cell.* 1988;20(4):493–503.

55. Van der Heiden K, Hierck BP, Krams R. Endothelial primary cilia in areas of disturbed flow are at the base of atherosclerosis. *Atherosclerosis.* 2008;196(2): 542–550.

56. Van der Heiden K, Groenendijk BC, Hierck BP, et al. Monocilia on chicken embryonic endocardium in low shear stress areas. *Dev Dyn.* 2006;235(1):19–28.

57. Dinsmore C, Reiter JF. Endothelial primary cilia inhibit atherosclerosis. *EMBO Rep.* 2016;17(2):156–166.

58. Iomini C, Tejada K, Mo W, Vaananen H, Piperno G. Primary cilia of human endothelial cells disassemble under laminar shear stress. *J Cell Biol.* 2004;164(6):811–817.

59. AbouAlaiwi WA, Takahashi M, Mell BR, et al. Ciliary polycystin-2 is a mechanosensitive calcium channel involved in nitric oxide signaling cascades. *Circ Res.* 2009;104(7):860–869.

60. Egorova AD, Khedoe PP, Goumans MJT, et al. Lack of primary cilia primes shear-induced endothelial-to-mesenchymal transition. *Circ Res.* 2011;108(9):1093–1101.

61. Nauli SM, Mohieldin AM, Alanazi M, Nauli AM. Mechanobiology of primary cilia in the vascular and renal systems. In: *Mechanobiology in Health and Disease.* Elsevier; 2018:305–326.

62. Abdul-Majeed S, Nauli SM. Dopamine receptor type 5 in the primary cilia has dual chemo-and mechano-sensory roles. *Hypertension.* 2011;58(2):325–331.

63. Pala R, Mohieldin AM, Sherpa RT, et al. Ciliotherapy: remote control of primary cilia movement and function by magnetic nanoparticles. *ACS Nano.* 2019;13(3): 3555–3572.

64. Poole CA. Articular cartilage chondrons: form, function and failure. *J Anat.* 1997;191(Pt 1):1–13.

65. Urban JP. The chondrocyte: a cell under pressure. *Br J Rheumatol.* 1994;33(10):901–908.

66. Kim E, Guilak F, Haider MA. The dynamic mechanical environment of the chondrocyte: a biphasic finite element model of cell-matrix interactions under cyclic compressive loading. *J Biomech Eng.* 2008;130(6):061009.

67. Wislman NJ, Fletcher TF. Cilia of neonatal articular chondrocytes: incidence and morphology. *Anat Rec.* 1978; 190(4):871–889.

68. Meier-Vismara E, Walker N, Vogel A. Single cilia in the articular cartilage of the cat. *Exp Cell Biol.* 1979;47(3): 161–171.

69. Poole CA, Flint MH, Beaumont BW. Analysis of the morphology and function of primary cilia in connective tissues: a cellular cybernetic probe? *Cell Motil.* 1985; 5(3):175–193.

70. Poole CA, Zhang ZJ, Ross JM. The differential distribution of acetylated and detyrosinated alpha-tubulin in the microtubular cytoskeleton and primary cilia of hyaline cartilage chondrocytes. *J Anat.* 2001;199(Pt 4): 393–405.

71. McGlashan SR, Haycraft CJ, Jensen CG, Yoder BK, Poole CA. Articular cartilage and growth plate defects are associated with chondrocyte cytoskeletal abnormalities in Tg737orpk mice lacking the primary cilia protein polaris. *Matrix Biol.* 2007;26(4):234–246.

72. McGlashan SR, Cluett EC, Jensen CG, Poole CA. Primary cilia in osteoarthritic chondrocytes: from chondrons to clusters. *Dev Dyn.* 2008;237(8):2013–2020.

73. McGlashan SR, Knight MM, Chowdhury TT, et al. Mechanical loading modulates chondrocyte primary cilia incidence and length. *Cell Biol Int.* 2010;34(5): 441–446.

74. Wann AK, Knight MM. Primary cilia elongation in response to interleukin-1 mediates the inflammatory response. *Cell Mol Life Sci.* 2012;69(17):2967–2977.

75. Farnum CE, Wilsman NJ. Orientation of primary cilia of articular chondrocytes in three-dimensional space. *Anat Rec.* 2011;294(3):533–549.

76. Guilak F, Ratcliffe A, Mow VC. Chondrocyte deformation and local tissue strain in articular cartilage: a confocal microscopy study. *J Orthop Res.* 1995;13(3):410–421.

77. Chang CF, Ramaswamy G, Serra R. Depletion of primary cilia in articular chondrocytes results in reduced Gli3 repressor to activator ratio, increased Hedgehog signaling, and symptoms of early osteoarthritis. *Osteoarthr Cartil.* 2012;20(2):152–161.

78. Kaushik AP, Martin JA, Zhang Q, Sheffield VC, Morcuende JA. Cartilage abnormalities associated with defects of chondrocytic primary cilia in Bardet-Biedl syndrome mutant mice. *J Orthop Res.* 2009;27(8): 1093–1099.

79. Millward-Sadler SJ, Wright MO, Lee HS, Caldwell H, Nuki G, Salter DM. Altered electrophysiological responses to mechanical stimulation and abnormal signalling through alpha5beta1 integrin in chondrocytes from osteoarthritic cartilage. *Osteoarthr Cartil.* 2000;8(4): 272–278.

80. Huang J, Ballou LR, Hasty KA. Cyclic equibiaxial tensile strain induces both anabolic and catabolic responses in articular chondrocytes. *Gene.* 2007;404(1–2):101–109.

81. Doi H, Nishida K, Yorimitsu M, et al. Interleukin-4 downregulates the cyclic tensile stress-induced matrix metalloproteinases-13 and cathepsin B expression by rat normal chondrocytes. *Acta Med Okayama.* 2008; 62(2):119–126.

82. Yellowley CE, Jacobs CR, Li Z, Zhou Z, Donahue HJ. Effects of fluid flow on intracellular calcium in bovine articular chondrocytes. *Am J Physiol.* 1997;273(1 Pt 1): C30–C36.

83. Degala S, Williams R, Zipfel W, Bonassar LJ. Calcium signaling in response to fluid flow by chondrocytes in 3D alginate culture. *J Orthop Res.* 2012;30(5):793−799.

84. Pingguan-Murphy B, El-Azzeh M, Bader DL, Knight MM. Cyclic compression of chondrocytes modulates a purinergic calcium signalling pathway in a strain rate- and frequency-dependent manner. *J Cell Physiol.* 2006; 209(2):389−397.

85. Tanaka N, Ohno S, Honda K, et al. Cyclic mechanical strain regulates the PTHrP expression in cultured chondrocytes via activation of the Ca2+ channel. *J Dent Res.* 2005;84(1):64−68.

86. Browning JA, Saunders K, Urban JPG, Wilkins RJ. The influence and interactions of hydrostatic and osmotic pressures on the intracellular milieu of chondrocytes. *Biorheology.* 2004;41(3−4):299−308.

87. Mizuno S. A novel method for assessing effects of hydrostatic fluid pressure on intracellular calcium: a study with bovine articular chondrocytes. *Am J Physiol Cell Physiol.* 2005;288(2):C329−C337.

88. Wann AK, Zuo N, Haycraft CJ, et al. Primary cilia mediate mechanotransduction through control of ATP-induced Ca2+ signaling in compressed chondrocytes. *FASEB J.* 2012;26(4):1663−1671.

89. Rais Y, Reich A, Simsa-Maziel S, et al. The growth plate's response to load is partially mediated by mechanosensing via the chondrocytic primary cilium. *Cell Mol Life Sci.* 2015;72(3):597−615.

90. Bonewald LF. The amazing osteocyte. *J Bone Miner Res.* 2011;26(2):229−238.

91. Schaffler MB, Cheung WY, Majeska R, Kennedy O. Osteocytes: master orchestrators of bone. *Calcif Tissue Int.* 2014;94(1):5−24.

92. Udagawa N, Takahashi N, Akatsu T, et al. Origin of osteoclasts: mature monocytes and macrophages are capable of differentiating into osteoclasts under a suitable microenvironment prepared by bone marrow-derived stromal cells. *Proc Natl Acad Sci U S A.* 1990;87(18):7260−7264.

93. Djavan B, Partin AW, Hoey MF, Roehrborn CG, Dixon CM, Marberger M. Transurethral radiofrequency therapy for benign prostatic hyperplasia using a novel saline-liquid conductor: the virtual electrode. *Urology.* 2000;55(1):13−16.

94. Tonna EA, Lampen NM. Electron microscopy of aging skeletal cells. I. Centrioles and solitary cilia. *J Gerontol.* 1972;27(3):316−324.

95. Uzbekov RE, Maurel DB, Aveline PC, Pallu S, Benhamou CL, Rochefort GY. Centrosome fine ultrastructure of the osteocyte mechanosensitive primary cilium. *Microsc Microanal.* 2012;18(6):1430−1441.

96. Coughlin TR, Voisin M, Schaffler MB, Niebur GL, McNamara LM. Primary cilia exist in a small fraction of cells in trabecular bone and marrow. *Calcif Tissue Int.* 2015;96(1):65−72.

97. Corrigan MA, Johnson GP, Stavenschi E, Riffault M, Labour MN, Hoey DA. TRPV4-mediates oscillatory fluid shear mechanotransduction in mesenchymal stem cells in part via the primary cilium. *Sci Rep.* 2018;8(1):3824.

98. Johnson GP, Stavenschi E, Eichholz KF, Corrigan MA, Fair S, Hoey DA. Mesenchymal stem cell mechanotransduction is cAMP dependent and regulated by adenylyl cyclase 6 and the primary cilium. *J Cell Sci.* 2018; 131(21):jcs222737.

99. Hoey DA, Tormey S, Ramcharan S, O'Brien FJ, Jacobs CR. Primary cilia-mediated mechanotransduction in human mesenchymal stem cells. *Stem Cells.* 2012;30(11): 2561−2570.

100. Tummala P, Arnsdorf EJ, Jacobs CR. The role of primary cilia in mesenchymal stem cell differentiation: a pivotal switch in guiding lineage commitment. *Cell Mol Bioeng.* 2010;3(3):207−212.

101. Brown JA, Santra T, Owens P, Morrison AM, Barry F. Primary cilium-associated genes mediate bone marrow stromal cell response to hypoxia. *Stem Cell Res.* 2014;13(2): 284−299.

102. Yuan X, Cao J, He X, et al. Ciliary IFT80 balances canonical versus non-canonical hedgehog signalling for osteoblast differentiation. *Nat Commun.* 2016;7:11024.

103. Oliazadeh N, Gorman KF, Eveleigh R, Bourque G, Moreau A. Identification of elongated primary cilia with impaired mechanotransduction in idiopathic scoliosis patients. *Sci Rep.* 2017;7:44260.

104. Teves ME, Sundaresan G, Cohen DJ, et al. Spag17 deficiency results in skeletal malformations and bone abnormalities. *PLoS One.* 2015;10(5):e0125936.

105. Malone AM, Anderson CT, Tummala P, et al. Primary cilia mediate mechanosensing in bone cells by a calcium-independent mechanism. *Proc Natl Acad Sci U S A.* 2007;104(33):13325−13330.

106. Moore ER, Ryu HS, Zhu YX, Jacobs CR. Adenylyl cyclases and TRPV4 mediate Ca(2+)/cAMP dynamics to enhance fluid flow-induced osteogenesis in osteocytes. *J Mol Biochem.* 2018;7:48−59.

107. Spasic M, Jacobs CR. Lengthening primary cilia enhances cellular mechanosensitivity. *Eur Cell Mater.* 2017;33: 158−168.

108. Kurata K, Uemura T, Nemoto A, et al. Mechanical strain effect on bone-resorbing activity and messenger RNA expressions of marker enzymes in isolated osteoclast culture. *J Bone Miner Res.* 2001;16(4):722−730.

109. Finetti F, Paccani SR, Rosenbaum J, Baldari CT. Intraflagellar transport: a new player at the immune synapse. *Trends Immunol.* 2011;32(4):139−145.

110. Singh M, Chaudhry P, Merchant AA. Primary cilia are present on human blood and bone marrow cells and mediate Hedgehog signaling. *Exp Hematol.* 2016; 44(12):1181−1187e2.

111. Chen JC, Hoey DA, Chua M, Bellon R, Jacobs CR. Mechanical signals promote osteogenic fate through a primary cilia-mediated mechanism. *FASEB J.* 2016;30(4): 1504−1511.

112. Temiyasathit S, Tang WJ, Leucht P, et al. Mechanosensing by the primary cilium: deletion of Kif3A reduces bone formation due to loading. *PLoS One.* 2012;7(3):e33368.

113. Corbit KC, Shyer AE, Dowdle WE, Gaulden J, Singla V, Reiter JF. Kif3a constrains beta-catenin-dependent Wnt

signalling through dual ciliary and non-ciliary mechanisms. *Nat Cell Biol.* 2008;10(1):70—76.

114. Haycraft CJ, Zhang Q, Song B, et al. Intraflagellar transport is essential for endochondral bone formation. *Development.* 2007;134(2):307—316.

115. Moore ER, Zhu YX, Ryu HS, Jacobs CR. Periosteal progenitors contribute to load-induced bone formation in adult mice and require primary cilia to sense mechanical stimulation. *Stem Cell Res Ther.* 2018;9(1):190.

116. Kwon RY, Temiyasathit S, Tummala P, Quah CC, Jacobs CR. Primary cilium-dependent mechanosensing is mediated by adenylyl cyclase 6 and cyclic AMP in bone cells. *FASEB J.* 2010;24(8):2859—2868.

117. Lee KL, Hoey DA, Spasic M, Tang T, Hammond HK, Jacobs CR. Adenylyl cyclase 6 mediates loading-induced bone adaptation in vivo. *FASEB J.* 2014;28(3):1157—1165.

118. Xiao ZS, Dallas M, Qiu N, et al. Conditional deletion of Pkd1 in osteocytes disrupts skeletal mechanosensing in mice. *FASEB J.* 2011;25(7):2418—2432.

119. Delaine-Smith RM, Sittichokechaiwut A, Reilly GC. Primary cilia respond to fluid shear stress and mediate flow-induced calcium deposition in osteoblasts. *FASEB J.* 2014;28(1):430—439.

120. Xie YF, Shi WG, Zhou J, et al. Pulsed electromagnetic fields stimulate osteogenic differentiation and maturation of osteoblasts by upregulating the expression of BMPRII localized at the base of primary cilium. *Bone.* 2016;93:22—32.

121. Yan JL, Zhou J, Ma HP, et al. Pulsed electromagnetic fields promote osteoblast mineralization and maturation needing the existence of primary cilia. *Mol Cell Endocrinol.* 2015;404:132—140.

122. Corrigan MA, Ferradaes TM, Riffault M, Hoey DA. Ciliotherapy treatments to enhance biochemically- and biophysically-induced mesenchymal stem cell osteogenesis: a comparison study. *Cell Mol Bioeng.* 2019;12(1):53—67.

123. Shi WG, Xie Y, He J, Wang J. Microgravity induces inhibition of osteoblastic differentiation and mineralization through abrogating primary cilia. *Sci Rep.* 2017;7(1):1866.

124. Besschetnova TY, Kolpakova-Hart E, Guan Y, Zhou J, Olsen BR, Shah JV. Identification of signaling pathways regulating primary cilium length and flow-mediated adaptation. *Curr Biol.* 2010;20(2):182—187.

125. Resnick A, Hopfer U. Force-response considerations in ciliary mechanosensation. *Biophys J.* 2007;93(4):1380—1390.

126. Espinha LC, Hoey DA, Fernandes PR, Rodrigues HC, Jacobs CR. Oscillatory fluid flow influences primary cilia and microtubule mechanics. *Cytoskeleton (Hoboken).* 2014;71(7):435—445.

127. Jeon OH, Yoo YM, Kim KH, Jacobs CR, Kim CH. Primary cilia-mediated osteogenic response to fluid flow occurs via increases in focal adhesion and Akt signaling pathway in MC3T3-E1 osteoblastic cells. *Cell Mol Bioeng.* 2011;4(3):379—388.

128. Hoey DA, Kelly DJ, Jacobs CR. A role for the primary cilium in paracrine signaling between mechanically stimulated osteocytes and mesenchymal stem cells. *Biochem Biophys Res Commun.* 2011;412(1):182—187.

129. Delling M, Indzhykulian AA, Liu X, et al. Primary cilia are not calcium-responsive mechanosensors. *Nature.* 2016;531(7596):656—660.

130. Lee KL, Guevarra MD, Nguyen AM, Chua MC, Wang Y, Jacobs CR. The primary cilium functions as a mechanical and calcium signaling nexus. *Cilia.* 2015;4:7.

131. Hwang SH, White KA, Somatilaka BN, Shelton JM, Richardson JA, Mukhopadhyay S. The G protein-coupled receptor Gpr161 regulates forelimb formation, limb patterning and skeletal morphogenesis in a primary cilium-dependent manner. *Development.* 2018;145(1).dev154054.

132. Mukhopadhyay S, Wen X, Ratti N, et al. The ciliary G-protein-coupled receptor Gpr161 negatively regulates the sonic hedgehog pathway via cAMP signaling. *Cell.* 2013;152(1—2):210—223.

133. Serra R. Role of intraflagellar transport and primary cilia in skeletal development. *Anat Rec.* 2008;291(9):1049—1061.

134. Nguyen AM, Jacobs CR. Emerging role of primary cilia as mechanosensors in osteocytes. *Bone.* 2013;54(2):196—204.

135. Halbritter J, Bizet AA, Schmidts M, et al. Defects in the IFT-B component IFT172 cause Jeune and Mainzer-Saldino syndromes in humans. *Am J Hum Genet.* 2013;93(5):915—925.

136. Tian H, Feng J, Li J, et al. Intraflagellar transport 88 (IFT88) is crucial for craniofacial development in mice and is a candidate gene for human cleft lip and palate. *Hum Mol Genet.* 2017;26(5):860—872.

137. Girisha KM, Shukla A, Trujillano D, et al. A homozygous nonsense variant in IFT52 is associated with a human skeletal ciliopathy. *Clin Genet.* 2016;90(6):536—539.

138. Mukhopadhyay S. TCTEX1D2, a potential link to skeletal ciliopathies. *Cell Cycle.* 2015;14(3):293—294.

139. Schmidts M, Frank V, Eisenberger T, et al. Combined NGS approaches identify mutations in the intraflagellar transport gene IFT140 in skeletal ciliopathies with early progressive kidney disease. *Hum Mutat.* 2013;34(5):714—724.

140. Nguyen AM, Young YN, Jacobs CR. The primary cilium is a self-adaptable, integrating nexus for mechanical stimuli and cellular signaling. *Biol Open.* 2015;4(12):1733—1738.

141. Thompson CL, Chapple JP, Knight MM. Primary cilia disassembly down regulates mechanosensitive hedgehog signalling: a feedback mechanism controlling ADAMTS-5 expression in chondrocytes. *Int J Exp Pathol.* 2014;95(3).A32—A32.

142. Rowson DT, Shelton JC, Screen HR, Knight MM. Mechanical loading induces primary cilia disassembly in tendon cells via TGF beta and HDAC6. *Sci Rep.* 2018;8(1):11107.

143. Luo N, Conwell MD, Chen X, et al. Primary cilia signaling mediates intraocular pressure sensation. *Proc Natl Acad Sci U S A.* 2014;111(35):12871—12876.

144. Thompson CL, Chapple JP, Knight MM. Primary cilia disassembly down-regulates mechanosensitive hedgehog signalling: a feedback mechanism controlling ADAMTS-5 expression in chondrocytes. *Osteoarthr Cartil.* 2014;22(3): 490–498.

145. Iomini C, Tejada K, Mo W, Vaananen H, Piperno G. Primary cilia of human endothelial cells disassemble under laminar shear stress. *JCB (J Cell Biol).* 2004;164(6):811–817.

146. Bershteyn M, Atwood SX, Woo WM, Li M, Oro AE. MIM and cortactin antagonism regulates ciliogenesis and hedgehog signaling. *Dev Cell.* 2010;19(2):270–283.

147. Sharma N, Kosan ZA, Stallworth JE, Berbari NF, Yoder BK. Soluble levels of cytosolic tubulin regulate ciliary length control. *Mol Biol Cell.* 2011;22(6):806–816.

148. Hernandez-Hernandez V, Pravincumar P, Diaz-Font A, et al. Bardet-Biedl syndrome proteins control the cilia length through regulation of actin polymerization. *Hum Mol Genet.* 2013;22(19):3858–3868.

149. Parkkinen JJ, Lammi MJ, Inkinen R, et al. Influence of short-term hydrostatic-pressure on organization of stress fibers in cultured chondrocytes. *J Orthop Res.* 1995;13(4): 495–502.

150. Knight MM, Toyoda T, Lee DA, Bader DL. Mechanical compression and hydrostatic pressure induce reversible changes in actin cytoskeletal organisation in chondrocytes in agarose. *J Biomech.* 2006;39(8):1547–1551.

151. Campbell JJ, Blain EJ, Chowdhury TT, Knight MM. Loading alters actin dynamics and up-regulates cofilin gene expression in chondrocytes. *Biochem Biophys Res Commun.* 2007;361(2):329–334.

152. Kim J, Jo H, Hong H. Actin remodelling factors control ciliogenesis by regulating YAP/TAZ activity and vesicle trafficking. *Nat Commun.* 2015;6, 6781.

153. Rich DR, Clark AL. Chondrocyte primary cilia shorten in response to osmotic challenge and are sites for endocytosis. *Osteoarthr Cartil.* 2012;20(8):923–930.

154. Zhang JW, Dalbay MT, Luo X, et al. Topography of calcium phosphate ceramics regulates primary cilia length and TGF receptor recruitment associated with osteogenesis. *Acta Biomater.* 2017;57:487–497.

155. Badano JL, Mitsuma N, Beales PL, Katsanis N. The ciliopathies: an emerging class of human genetic disorders. *Annu Rev Genom Hum Genet.* 2006;7:125–148.

156. Gambassi S, Geminiani M, Thorpe SD. Smoothened-antagonists reverse homogentisic acid-induced alterations of Hedgehog signaling and primary cilium length in alkaptonuria. *J Cell Physiol.* 2017;232(11):3103–3111.

157. Rydholm S, Zwartz G, Kowalewski JM, Kamali-Zare P, Frisk T, Brismar H. Mechanical properties of primary cilia regulate the response to fluid flow. *Am J Physiol Renal Physiol.* 2010;298(5):F1096–F1102.

158. Engel BD, Ludington WB, Marshall WF. Intraflagellar transport particle size scales inversely with flagellar length: revisiting the balance-point length control model. *JCB (J Cell Biol).* 2009;187(1):81–89.

159. Marshall WF, Rosenbaum JL. Intraflagellar transport balances continuous turnover of outer doublet microtubules: implications for flagellar length control. *JCB (J Cell Biol).* 2001;155(3):405–414.

160. Kukic I, Rivera-Molina F, Toomre D. The IN/OUT assay: a new tool to study ciliogenesis. *Cilia.* 2016;5:23.

161. Benmerah A. The ciliary pocket. *Curr Opin Cell Biol.* 2013; 25(1):78–84.

162. Donnelly E, Ascenzi MG, Farnum C. Primary cilia are highly oriented with respect to collagen direction and long axis of extensor tendon. *J Orthop Res.* 2010;28(1): 77–82.

163. Rowson DT, Shelton JC, Screen HR, Knight MM. Mechanical loading induces primary cilia disassembly in tendon cells via TGFbeta and HDAC6. *Sci Rep.* 2018;8(1):11107.

164. Corrigan MA, Johnson GP, Stavenschi E, Riffault M, Labour MN, Hoey DA. TRPV4-mediates oscillatory fluid shear mechanotransduction in mesenchymal stem cells in part via the primary cilium. *Sci Rep.* 2018;8(1):3824.

165. Masyuk AI, Masyuk TV, Splinter PL, Huang BQ, Stroope AJ, LaRusso NF. Cholangiocyte cilia detect changes in luminal fluid flow and transmit them into intracellular Ca2+ and cAMP signaling. *Gastroenterology.* 2006;131(3):911–920.

166. Mick DU, Rodrigues RB, Leib RD, et al. Proteomics of primary cilia by proximity labeling. *Dev Cell.* 2015;35(4): 497–512.

167. Kathem SH, Mohieldin AM, Abdul-Majeed S, et al. Ciliotherapy: a novel intervention in polycystic kidney disease. *J Geriatr Cardiol.* 2014;11(1):63–73.

# In Vivo Models of Muscle Stimulation and Mechanical Loading in Bone Mechanobiology

YI-XIAN QIN, PHD • MINYI HU, PHD

## 1. BACKGROUND

Bone fracture commonly occurs across the entire spectrum of the human population, including male, female, young and/or old populations, and all ethnic backgrounds. While sports- and trauma-related fractures occur among young adults, aging and a sedentary lifestyle can conspire to reduce bone quantity and quality, decrease muscle mass and strength, and undermine postural stability, culminating in an elevated risk of skeletal fracture. Concurrently, a marked reduction in the available bone-marrow-derived population of mesenchymal stem cells (MSCs) jeopardizes the regenerative potential that is critical to recovery from musculoskeletal injury and diseases. A potential way to combat the deterioration involves harnessing the sensitivity of bone to mechanical signals, which is crucial in defining, maintaining, and recovering bone mass. Bone mineral density (BMD) and muscle strength are highly biomechanically related to each other.[1,2] High physical activity level has been associated with high bone mass and low fracture risk and is therefore recommended to reduce fractures in old age.[3-7] As a direct consequence of exposure to microgravity, astronauts experience several physiologic changes, which can have serious medical complications. Most immediate and significant are the musculoskeletal implications in bone and muscles.[8-11] Results of the joint Russian/US studies of the effect of microgravity on bone tissue from 4.5- to 14.5-month-long missions have demonstrated that BMD (g/cm$^2$) and bone mineral content (g) have decayed in the whole body of the astronauts.[12] The greatest BMD losses have been observed in the skeleton of the lower body, i.e., in pelvic bones ($-11.99\% \pm 1.22\%$) and in the femoral neck ($-8.17\% \pm 1.24\%$), while there was no evitable decay found in the skull region. Overall changes in bone mass of the whole skeleton of male cosmonauts during the period of about 6 months on a mission made up $-1.41\% \pm 0.41\%$ and suggest the mean balance of calcium over flight equal to $-227 \pm 62.8$ mg/day. On average, the magnitude and rate of the loss are staggering; astronauts lose bone mineral in the lower appendicular skeleton at a rate approaching 2% per month with muscle atrophy.[5,6,13-15] In simulated or actual microgravity, human postural muscles undergo substantial atrophy: after about 270 days the muscle mass asymptotically approaches a constant value of about 70% of the initial one. Most animal studies reported preferential atrophy of slow-twitch fibers whose mechanical properties change toward the fast type. Following exposure to microgravity, the maximal force of several muscle groups showed a substantial decrease (6%–25% of preflight values).[8,10,16-19] The mechanism that explains both muscle and bone decays in the functional disuse environment is still unclear. In the recent years, considerable attention has focused on identifying particular parameters and exercise paradigms to ameliorate the deficits of muscle atrophy and bone density. Perhaps, microcirculation and interstitial fluid flow that are linked with exercise and muscle contraction can identify the interrelationship between muscle and bone fluid flow in response to loading and disuse environment. The headward shift of body fluids and the removal of gravitational loading from bone and muscles lead to progressive changes in the musculoskeletal systems. The underlying factors producing these changes may primarily be the fluid flow and circulations in both muscular and bone tissues.

The ability of musculoskeletal tissues to respond to changes in their functional milieu is one of the most intriguing aspects of such living tissues, and it certainly contributes to their success as a structure. The ability of

Mechanobiology. https://doi.org/10.1016/B978-0-12-817931-4.00007-8

bone and muscle to rapidly accommodate changes in their functional environments ensures that sufficient skeletal mass is appropriately placed to withstand the rigors of functional activity, an attribute described as Wolff's law.[20,21] This adaptive capability of musculoskeletal tissues suggests that biophysical stimuli may be able to provide a site-specific, exogenous treatment for controlling both bone mass and morphology. The premise of a mechanical influence on bone morphology has become a basic tenet of bone physiology.[22–26] Absence of functional loading results in the loss of bone mass,[22,27–29,127,144] whereas exercise or increased activity results in increased bone mass.[30–34] Similarly, the increasing exercise of muscle tissues can significantly increase blood flow, oxygen, and exchange fluid in muscles. During muscle contraction, several mechanisms regulate blood flow to ensure a close coupling between muscle oxygen delivery and metabolic demand.[8,10,16–19,35–42] Defining the formal relationship between the mechanical milieu and the adaptive response and the relationship between muscle pump and interstitial fluid flow will prove instrumental in devising a mechanical intervention for musculoskeletal disorders such as osteoporosis, muscle fatigue, and atrophy, designing biomechanical means to accelerate fracture healing, and promoting bony ingrowth.

## 2. BONE HOMEOSTASIS, STRUCTURE, PHYSIOLOGY, AND BASIC BIOMECHANICS

The skeleton carries out a diverse range of developmental and metabolic functions: it serves as a protective cage for internal organs and a safe niche for marrow, facilitates locomotion, and is a principal reservoir of minerals.[43,44] The constant remodeling cycle of formation and resorption facilitates both the rapid repair of bone microdamage, the replacement of dead osteocytes and orchestrates changes in mass and morphology to meet any changing demands of mechanical loads or metabolic need. Like most biological tissues, bone has its unique structure and serves as a functional unit[43,45–47] (Fig. 2.3.1). In general, the basic structure of a mature long bone includes both cortical (or compact) bone and trabecular (or cancellous) bone. Within the mature cortical bone, the skeleton exhibits a lamellar structure. Such a plate structure is centered with the haversian canal mostly aligned longitudinally and connected with the Volkmann canals running horizontally in the cortex. These two conduits are rich in vascular tissue; two-thirds of the vascular supply is provided through the medullary cavity and the other

one-third is provided through the periosteum. Bone cells present in the bone lacunae are interconnected by the canaliculi, tubular structures, and communicate with each other through gap junctions. Each concentric structure consists of a capillary tube, a concentric plate structure, lacuna, and canaliculi. Such a unique bone unit is called an osteon.

The haversian canals, canaliculi, and lacunae occupy 13.3% of the volume of the cortical bone.[48,49] The remaining solid portion is the matrix, which is occupied by mineralized material containing hydroxyapatite crystals and collagen fibers. The solid bone matrix of the bone contains pores on the order of 0.01–0.1 μm in diameter.[48–52] The primary constituents of an osteon are collagen (organic), hydroxyapatite (inorganic), and fluid. Fluid flows through the various microstructural spaces to transfer metabolites to the osteocytes, ensuring that bone tissue remains viable. This continuous perfusion allows the remodeling processes to be perpetual. The requisite nutrition of the cells and essential disposal of their waste products are carried by this dynamic fluid stream. Because of the intricate microstructure of the cortical bone, transportation between cells would be very poor if it relied solely on fluid diffusion. Indeed, a complex network of very thin (∼1 μm) and long (up to 100 μm) canaliculi must employ a more active policy of nutrient allocation and signal dispersion.[48–52]

Regarding the potential mechanotransduction in bone induced by fluid flow, Cowin[50,51] summarized three primary levels of bone porosities within cortical microstructure, which might mediate fluid flow. First, an osteon, the basic structure of bone with dominant cylindric structures of 100–150 μm radii running primarily along the long axes of bone, contains at its center an *osteonal canal* (OC), including blood vessels, a nerve, and surrounding fluid. There are cells attached on the walls of the OC. At the level of the osteon, fluid transduction from the haversian canal to the cement line or its surrounding micropores, via either perfusion or convection, appears to be significantly influenced by the pathway of the fluid flow. Second, an osteon contains lacunae structure of 3–10 μm in radius surrounding the OC, which connects to the OC via canaliculi. Osteocytes (∼2 μm radii) are contained within the lacunae. Third, canaliculi, capillaries approximately 0.1–0.5 μm in radius that run radially and surround and connect the lacunae, OC, and cement lines together. In addition, the microstructures of collagen-apatite porosity (100–300 Å) contain fluid in the collagen matrix. The cortical vascular supply begins primarily at the marrow cavity and passes through the

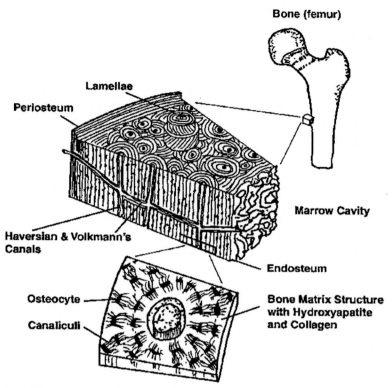

FIG. 2.3.1 Cortical bone and osteon structure. The general structure of cortical bone includes marrow cavity, periosteum, and endosteum. The microstructure includes haversian canals, Volkmann canal, osteocyte, and lacuna and canaliculi system.

endosteum, accommodating approximately two-thirds of the vascular supply in the cortex. These levels of porosities participate and interact with each other in fluid diffusion/perfusion, and fluid pathways may include vascular canals, lacunocanalicular spaces, and collagen-apatite spaces.

## 3. MECHANICAL PROPERTIES AND CHARACTERIZATION OF BIOLOGICAL TISSUES

The mechanical strength of biological tissue or a sample can be defined and quantified by its mechanical parameters, such as stress, strain, and modulus. These mechanical parameters are defined as follows.

*Stress*: In uniaxial loading, the stress is defined as the axial force applied normalized by the perpendicular cross-sectional area (CSA). The load and force per unit area is the response to externally applied loads on a structural section. In general, a unit point can experience three-dimensional loads. External forces and

moments can be applied to the structure in different directions, resulting in tensile, compressive, bending, torsional, shear, and composite stresses. Fractures often occur clinically at the location where the principal stress exceeds the mechanical strength of the tissue. The unit of stress in the SI system is defined as newtons per square meter ($N/m^2$) or pascals (Pa), megapascals (MPa) ($10^6$Pa), and gigapascals (GPa) ($10^9$Pa).

*Strain*: External loading on a structure induces a change in dimension of the structure, which is called deformation. The strain is defined as the structural deformation normalized by the original dimensions of the structure caused by an applied load, or more technically the spatial gradient of the deformation. There are two types of strain: one is the normal strain, which causes a change in the length of the specimen, and the other is the shear strain, which causes a change in the angular relationship within the structure. Strain has no unit quantity but is usually expressed in percentage, $\varepsilon$, or $\mu\varepsilon$.

*Elastic Modulus*: The modulus is the relation between strain and stress, such as the ratio of stress to strain, and is the basic material constant that reflects the mechanical properties of a material. The relationship between stress and strain can be defined as stress/strain = elastic modulus. The unit of elastic modulus is also pascals (Pa), megapascals (MPa), or gigapascals (GPa). Typically, but not always, a larger value of elastic modulus is also associated with higher strength of the material (Fig. 2.3.2).

Bone intensity is referred to as the ability of bone tissue to resist structural damage after being loaded. It is a comprehensive index that integrates bone structure, bone mass, and material properties.

The mechanical properties of the tissue can be quantified by various mechanical loading and evaluations. Mechanical strength testing instruments can be used to obtain measurements of the tissue strength, stiffness, hardness, and toughness. The mechanical properties of bone are usually assessed by mechanical testing of the bone using a 3- or 4-point bending device (Fig. 2.3.3). The shear mechanical test of bone can be achieved by torsional testing (Fig. 2.3.3). The choice of the test type is determined by various technical and physiological factors. For example, in the study of healing of long bone fractures, bending and torsion tests are a logical choice because they can test the bending and torsional strength of the tissue. The torsional test subjects each cross section of the bone to the same torque, whereas the 4-point bending test creates a uniform bending moment[53] for the entire epiphysis.

For the torsional test, an additional parameter, twisting to failure (fracture), can be used as a measure of the callus ductility. Although it can only be measured once for a given callus, it is possible to obtain multiple measurements on hardness and stiffness before reaching failure. Multilevel testing methods have been reported to test the loading model by performing a

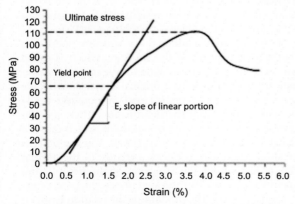

FIG. 2.3.2  Strain-stress relation under mechanical loading.

noninvasive load on the bone. Within these methods, bending strength or torsional and compressive stiffness can be quantified in multiple planes.

## 4. MUSCULOSKELETAL TISSUE RESPONSE TO DYNAMIC MECHANICAL SIGNALS

Elevated levels of physical activity have been associated with high bone mass and low fracture risk and is therefore recommended to reduce fractures.[3,4,6] The ability of musculoskeletal tissues to respond to changes in the functional loading milieu is one of their most intriguing aspects, and it certainly contributes to their longevity as a load-bearing structure. Bone and muscle rapidly accommodate changes in their functional environment to ensure that sufficient skeletal mass is appropriately placed in appropriate regions to withstand functional activity.[20,21] This adaptive capability of musculoskeletal tissues suggests that biophysical stimuli may be able to provide a site-specific, exogenous treatment to control both bone mass and morphology. The premise of mechanical influence on bone morphology has become a basic tenet of bone physiology.[24–26] Absence of functional loading results in loss of bone mass,[28,29,54,55] whereas exercise or increased activity results in increased bone mass.[30,32,33] Similarly, the increasing exercise of musculoskeletal tissues can significantly increase blood flow, oxygen, and the exchange of fluid in muscle. Several mechanisms of muscle contraction regulate blood flow, which couple oxygen delivery and metabolic demand in the muscles.[16,18,19,35,37,41] Based on the muscle pump theory, vascular arteries and veins within skeletal muscles are compressed upon muscle contraction, thereby increasing the arteriovenous pressure gradient and promoting capillary infiltration.[56–58] Mapping the pathways between mechanical loading and the adaptive responses of muscle pump and interstitial fluid may lead to great insights in developing mechanical intervention for musculoskeletal disorders.

Modeling (formation) and remodeling (coupling of resorption and formation) responses in bone are sensitive only to dynamic (time-varying) strains; static strains are ignored as a source of osteogenic stimuli.[59] With that said, it is also clear the osteogenic potential of mechanical signals is defined by a strong interdependence among cycle number, strain magnitude, and frequency. In cortical bone, 2000 με induced at 0.5 Hz (one cycle every 2 seconds) maintains bone mass and achieves this with just four cycles of loading encompassing 8 seconds per day.[54] Reducing this strain to 1000 με at 1 Hz requires 100 cycles, and 100 seconds, to maintain bone mass.[60] Raising the loading frequency to 3 Hz, bone

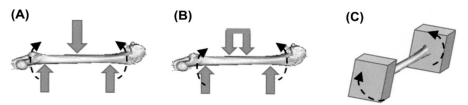

FIG. 2.3.3 Long bone under loading of **(A)** 3-point bending, **(B)** 4-point bending, and **(C)** torsion tests.

mass can be retained with 1800 cycles of only 800 $\mu\varepsilon$,[53] whereas at 30 Hz, the same 600 seconds requires only 200 $\mu\varepsilon$ to maintain cortical bone, a protocol employing 18,000 cycles of loading. Increasing the 30 Hz signal to 1h per day (108,000 cycles), only 70 $\mu\varepsilon$ is necessary to inhibit bone loss. These data demonstrate that the sensitivity of bone to mechanical loading goes up quickly with the total number of loading cycles, which is closely associated with the loading frequency, i.e., extremely low levels of strain will maintain bone mass if sufficient displacement cycles at a relatively higher frequency are applied. This relation between the loading cycle and resultant bone strain can be plotted in a nonlinear response curve[53] (Fig. 2.3.4). It is suggested that a nonlinear governing relationship between the daily loading cycles and the peak bone strain magnitude generated in the body can be generated using curve fitting analysis.[53]

$$S = 10^{2.28}(5.6 - \log_{10} C)^{1.5} \qquad (2.3.1)$$

where S represents the peak bone strain and C represents the total daily loading cycles. Such a high level of loading cycles can be achieved using higher frequency daily loading regimen.

It is well accepted that overloading will damage the bone, leading to failure—just as too much light, noise, and pressure will overwhelm sight, hearing, and touch. Although the skeleton's primary responsibility is structural in nature, its overall responsibilities are broader than first presumed and include, even, a critical role of the acoustic sensory organ in elephants.[61] Emphasizing this point, bone's adaptation to mechanical signals is nonlinear, such that it can be influenced by a very few high-magnitude strain events or by many thousands of low-magnitude strain events.

To adapt to the changing demands of mechanics, bone mass and bone morphology can be regulated via bone remodeling at specific sites. This crucial process of structural remodeling of the bone involves bone resorption and the subsequent bone formation. However, difficulties in determining specific mechanical components will hamper our understanding of bone-remodeling-related diseases, as well as limit our judgments on bone fractures and healing capacity. Therefore continuous studies of the bone remodeling process, for example, to determine the mechanical model of this remodeling process, should ultimately lead to interventions for prevention and treatment of musculoskeletal disorders.

## 5. BONE TISSUE ADAPTATION RESPONSE TO HIGH-RATE, BUT LOW-INTENSITY, MECHANICAL STIMULATION

Static loading has a lesser effect on bone healing and remodeling,[62–64] whereas dynamic stimulation has been shown effective for bone formation through in vivo animal models and preliminary clinical studies.[65,66] The potential of using high-frequency, low-magnitude mechanical stimulation to improve the quantity and quality of skeletal tissue in animal and human studies has been demonstrated in the literature. In a "proof-of-principle" assessment of dynamic loading, mature female sheep that were subjected to a 1-year treatment of brief (20 minutes per day), low-magnitude (0.3 g), high-frequency (30 Hz) mechanical signals attained 30% increases in the trabecular density and volume of the femur compared with controls, paralleled by an increase in bone stiffness and strength (Fig. 2.3.5).[66,67] These studies provided evidence that extremely small strains ($<10$ $\mu\varepsilon$), far below those generated during strenuous activity, could readily serve as anabolic agents to bone.

## 6. FUNCTIONAL DISUSE-INDUCED BONE LOSS AND MUSCLE ATROPHY

Disuse osteoporosis is a common skeletal disorder in patients subjected to prolonged immobility or bed rest, e.g., fracture and spinal injury. In addition to bone loss, functional disuse and microgravity can cause muscle atrophy. These physiologic changes generate additional health complications, including increased risk of falls and fracture and poor long-term recovery. Analyses of patients with spinal cord injury (SCI) showed a reduction in BMD in disused limbs[68–70]

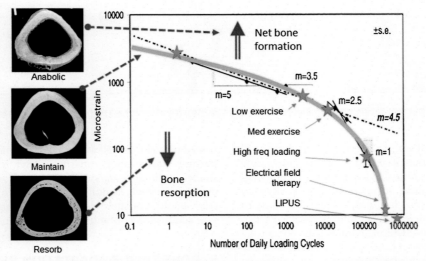

FIG. 2.3.4 Nonlinear interrelationship of cycle number and daily strain magnitude following a governing equation (Eq. 2.3.1). It appears that bone mass can be retained through a number of distinct strategies: (1) bone is preserved with extremely low daily loading cycles and extremely high frequency, i.e., four cycles per day of 2000 µε; (2) bone mass is maintained with 100 cycles per day of 1000 µε, or 10,000s of cycles of signals well below 100 µε (each represented as a star); and (3) bone is responded to extremely high frequency and extremely low strain magnitude, such as electric field-induced muscle contraction and low-intensity pulsed ultrasound (LIPUS). It is proposed that falling below this "daily strain energy" would stimulate bone loss or any combination exceeding this relationship would stimulate net bone gain.[53]

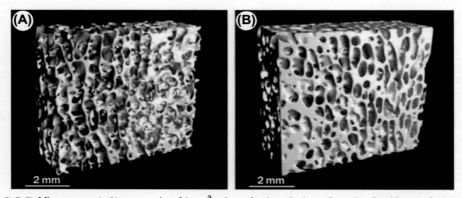

FIG. 2.3.5 Micro computed tomography of 1-cm³ cubes of trabecular bone from the distal femur of adult sheep **(A)** under aging control (8 year old) and **(B)** subjected to 20 min/day of 30 Hz, 0.3 g mechanical stimulation for 1 year. These images are supportive of the changes in both the density and architecture of bone.[67]

and a higher incidence of fracture.[71–74] More than 1 year after SCI, as much as 30%–40% demineralization was observed in the femoral neck, distal femur, and proximal tibia.[69] It has been reported that SCI-induced osteopenia/osteoporosis reaches the fracture threshold (BMD of 1 g/cm²) 1–5 years after the injury, with a fracture frequency of 5%–34%.[71,75–77] These patients were also observed to have significant reductions in muscle mass.[78] Analyses from space missions of 4–12 months duration have shown that weightlessness can induce 1%–2% BMD loss per month in the spine, hip, and lower extremities.[5,79] The reduction in trabecular BMD in both hip and femur regions was greater than 2% per month, whereas there was only a minimal decrease in cortical bone.[80] Similar results were observed in animal studies. Burr et al.[81]

showed an increase in the bone turnover rate in cast-immobilized animals that received muscle stimulation (MS) for 17 days.

Lower extremity muscle volume was similarly altered by disuse. Exposure to a 6-month space mission resulted in a decrease in muscle volume of 10% in the quadriceps and 19% in the gastrocnemius and soleus.[6,82] Computed tomographic measurements of the muscle CSA indicated a decrease of 10% in the gastrocnemius and 10%−15% in the quadriceps after short-term missions.[14,83] Similar results were observed following SCI, where patients suffered significant 21%, 28%, and 39% reductions in CSA at the quadriceps femoris, soleus, and gastrocnemius muscles, respectively.[61,84] In addition to the effects on whole muscle volume, muscle fiber characteristics were also modified due to inactivity.[85−87] There are two primary muscle fibers: slow (type I) fibers play an important role in maintaining body posture, while fast (type II) fibers are responsive during more intense physical activity. Under disuse conditions, all fiber types were decreased in size, 16% for type I and 23%−36% for type II.[86−88] The atrophied soleus muscles also underwent a shift from type I (−8% in fiber numbers) to type II fibers.[86,88,89]

Clinical MS has been examined extensively in patients with SCI to strengthen skeletal muscle and alleviate muscle atrophy with promising outcomes.[90−93] A few physical training studies further investigated electric stimulation to determine its effect on osteopenia. These studies showed mixed results with respect to bone density data.[90,94−97] Using dual-energy X-ray absorptiometry, BeDell et al. found no change in BMD of the lumbar spine and femoral neck regions after functional electric stimulation-induced cycling exercise, whereas Mohr et al. showed a 10% increase in BMD in the proximal tibia following 12 months of similar training.[98−100] In a 24-week study of patients with SCI in whom 25-Hz electric stimulation was applied to the quadriceps muscles daily, Belanger and colleagues[99] reported a 28% recovery of BMD in the distal femur and proximal tibia, along with increased muscle strength. A number of reported animal studies also indicated that MS can enhance not only muscle mass but also BMD.[101,102] Both animal and human studies seem to strongly support that functional disuse can result in significant bone loss and muscle atrophy.

# 7. FREQUENCY-DEPENDENT MARROW PRESSURE AND BONE STRAIN GENERATED BY MUSCLE STIMULATION

A recent study has revealed that marrow fluid pressure and bone strain induced by MS were dependent on dynamic loading parameters and optimized at certain loading frequencies.[103] Adult Sprague-Dawley retired breeder rats with a mean body weight of $387 \pm 41$ g (Taconic, NY) were used to measure the relationships among intramedullary pressure (ImP), bone strain, and induced muscle contraction. Rats were anesthetized using standard isoflurane inhalation. A microcardiovascular pressure transducer (Millar Instruments, SPR-524, Houston, TX) was inserted into the femoral marrow cavity, guided via a 16-gauge catheter. A single element strain gauge ($120\Omega$, factor 2.06, Kenkyujo Co., Tokyo) was firmly attached onto the lateral surface of the same femur at the mid-diaphyseal region, with minimal disruption to the quadriceps, in order to measure bone strains during the loading. The MS was applied at various frequencies (1, 2.5, 5, 10, 15, 20, 30, 40, 50, 60, and 100 Hz).

Normal heartbeat generated approximately 5 mm Hg of ImP in the femur at a frequency of $5.37 \pm 0.35$ Hz. The ImP value (peak-peak) increased significantly during dynamic MS at 5, 10, 15, 20, 30, and 40 Hz ($P < .05$ for 5, 10, 30, and 40 Hz; $P < .01$ for 15 and 20 Hz). The response trend of the ImP against frequency was nonlinear; the ImP reached a maximum value of $45 \pm 9.3$ mm Hg (peak-peak) at 20 Hz (Fig. 2.3.6A), although there was no significant difference between 10, 20, and 30 Hz. In the range from 5 to 40 Hz, the ImP was approximately 60% of the maximum ImP. MS generated ImP in the marrow cavity with values of $17.4 \pm 6.2$, $24 \pm 5.4$, $37.5 \pm 11.0$, $26.3 \pm 11.1$, and $3.7 \pm 1.5$ mm Hg at frequencies of 1, 5, 10, 40, and 100 Hz, respectively.

The response of the mineralized bone matrix strain to various MS frequencies was also nonlinear (Fig. 2.3.6B). Dynamic MS generated femoral matrix strains of $61.8 \pm 6.2$, $87.5 \pm 5.1$, $128.4 \pm 19.2$, $78.3 \pm 6.8$, $18.7 \pm 1.3$, and $10.1 \pm 1.8$ $\mu\varepsilon$ at frequencies of 1, 5, 10, 20, 40, and 100 Hz, respectively. Although the ImP trend indicated that the peak ImP value was observed at 20 Hz, the maximum matrix strain was measured at 10 Hz. Bone strain induced by MS at 10 Hz was significantly different ($P < .01$) from all other stimulation frequencies, with the exception of 15 Hz. In addition, the strains generated by MS above 30 Hz were significantly lower than those measured for stimulation at and below 20 Hz ($P < .005$), in which matrix strains, when loaded above 30 Hz, decreased by more than 75% of the peak strain measured at 10 Hz. For frequencies from 40 to 100 Hz, the MS-induced matrix strain was less than 20 $\mu\varepsilon$. These results suggest that MS with a relatively

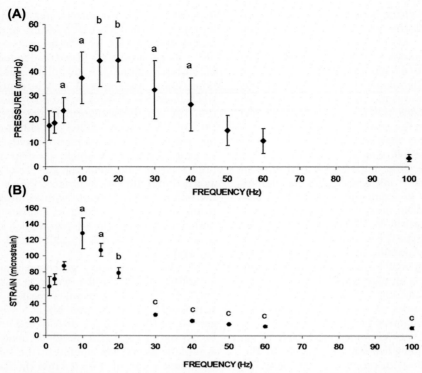

FIG. 2.3.6 **(A)** Intramedullary pressure (ImP) in rat femur increased significantly with electric frequency at 5, 10, 15, 20, 30, and 40 Hz. In the loading spectrum from 1 to 100 Hz, stimulation at 1 Hz generated an ImP of 18 mm Hg. A maximum ImP of 45 mm Hg was measured at 20 Hz, which was 2.5-fold higher than 1 Hz a, $P < .05$ versus baseline ImP; b, $P < .01$ versus baseline ImP. **(B)** Bone surface strain measurement. Dynamic muscle stimulation applied at various frequencies significantly increased bone strain. In the loading spectrum from 1 to 100 Hz, stimulation at 1 Hz produced a strain of 62 $\mu\varepsilon$. A peak strain of 128 $\mu\varepsilon$ was recorded at 10 Hz stimulation. The strain magnitude was reduced by >75% of the peak strain for stimulation frequencies greater than 30 Hz a, $P < .01$ versus 1, 2.5, and 5 Hz; b, $P < .01$ versus 10 Hz; and c, $P < .001$ versus stimulation at 20 Hz and below.

high rate and small magnitude can trigger significant fluid pressure in the skeleton.

## 8. DYNAMIC MUSCLE-STIMULATION-INDUCED ATTENUATION OF BONE LOSS

These findings were verified in an in vivo experiment under functional disuse conditions.[104] A total of 56 6-month-old female Sprague-Dawley retired breeder rats (Taconic, NY) were used to investigate the effects of frequency-dependent dynamic MS on skeletal adaptation under disuse conditions. Animals were randomly assigned to seven groups with n = 8 per group: (1) baseline control, (2) age-matched control, (3) hind limb suspension (HLS), (4) HLS + 1 Hz MS, (5) HLS + 20 Hz MS, (6) HLS + 50 Hz MS, and (7)

HLS + 100 Hz MS. Functional disuse was induced by an HLS setup modified from Morey-Holton and Globus.[105,106] An approximately 30 degrees head-down tilt was set to prevent contact of the animal's hind limbs with the cage bottom. The body weight of each animal was weighed three times per week throughout the study.

Throughout the experimental period, the body weights were not significantly different between groups at the beginning of the study, with an average of $320 \pm 47$ g. Age-matched control animals were able to maintain a steady body weight throughout the study, with only a $-0.15\%$ difference between the start and end dates. Animals subjected to 4-week functional disuse lost a significant amount of body mass. These weight reductions were similar in HLS and HLS + MS groups, with −10% for HLS ($P < .05$), −8% for 1 Hz ($P = .07$), −9% for 20 Hz ($P < .05$), −11% for 50 Hz ($P < .01$), and −8% for ($P = .09$).

Trabecular bone structure changes by MS seem to be sensitive to the fluid pressure magnitude experienced by the tissue, where a larger response occurred at the region near the marrow cavity, and attenuated at the region near the growth plate. For example, M1 is the distal metaphyseal region 1.5 mm above the growth plate. The lack of weight-bearing activity for 4 weeks significantly reduced trabecular bone quantity and quality, demonstrated by a 70% decrease in bone volume to total volume ratio (BV/TV), an 86% decrease in connectivity density (Conn.D), a 28% decrease in trabecular number (Tb.N), a 57% increase in structural model index (SMI), and a 43% increase in trabecular spacing (Tb.Sp) compared with baseline ($P < .001$). All these changes are associated with lower modulus and strength in trabecular bone. Similar results were observed when compared with age-matched controls ($P < .001$); decreases in BV/TV (66%), Conn.D (86%), and Tb.N (26%) and increases in SMI (39%) and Tb.Sp (39%) were observed. Trabecular BV/TV in electrically stimulated animals, with the exception of 1 Hz, was significantly greater than that of disused bone. Animals with MS at 20 and 50 Hz exhibited an increase in BV/TV of 143% ($P < .05$) and 147% ($P < .01$), respectively. Stimulation at 100 Hz resulted in an 86% increase in BV/TV, but this change was not statistically different from the HLS group. The other outcome measures of Conn.D, Tb.N, and Tb.Sp were also significantly affected by MS at 20, 50, and 100 Hz frequencies. There were up to 600% and 38% increases for Conn.D and Tb.N and up to a 36% decrease for trabecular separation (20 Hz, $P < .01$; 50 Hz, $P < .001$; and 100 Hz, $P < .05$). SMI and trabecular thickness (Tb.Th) were not affected by the stimulus, regardless of its frequency. The animals subjected to 4 weeks of 1 Hz MS showed the same level of bone loss and structural deterioration as did the HLS animals without MS, and they were significantly different compared with stimulation at higher frequencies. M3, the distal metaphyseal portion directly above the growth plate, is a region with the most abundant trabecular network with $0.3 \pm 0.05$ BV/TV and $4.72 \pm 0.64$ Tb.N (Fig. 2.3.7). Disuse induced a 38% bone loss, 75% decrease in Conn.D, 30% reduction in Tb.N, and 43% increase in Tb.Sp within this region. Similar to the results reported for the M2 region, 50 Hz MS resulted in the greatest preventive effects against disuse osteopenia, with increased BV/TV (40%; $P < .05$), Conn.D (305%; $P < .001$), and Tb.N (41%; $P < .001$) and reduced Tb.Sp (31%; $P < .001$). BV/TV

was not significantly altered by MS at 20 Hz (+26%) and 100 Hz (+20%), whereas trabecular qualities, i.e., Conn.D, Tb.N, and Tb.Sp, were improved (up to 226%, 28%, and 24% respectively; $P < .001$). Like the other metaphyseal regions, SMI and Tb.Th were not affected by the stimulation. With the exception of 1 Hz, stimulation frequencies at 20, 50, and 100 Hz had greater effects on the trabecular bone 2.25 cm away from the growth plate, closer to the diaphysis. Also, BV/TV inhibition at M1 was significantly higher ($P < .05$) than that of M3 with 20 Hz MS. Although following a trend similar to that of these indices, at 50 and 20 Hz MS the percentage changes of the micro computed tomographic (µCT) measurements were not statistically significant between the three metaphyseal regions.

Trabecular bone responded to MS showing structural property changes, whereas the epiphyseal trabecular bone was not significantly affected by the 4-week HLS. The relative changes in the trabecular morphometric measures were minor in comparison to the metaphyseal regions, with a 5% decrease in BV/TV, 44% decrease in Conn.D, 7% decrease in Tb.N, and a 4% increase in Tb.Sp. In this region, MS did not induce any measurable effect on the bone volume and trabecular integrity at any stimulation frequency. All stimulated values were comparable to age-matched and HLS animals, with up to 10% higher BV/TV, 8% greater Tb.N, and a 9% reduction in Tb.Sp. These changes were not statistically significant.

The µCT data were correlated with the histomorphometric analyses. In the metaphyseal trabecular bone, BV/TV measured by the two-dimensional histomorphometric method was 43% lower in the HLS group than in age-matched controls ($P < .001$). Animals subjected to MS also experienced 22%−29% bone loss ($P < .01$). The result was correlated with the BV/TV values from the µCT analysis, with an $R^2$ value of 0.84 ($P < .05$). In other bone formation indices, HLS animals also showed significant decline in mineralized surface/bone surface (MS/BS) (76%, $P < .001$), mineral apposition rate (MAR) (80%, $P < .001$), and bone formation rate/bone surface (BFR/BS) (92%, $P < .001$). Disuse had an insignificant effect on the trabecular BV/TV (−10%) at the epiphyseal region, similar to the results of the µCT analysis. Bone formation indices in the epiphysis were reduced due to HLS (52% for MS/BS, 147% for MAR, and 59% for BFR/BS), and daily MS failed to prevent this reduction in bone formation activity.

These data imply that MS, applied at a high frequency with low magnitude and for a short duration, is able to at least partially mitigate bone loss induced

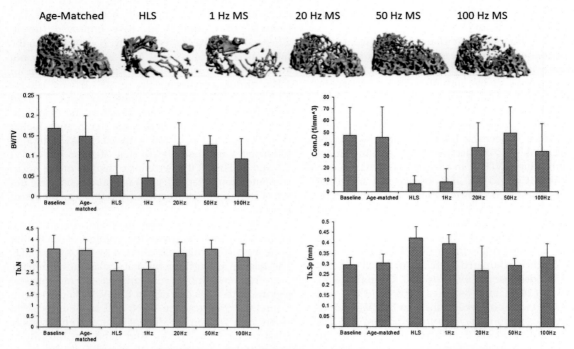

FIG. 2.3.7 Muscle stimulation (MS)-induced intramedullary pressure maintained bone mass at higher loading frequency, indicated by micro computed tomographic images of trabecular bone of the distal femur. Graphs show mean $\pm$SD values for bone volume fraction (BV/TV, %), connectivity density (Conn.D, 1/mm$^3$), trabecular number (Tb.N, 1/mm), and trabecular spacing (Tb.Sp, mm). MS at 50 Hz produced a significant difference in all indices, compared with hind limb suspension (HLS).

by functional disuse. There was, however, no evidence to suggest that such loading would enhance overall new bone formation, e.g., the total bone mass was less than that of age-matched animals. However, further studies related to cellular activities, e.g., osteoclast and osteoblast, linked to bone resorption and bone formation, may be necessary to further explore the balance of resorption and formation in such functional disuse model.

# 9. CELLULAR AND MOLECULAR PATHWAYS OF BONE IN RESPONSE TO MECHANICAL LOADING

Bone remodeling involves a number of related cell types, i.e., osteoblast, osteoclast, osteocyte, T cells, B cells, megakaryocytes, and lining cells. Thus all these cells are potentially mechanosensitive and even interrelated. These cells respond to mechanical loading by expressing specific molecular pathways. This section will discuss several potential pathways involved in mechanical stimulation-induced adaptation.

## 9.1. Basic Multicellular Units

To explore the interrelation among overall bone cells, a cluster of bone-forming and bone-resorbing cells is used which contribute to dynamic and temporal adaptation structures and are referred to as "basic multicellular units" (BMUs).[107] Bone adaptation occurs constantly and each cycle may take over several weeks, involving a continuous combination of resorption and formation. Each phase can involve targeted molecular and gene activation. An active BMU consists of a leading front of bone-resorbing osteoclasts. Reversal cells, of unclear phenotype, follow the osteoclasts, covering the newly exposed bone surface, and prepare it for deposition of replacement bone, followed by deposition of an unmineralized bone matrix known as osteoid. Related molecular factors are represented in this temporal sequence (Fig. 2.3.8).

In response to mechanical loading, the first stage of remodeling reflects the detection of initiating triggering signals such as fluid flow or other of physical stimulation, such as pressure, electric, and acoustic waves. Before activation, the resting bone surface is covered with bone-lining cells, including preosteoblasts. B cells are present in the bone marrow and secrete osteoprotegerin (OPG), which suppresses osteoclastogenesis.

During the *activation* phase, the endocrine bone-remodeling signal parathyroid hormone (PTH) binds to the PTH receptor on preosteoblasts. Damage to the mineralized bone matrix results in localized osteocyte apoptosis, reducing the local transforming growth factor β concentration and its inhibition of osteoclastogenesis.

In the *resorption* phase, in response to PTH signaling, preosteoclasts are recruited to the bone surface. Additionally, osteoblast expression of OPG is decreased and the production of colony-stimulating factor (CSF) 1 and RANKL is increased to promote the proliferation of osteoclast precursors and differentiation of mature osteoclasts. Mature osteoclasts anchor to RGD-binding sites, creating a localized microenvironment (sealed zone) that facilitates degradation of the mineralized bone matrix.

In the *reversal* phase, reversal cells engulf and remove demineralized undigested collagen from the bone surface. Transition signals are generated that halt bone resorption and stimulate the bone formation process.

During the *formation* phase, formation signals and molecules arise from the degraded bone matrix, mature osteoclasts, and potentially reversal cells. PTH and mechanical activation of osteocytes reduce sclerostin expression, allowing for Wnt-directed bone formation to occur.

Finally, in the *termination* phase, sclerostin expression likely returns and bone formation ceases. The newly deposited osteoid is mineralized, the bone surfaces return to a resting state with bone-lining cells intercalated with osteomacs, and the remodeling cycle concludes.

Mechanical stimulation is likely involved in each of these phases and eventually regulates related molecular and genetic factors. This unique spatial and temporal arrangement of cells within the BMU is critical to bone remodeling, ensuring coordination of the distinct and sequential phases of this process: activation, resorption, reversal, formation, and termination.

## 10. MECHANICAL SIGNAL-INDUCED MARROW STEM CELL ELEVATION AND ADIPOSE CELL SUPPRESSION

The data from disuse osteopenia and clinical osteoporosis have shown significant reduction of bone density and structural integrity, culminating in an elevated risk of skeletal fracture. Concurrently, a marked reduction in the available bone-marrow-derived population of MSCs[66] jeopardizes the regenerative potential that is critical to recovery from bone loss, musculoskeletal injury, and diseases. A potential way to combat the deterioration involves harnessing the sensitivity of bone to mechanical signals, which is crucial in defining, maintaining, and recovering bone mass. As discussed earlier, osteoblasts, osteoclasts, and osteocytes may sense external mechanical loading directly to maintain the balance of formation and resorption in the remodeling process; specific mechanotransductive signals may also bias MSC differentiation toward osteoblastogenesis and away from adipogenesis. Mechanical targeting of the bone marrow stem cell pool might, therefore, represents a novel, drug-free means of slowing the age-related decline of the musculoskeletal system.

Considering the importance of exercise in stemming both osteoporosis and obesity, combined with the fact that MSCs are progenitors of both osteoblasts and adipocytes (fat cells), as well as the anabolic response of the skeletal system to mechanical loadings, it was hypothesized that mechanical signals anabolic to bone would invariably cause a parallel decrease in fat production. In an in vivo setting, 7-week-old C57BL/6J mice on a normal chow diet were randomized to undergo low-magnitude mechanical stimulation (LMMS) (90 Hz at 0.2 g for 15 minutes per day) or placebo treatment.[108] At 15 weeks, with no differences in food consumption between groups, in vivo computed tomographic scans showed that the abdominal fat volume of mice subjected to loading was 27% lower than that of controls ($P < .01$).[109,110] Wet weights of visceral and subcutaneous fat deposits in the loading group were correspondingly lower. Confirmed by fluorescent labeling and flow cytometry studies,[109,110] these data indicated that mechanical signals influence not only the resident bone cell (osteoblast/osteocyte) population but also their progenitors, biasing MSC differentiation toward bone (osteoblastogenesis) and away from fat (adipogenesis). In a follow-up test of this hypothesis, mice fed a high-fat diet were subjected to low—magnitude loading or placebo treatment.[109,110] Suppression of adiposity by the mechanical signals was accompanied by a "mechanistic response" at the molecular level showing that loading significantly influenced MSC commitment to either an osteogenic (indicated by expression of Runx2, a transcription factor central to osteoblastogenesis) or adipogenic (indicated by expression of peroxisome proliferator-activated receptor [PPAR] γ, a transcription factor central to adipogenesis)

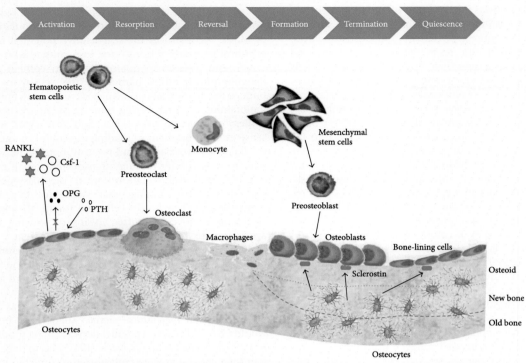

FIG. 2.3.8 Bone remodeling cycle and its associated molecular pathways. Three primary cell types are involved in bone modeling and remodeling: osteoblasts, osteoclasts, and osteocytes. The remodeling cycle of bone is composed of sequential phases of resorption and formation, including the activation of precursor cells, osteoclast activation generating bone resorption, bone formation by osteoblasts after mineral removal and reversal, and mineralization. The osteoblasts that are buried within the newly formed mineral matrix become osteocytes. Other osteoblasts that rest on the bone surface become bone-lining cells.[46] *CSF-1*, colony-stimulating factor 1; *OPG*, osteoprotegerin; *PTH*, parathyroid hormone.

fate. Runx2 expression was greater and PPARγ expression was decreased in mice that underwent LMMS compared with controls. The PPARγ transcription factor, when absent or present as a single copy, facilitates osteogenesis at least partly through enhanced canonical Wnt signaling,[111,112] a pathway critically important to MSC entry into the osteogenic lineage and expansion of the osteoprogenitor pool. Notably, low-magnitude mechanical loading treatment also resulted in a 46% increase in the size of the MSC pool ($P < .05$).[109,110] These experiments, although not obviating a role for the osteoblast/osteocyte syncytium, provide evidence that bone marrow stem cells are capable of sensing exogenous mechanical signals and responding with an alteration in cell fate that ultimately influences both the bone and fat phenotype. The inverse correlation of bone and fat phenotype has increasing support in the clinical literature. Although controversial, and despite

the presumption that conditions such as obesity could inherently protect the skeleton through increased loading events, data in humans evaluating bone-fat interactions indicate that an ever-increasing adipose burden comes at the cost of bone structure and increased risk of fracture.[113]

## 11. OSTEOCYTES AND THEIR RESPONSE TO MECHANICAL SIGNALS COUPLED WITH WNT SIGNALING

Osteocytes, cells embedded within the mineralized matrix of bone, are the target of intensive investigation in the mechanical regulation of bone adaptation.[107,114–116] Osteoblasts are defined as cells that make the bone matrix and are thought to translate mechanical loading into biochemical signals that affect bone modeling and remodeling. The interrelationship between osteoblasts and osteocytes would be expected to have the same lineage, yet these cells also have distinct differences, particularly in their responses to mechanical loading and utilization of the various biochemical pathways to accomplish their respective functions. Among the many factors, Wnt/β-catenin signaling pathway may be recognized as an important regulator of bone mass and bone cell functions.[107,116] Although osteocytes are embedded within the mineral matrix, Wnt/β-catenin signaling pathway may serve as a transmitter to transfer mechanical signals sensed by osteocytes to the surface of bone. Furthermore, new data suggest that the Wnt/β-catenin pathway in osteocytes may be triggered by cross talk with the prostaglandin pathway in response to loading, which then leads to a decrease in expression of negative regulators of the pathway such as sclerostin (Sost) and Dickkopf-related protein (Dkk) 1.[116,117]

The Wnt pathway is a key element of bone cell differentiation, proliferation, and apoptosis.[116,118] Regulation of the Wnt/β-catenin signaling pathway is vested largely in proteins that either act as competitive binders of Wnts, notably the secreted frizzled-related protein (sFRP) family, or act at the level of low-density lipoprotein receptor–related protein 5 (LRP5), including the osteocyte-specific protein, sclerostin (the Sost gene product) and the Dkk proteins, particularly Dkk-1 and Dkk-2.[116,118–121] Sclerostin is expressed by mature osteocytes and inhibits Wnt/β-catenin signaling by binding to LRP5, thereby preventing the binding of Wnt. Dkk-1 is also highly expressed in osteocytes.[119–121] Clinical trials using antibodies to sclerostin that competitively bind to prevent interaction with LRP5 have resulted in increased bone mass, suggesting that targeting these negative regulators of the Wnt/β-catenin signaling pathway might be an anabolic treatment for diseases such as osteoporosis.[119] Mechanical loading has been shown to reduce sclerostin levels in bone,[119] suggesting that this treatment targets the pathways activated by the early events after mechanical loading, which are normally regulated by the transcription of genes encoding these negative modulators of the Wnt/β-catenin signaling pathway.

There is still much to be learned regarding how the bone cells, i.e., osteocytes, sense and transmit signals in response to or in the absence of loading and further elevate the activity of other cells. Although fluid shear stress is proposed as a triggering force, identification of such particular mechanical signals is still a challenging area to study. The osteocyte is joining the osteoblast and osteoclast as targets for therapeutics to treat or prevent bone disease. Clearly, targeting the Wnt/β-catenin pathway in osteocytes because of its central

role in bone mass regulation and bone formation in response to mechanical loading may prove useful for designing new paradigms and pharmaceutical products to treat bone diseases in the future.

## 11.1. The Role of LRP5 in Bone Response to Mechanical Loading

LRP5 has been shown to have important functions in the mammalian skeleton. Experimental evidence has pointed to LRP5 as a critical factor in translating mechanical signals into the proper skeletal response. For example, loss-of-function mutations in LRP5 have been reported to cause the autosomal recessive human disease osteoporosis pseudoglioma syndrome, which leads to a significant reduction of BMD and patients are more susceptible to skeletal fracture and deformity.[122–124] Moreover, the mechanical importance of LRP5 have been demonstrated in LRP5−/− mice, which have an almost complete ablation in ulnar loading-induced bone formation compared with wild-type controls.[124,125] Multiple single nucleotide polymorphisms, located in exons 18 and 10, have been reported, which can significantly affect the interconnection between physical activity and bone mass.[124,126] A high-bone-mass phenotype in humans was reported to be caused by certain missense mutations near the N terminus of LRP5.[127,128] An LRP5 overexpression mutation is, on the other hand, associated with high bone mass and induced osteoblast proliferation.[127] Increased sensitivity to load due to a lower threshold for initiating bone formation was also reported in this mouse.[128] A study done by Zhong et al.[129] showed that in vitro tension on MC3T3-E1 cells increased LRP5 gene expression at 1, 3, and 5 hours of loading.

## 11.2. MicroRNAs and Their Role in Mechanotransduction in Tissue

The newly discovered microRNAs (miRNAs) are short noncoding RNAs, which can be complementary to messenger RNA sequences to silence gene expression by either degradation or inhibitory translation of target transcripts.[130,131] Regulation of Runx2, bone morphogenetic protein (BMP), and Wnt signaling pathways is by far the well-studied miRNA related to osteoblast function. Positive and negative regulation of miRNAs on Runx2 expression has been shown to affect skeletal morphogenesis and osteoblastogenesis.[132] Inhibition of osteoblastogenesis can result from miRNA-135 and miRNA-26a regulation of the BMP-2/Smad signaling pathway.[133] Activation of Wnt signaling and β-catenin expression are upregulated by miRNA-29a targeting of Wnt inhibitors during osteoblast differentiation.[134] In addition, studies have been performed to investigate the miRNA function on self-renewal and lineage determination for tissue regeneration via human stromal stem cells.[135,136] Moreover, extensive studies have also been done to access the effects of miRNAs on osteogenic functions in committed cell lines, including osteoprogenitors, osteoblasts, and osteocytic cell lines. In general, actions of miRNA may affect bone cell differentiation in either positive or negative way.[130,136]

Recent research has been aimed at transcription and miRNA regulation to better understand gene expression regulation in a mechanical loading model. Transcription factors can bind to motifs in the promoter of genes and directly affect their expression; therefore, mechanotransduction in bone may result in alteration of transcription factors for regulation. Using a predictive bioinformatics algorithm, a recent study investigated the time-dependent regulatory mechanisms that governed mechanical loading-induced gene expression in bone. Axial loading was performed on the right forelimb in rodents. A linear model of gene expression was created, and 44 transcription-factor-binding motifs and 29 miRNA-binding sites were identified to predict the regulated gene expression across the time course of the experiment.

## 12. MECHANOTRANSDUCTIVE IMPLICATIONS IN BONE TISSUE ENGINEERING

Development of bone tissue from cell-seeded artificial scaffolds for musculoskeletal applications could take advantage of mechanotransduction phenomena to achieve integrity and function. Mechanical signals delivered to bone cells may be interfered by the scaffold deformation and should be taken into account. Fortunately, mechanotransduction could be used to control the proliferation and differentiation of bone cells.[137–141] Fluid flow has also been proposed as an important mechanical aspect when developing bone scaffolds.[137,138,141–143] Studies using bioreactors have helped us understand the phenomenon of mechanotransduction used in scaffold design.[139] For example, rotating bioreactors, flow perfusion bioreactors, and other mechanical stimuli such as strain have been designed to increase mass transfer by inducing dynamic flow conditions in culture, to create osteoinductive factors on MSCs by generating fluid shear stress,[144] and to induce the osteogenic differentiation of MSCs,[145] respectively. Among all, mimicking the natural bone strain to favor osteogenesis is one of the most rational aims of scaffold development.

Matching of the strain histograms of a scaffold and the actual bone can be performed using μCT measurements and the finite element method.[137,146,147]

## 13. DISCUSSION

These data have suggested that dynamic stimulation can generate fluid pressure in bone with simultaneously low-level bone strain. MS adjacent to the rat femur induces a peak ImP at 20–30 Hz. The increase in bone fluid pressure suggests that hypertension in the skeletal nutrient vessels may increase ImP and regulate fluid flow in bone.[58] Similarly, the bone strain was highest at approximately 10 Hz. Stimulus-induced bone fluid and matrix deformation is dependent on the stimulation frequency. It appears that the oscillatory MS stimulates relatively high fluid pressure at the frequency range between 20 and 50 Hz in the tested frequencies up to 100 Hz. At such an optimized loading rate (e.g., 20–50 Hz), relatively high ImP and relatively low bone strain were observed in response to the MS loading, which may be critical to regulate fluid flow and adaptation in bone in a functional loading frequency-dependent manner. It is also noted that both ImP and strain are higher at 10 and 20 Hz than at lower frequencies, i.e., 1 Hz. There is a significant strain difference between 10 and 20 Hz, whereas no significant ImP difference was observed between 10 and 20 Hz. At an optimal frequency, MS can produce high fluid pressure gradients within the femoral marrow cavity and a high strain value in bone. It is possible that loading-generated matrix strain and fluid pressure in bone may have combined effects on adaptation if loading occurred at relatively high frequencies.

Dynamic MS was able to inhibit bone loss and trabecular architectural deterioration caused by a lack of daily weight-bearing activities. The importance of selecting an effective loading regimen was investigated in this chapter, in which a wide range of stimulation frequencies was tested to determine their effects on skeletal adaptive responses. Throughout this chapter, we have referred to 1 Hz as low frequency, 20 and 50 Hz as mid-frequency, and 100 Hz as high frequency. From our results, we concluded that the effectiveness of dynamic MS was greatly dependent on the stimulation frequency. The optimized frequency, which resulted in a strong adaptive response in the disuse osteopenia model was in the range of 50 Hz. While the strain level generated at 50 Hz by MS was relatively low, e.g., approximately 10 με, the adaptive response may be mainly contributed by induced fluid flow and uncoupled to strain.[148] The degree of effectiveness of MS in attenuating bone loss varied in different regions of the distal femur. Such spatial response may also depend on the fluid pressure generated in the local region in a dose-dependent manner. Low-frequency MS was unsuccessful in preventing osteopenia, whereas mid-frequency MS applied to the quadriceps was able to maintain trabecular bone mass.

These results were consistent with the previous in vivo results in which mechanical loading at frequencies between 20 and 50 Hz was shown to be anti-catabolic to bone.[53,149–151] This sensitivity was even more apparent in trabecular bone, perhaps because of the increased surface area in the trabecular network, which exposes it to rapid changes in fluid pressure.[152] For example, trabecular osteoblast surface in the tibia was increased by 26% when an MS protocol at 10 Hz was applied for 3 weeks.[151] Likewise, whole-body vibration at 45 Hz increased the rate of formation of the growing skeleton by 30%.[153]

Both ImP and matrix strain have indicated a nonlinear response in the MS spectrum between 1 and 100 Hz, but peaked differently at 20 Hz (ImP) and 10 Hz (strain). While no obvious muscle fatigue was observed, perhaps because of the rest period during the stimulation, the mechanism behind such a nonlinear response is not clear. From the characteristics of tissue material point of view, e.g., the viscoelastic property of the tissues, both muscle and bone could quickly dampen the response at high frequencies through the MS loading. But owing to the difference in densities and viscosities between muscle and bone, MS-induced ImP and matrix strain could result in different frequency responses or optimized/resonant patterns for different tissues against the loading. In addition, mechanotransduction of MS through different connective tissues may attenuate the high-frequency response in bone, e.g., via the connective pathway from muscle, tendon to bone, thus resulting in peak strain and peak ImP at varied frequencies. Future research on such complex interrelations between muscle kinematics, bone fluid flow, and matrix strain is necessary to elucidate the mechanism further.

Even in the absence of bone matrix strain, previous data have shown that ImP alone can induce bone adaptation.[150] Using a turkey ulna osteotomy model, disuse alone resulted in a 5.7% loss of cortical bone. Direct fluid loading at 20 Hz for 4 weeks increased cortical bone mass by 18% by enhancing the formation of bone at both periosteal and endosteal surfaces.[150] Transcortical fluid pressure gradient and total bone formation were strongly positively correlated. Bone fluid flow plays an important role in triggering bone remodeling.[154–156] Strong evidence suggests that interstitial fluid flow in bone interacts strongly with external muscular activities via various mechanisms.[157,158] According to a muscle pump hypothesis, an arteriovenous pressure gradient enhances muscle perfusion.[56,159] This process may, in turn, increase the hydraulic pressure in skeletal nutrient vessels and amplify the capillary filtration in bone tissue.[56–58]

As a clinical application, MS in spinal cord-injured patients can cause a partial reversal of disuse osteopenia and recovery of muscular strength.[99] Other in vivo studies have also reported positive effects of using MS to inhibit muscle atrophy. Immobilization studies using MS at 50–100 Hz have shown to minimize the reduction of the CSA of muscle fiber and to restore mechanical properties.[160,161] Previous data showed that stimulation of distal nerve stumps had similar action potential response between normal and innervated muscles.[162] Although the response of ImP and bone mass by MS under such periphery nerve block conditions still remains unknown, MS could serve as a mitigating agent to retain bone mass under chronic nerve damage conditions, e.g., SCI. Taken together with the results from our current study, dynamic MS may be applied as both a skeletal therapy and a muscular therapy to prevent osteopenia and sarcopenia.

Loading-induced fluid flow in the musculoskeletal tissues may ultimately enhance interstitial flow and mechanotransduction in bone. Furthermore, dynamic loading, if applied at an optimal frequency, has been shown to have preventive potential in osteopenia in a functional disuse environment as a biomechanically based intervention for preventing and treating osteoporosis and muscle atrophy.

## ACKNOWLEDGMENTS

This research was supported by the National Institutes of Health (R01 AR52379, AR61821, and AR49286) and the National Space Biomedical Research Institute through a NASA Cooperative Agreement NCC 9–58. The authors are grateful to Ms. Alyssa Tuthill for her technical assistance and all the members in the Orthopaedic Bioengineering Research Lab, particularly for their tireless assistance in the study.

## REFERENCES

1. Ozdurak RH, Duz S, Arsal G, et al. Quantitative forearm muscle strength influences radial bone mineral density in osteoporotic and healthy males. *Technol Health Care*. 2003;11:253–261.

2. Srinivasan S, Weimer DA, Agans SC, Bain SD, Gross TS. Low-magnitude mechanical loading becomes osteogenic when rest is inserted between each load cycle. *J Bone Miner Res.* 2002;17:1613−1620.

3. Gerdhem P, Ringsberg KA, Akesson K, Obrant KJ. Influence of muscle strength, physical activity and weight on bone mass in a population-based sample of 1004 elderly women. *Osteoporos Int.* 2003;14:768−772.

4. Gerdhem P, Ringsberg KA, Magnusson H, Obrant KJ, Akesson K. Bone mass cannot be predicted by estimations of frailty in elderly ambulatory women. *Gerontology.* 2003;49:168−172.

5. LeBlanc A. Summary of research issues in human studies. *Bone.* 1998;22:117S−118S.

6. LeBlanc A, Lin C, Shackelford L, et al. Muscle volume, MRI relaxation times (T2), and body composition after spaceflight. *J Appl Physiol.* 2000;89:2158−2164.

7. LeBlanc AD, Evans HJ, Johnson PC, Jhingran S. Changes in total body calcium balance with exercise in the rat. *J Appl Physiol.* 1983;55:201−204.

8. Convertino VA, Sandler H. Exercise countermeasures for spaceflight. *Acta Astronaut.* 1995;35:253−270.

9. Fluckey JD, Dupont-Versteegden EE, Montague DC, et al. A rat resistance exercise regimen attenuates losses of musculoskeletal mass during hindlimb suspension. *Acta Physiol Scand.* 2002;176:293−300.

10. Keller TS, Strauss AM, Szpalski M. Prevention of bone loss and muscle atrophy during manned space flight. *Microgravity Q.* 1992;2:89−102.

11. McPhee JC, White RJ. Physiology, medicine, long-duration space flight and the NSBRI. *Acta Astronaut.* 2003;53:239−248.

12. Grigoriev A, Morukov B, Stupakov G, Bobrovnik E. Influence of bisphosphonates on calcium metabolism and bone tissue during simulation of the physiological effects of microgravity. *J. Gravit Physiol.* 1998;5:69−70.

13. LeBlanc A, Marsh C, Evans H, Johnson P, Schneider V, Jhingran S. Bone and muscle atrophy with suspension of the rat. *J Appl Physiol.* 1985;58:1669−1675.

14. LeBlanc A, Rowe R, Schneider V, Evans H, Hedrick T. Regional muscle loss after short duration spaceflight. *Aviat Space Environ Med.* 1995;66:1151−1154.

15. LeBlanc A, Schneider V. Countermeasures against space flight related bone loss. *Acta Astronaut.* 1992;27:89−92.

16. Convertino VA. Mechanisms of microgravity induced orthostatic intolerance: implications for effective countermeasures. *J Gravit Physiol.* 2002;9:1−13.

17. Larina IM, Tcheglova IA, Shenkman BS, Nemirovskaya TL. Muscle atrophy and hormonal regulation in women in 120 day bed rest. *J Gravit Physiol.* 1997;4:121−122.

18. Mayet-Sornay MH, Hoppeler H, Shenkman BS, Desplanches D. Structural changes in arm muscles after microgravity. *J Gravit Physiol.* 2000;7:S43−S44.

19. Serova LV. Microgravity and aging of animals. *J Gravit Physiol.* 2001;8:137−138.

20. Wolff J. *Das Gesetz Der Transformation Der Knochen. Berlin.* 1892.

21. Wolff J. *The Law of Bone Remodeling. Berlin.* 1986.

22. Evans FG. The mechanical properties of bone. *Artif Limbs.* 1969;13:37−48.

23. Evans FG. Factors affecting the mechanical properties of bone. *Bull N Y Acad Med.* 1973;49:751−764.

24. Evans FG, Vincentelli R. Relations of the compressive properties of human cortical bone to histological structure and calcification. *J Biomech.* 1974;7:1−10.

25. Martin RB, Burr DB. *Structure, Function and Adaptation of Compact Bone. New York.* 1989.

26. Stokes IA. Analysis of symmetry of vertebral body loading consequent to lateral spinal curvature. *Spine.* 1997;22:2495−2503.

27. Donaldson CL, Hulley SB, Vogel JM, Hattner RS, Bayers JH, McMillan DE. Effect of prolonged bed rest on bone mineral. *Metabolism.* 1970;19:1071−1084.

28. Gross TS, Edwards JL, McLeod KJ, Rubin CT. Strain gradients correlate with sites of periosteal bone formation. *J Bone Miner Res.* 1997;12:982−988.

29. Qin YX, Otter MW, Rubin CT, McLeod KJ. The influence of intramedullary hydrostatic pressure on transcortical fluid flow patterns in bone. *Trans Ortho Res Soc.* 1997;22:885.

30. Jones HH, Priest JD, Hayes WC, Tichenor CC, Nagel DA. Humeral hypertrophy in response to exercise. *J. Bone Jt Surg Am.* 1977;59:204−208.

31. Judex S, Zernicke RF. High-impact exercise and growing bone: relation between high strain rates and enhanced bone formation. *J Appl Physiol.* 2000;88:2183−2191.

32. Krolner B, Toft B, Pors NS, Tondevold E. Physical exercise as prophylaxis against involutional vertebral bone loss: a controlled trial. *Clin Sci.* 1983;64:541−546.

33. Nilsson BE, Westlin NE. Bone density in athletes. *Clin Orthop.* 1971;77:179−182.

34. Stokes IA, Aronsson DD, Spence H, Iatridis JC. Mechanical modulation of intervertebral disc thickness in growing rat tails. *J Spinal Disord.* 1998;11:261−265.

35. Hicks A, McGill S, Hughson RL. Tissue oxygenation by near-infrared spectroscopy and muscle blood flow during isometric contractions of the forearm. *Can J Appl Physiol.* 1999;24:216−230.

36. Joyner MJ. Does the pressor response to ischemic exercise improve blood flow to contracting muscles in humans? *J Appl Physiol.* 1991;71:1496−1501.

37. Joyner MJ. Blood pressure and exercise: failing the acid test. *J Physiol.* 12-1-2001;537:331.

38. Joyner MJ, Dietz NM. Nitric oxide and vasodilation in human limbs. *J Appl Physiol.* 1997;83:1785−1796.

39. Joyner MJ, Lennon RL, Wedel DJ, Rose SH, Shepherd JT. Blood flow to contracting human muscles: influence of increased sympathetic activity. *J Appl Physiol.* 1990;68:1453−1457.

40. Joyner MJ, Nauss LA, Warner MA, Warner DO. Sympathetic modulation of blood flow and $O_2$ uptake in rhythmically contracting human forearm muscles. *Am J Physiol.* 1992;263:H1078−H1083.

41. Joyner MJ, Proctor DN. Muscle blood flow during exercise: the limits of reductionism. *Med.Sci.Sports Exerc.* 1999;31:1036−1040.

42. Joyner MJ, Wieling W. Increased muscle perfusion reduces muscle sympathetic nerve activity during handgripping. *J Appl Physiol*. 1993;75:2450−2455.

43. Lee NK, Sowa H, Hinoi E, et al. Endocrine regulation of energy metabolism by the skeleton. *Cell*. 2007;130:456−469.

44. Wolf G. Energy regulation by the skeleton. *Nutr Rev*. 2008;66:229−233.

45. Mak AF, Huang L, Wang Q. A biphasic poroelastic analysis of the flow dependent subcutaneous tissue pressure and compaction due to epidermal loadings: issues in pressure sore. *J Biomech Eng*. 1994;116:421−429.

46. Qin YX, Hu M. Mechanotransduction in musculoskeletal tissue regeneration: effects of fluid flow, loading, and cellular-molecular pathways. *BioMed Res Int*. 2014;2014:863421.

47. Qin YX, McLeod KJ, Otter MW, Rubin CT. Patterns of loading induced fluid flow in cortical bone. *Ann Biomed Eng*. 2002;30(5):693−702.

48. Frost HM. *Intermediary Organization of the Skeleton. Boca Raton*. 1986.

49. Frost HM. Bone's mechanostat: a 2003 update. *Anat Rec*. 2003;275A:1081−1101.

50. Cowin SC. *Bone Mechanics. Boca Raton*. 1989.

51. Cowin SC. Bone poroelasticity. *J Biomech*. 1999;32:217−238.

52. Piekarski K, Demetriades D, Mackenzie A. Osteogenetic stimulation by externally applied dc current. *Acta Orthop Scand*. 1978;49:113−120.

53. Qin Y-X, Rubin CT, McLeod KJ. Non-linear dependence of loading intensity and cycle number in the maintence of bone mass and morphology. *J Orthop Res*. 1998;16:482−489.

54. Rubin CT, Lanyon LE. Regulation of bone formation by applied dynamic loads. *J.Bone Jt Surg Am*. 1984;66:397−402.

55. Turner CH. Site-specific skeletal effects of exercise: importance of interstitial fluid pressure. *Bone*. 1999;24:161−162.

56. Laughlin MH. The muscle pump: what question do we want to answer? *J Appl Physiol*. 2005;99:774.

57. Otter MW, Qin YX, Rubin CT, McLeod KJ. Does bone perfusion/reperfusion initiate bone remodeling and the stress fracture syndrome? *Med Hypotheses*. 1999;53:363−368.

58. Winet H. A bone fluid flow hypothesis for muscle pump-driven capillary filtration: II. Proposed role for exercise in erodible scaffold implant incorporation. *Eur.Cell Mater*. 2003;6:1−10.

59. Lanyon LE, Rubin CT. Static vs dynamic loads as an influence on bone remodelling. *J Biomech*. 1984;17:897−905.

60. Rubin CT, Lanyon LE. Kappa Delta Award paper. Osteoregulatory nature of mechanical stimuli: function as a determinant for adaptive remodeling in bone. *J Orthop Res*. 1987;5:300−310.

61. Gorgey AS, Dudley GA. Skeletal muscle atrophy and increased intramuscular fat after incomplete spinal cord injury. *Spinal Cord*. 2007;45:304−309.

62. Hu M, Cheng J, Bethel N, et al. Interrelation between external oscillatory muscle coupling amplitude and in vivo intramedullary pressure related bone adaptation. *Bone*. 2014;66:178−181.

63. Hu M, Serra-Hsu F, Bethel N, et al. Dynamic hydraulic fluid stimulation regulated intramedullary pressure. *Bone*. 2013;57:137−141.

64. Meakin, L. B., J. S. Price, and L. E. Lanyon. The contribution of experimental in vivo models to understanding the mechanisms of adaptation to mechanical loading in bone. Front Endocrinol 5:154-2014.

65. Leonard MB, Shults J, Long J, et al. Effect of low magnitude mechanical stimuli on bone density and structure in pediatric crohn's disease: a randomized placebo controlled trial. *J Bone Miner Res*. 2016;31(6):1177−1188.

66. Ozcivici E, Luu YK, Adler B, et al. Mechanical signals as anabolic agents in bone. *Nat Rev Rheumatol*. 2010;6:50−59.

67. Rubin C, Turner AS, Muller R, et al. Quantity and quality of trabecular bone in the femur are enhanced by a strongly anabolic, noninvasive mechanical intervention. *J Bone Miner Res*. 2002;17:349−357.

68. Biering-Sorensen F, Bohr HH, Schaadt OP. Longitudinal study of bone mineral content in the lumbar spine, the forearm and the lower extremities after spinal cord injury. *Eur J Clin Investig*. 1990;20:330−335.

69. Dauty M, Perrouin VB, Maugars Y, Dubois C, Mathe JF. Supralesional and sublesional bone mineral density in spinal cord-injured patients. *Bone*. 2000;27:305−309.

70. Garland DE, Adkins RH, Stewart CA, Ashford R, Vigil D. Regional osteoporosis in women who have a complete spinal cord injury. *J Bone Jt Surg Am*. 2001;83-A:1195−1200.

71. Lazo MG, Shirazi P, Sam M, Giobbie-Hurder A, Blacconiere MJ, Muppidi M. Osteoporosis and risk of fracture in men with spinal cord injury. *Spinal Cord*. 2001;39:208−214.

72. Ryg J, Rejnmark L, Overgaard S, Brixen K, Vestergaard P. Hip fracture patients at risk of second hip fracture: a nationwide population-based cohort study of 169,145 cases during 1977−2001. *J Bone Miner Res*. 2009;24:1299−1307.

73. Vestergaard P, Krogh K, Rejnmark L, Mosekilde L. Fracture rates and risk factors for fractures in patients with spinal cord injury. *Spinal Cord*. 1998;36:790−796.

74. Vestergaard P, Rejnmark L, Mosekilde L. Osteoarthritis and risk of fractures. *Calcif Tissue Int*. 2009;84:249−256.

75. Heaney RP. Pathophysiology of osteoporosis. *Endocrinol Metab Clin North Am*. 1998;27:255−265.

76. Ingram RR, Suman RK, Freeman PA. Lower limb fractures in the chronic spinal cord injured patient. *Paraplegia*. 1989;27:133−139.

77. Szollar SM, Martin EM, Sartoris DJ, Parthemore JG, Deftos LJ. Bone mineral density and indexes of bone metabolism in spinal cord injury. *Am J Phys Med Rehabil*. 1998;77:28−35.

78. Macdonald JH, Evans SF, Davie MW, Sharp CA. Muscle mass deficits are associated with bone mineral density

in men with idiopathic vertebral fracture. *Osteoporos Int.* 2007;18:1371−1378.

79. McCarthy I, Goodship A, Herzog R, Oganov V, Stussi E, Vahlensieck M. Investigation of bone changes in microgravity during long and short duration space flight: comparison of techniques. *Eur J Clin Investig.* 2000;30: 1044−1054.

80. Lang T, LeBlanc A, Evans H, Lu Y, Genant H, Yu A. Cortical and trabecular bone mineral loss from the spine and hip in long-duration spaceflight. *J Bone Miner Res.* 2004;19:1006−1012.

81. Burr DB, Frederickson RG, Pavlinch C, Sickles M, Burkart S. Intracast muscle stimulation prevents bone and cartilage deterioration in cast-immobilized rabbits. *Clin Orthop.* 1984:264−278.

82. LeBlanc A, Rowe R, Evans H, West S, Shackelford L, Schneider V. Muscle atrophy during long duration bed rest. *Int J Sports Med.* 1997;18(Suppl 4):S283−S285.

83. Narici M, Kayser B, Barattini P, Cerretelli P. Effects of 17-day spaceflight on electrically evoked torque and cross-sectional area of the human triceps surae. *Eur J Appl Physiol.* 2003;90:275−282.

84. Shah PK, Stevens JE, Gregory CM, et al. Lower-extremity muscle cross-sectional area after incomplete spinal cord injury. *Arch Phys Med Rehabil.* 2006;87:772−778.

85. Riley DA, Slocum GR, Bain JL, Sedlak FR, Sowa TE, Mellender JW. Rat hindlimb unloading: soleus histochemistry, ultrastructure, and electromyography. *J Appl Physiol.* 1990;69:58−66.

86. Roy RR, Pierotti DJ, Garfinkel A, Zhong H, Baldwin KM, Edgerton VR. Persistence of motor unit and muscle fiber types in the presence of inactivity. *J Exp Biol.* 2008;211: 1041−1049.

87. Zhong H, Roy RR, Siengthai B, Edgerton VR. Effects of inactivity on fiber size and myonuclear number in rat soleus muscle. *J Appl Physiol.* 2005;99:1494−1499.

88. Stewart BG, Tarnopolsky MA, Hicks AL, et al. Treadmill training-induced adaptations in muscle phenotype in persons with incomplete spinal cord injury. *Muscle Nerve.* 2004;30:61−68.

89. Zhou MY, Klitgaard H, Saltin B, Roy RR, Edgerton VR, Gollnick PD. Myosin heavy chain isoforms of human muscle after short-term spaceflight. *J Appl Physiol.* 1995; 78:1740−1744.

90. Griffin L, Decker MJ, Hwang JY, et al. Functional electrical stimulation cycling improves body composition, metabolic and neural factors in persons with spinal cord injury. *J Electromyogr Kinesiol.* 2009;19:614−622.

91. Lim PA, Tow AM. Recovery and regeneration after spinal cord injury: a review and summary of recent literature. *Ann Acad Med Singapore.* 2007;36:49−57.

92. Rodgers MM, Glaser RM, Figoni SF, et al. Musculoskeletal responses of spinal cord injured individuals to functional neuromuscular stimulation-induced knee extension exercise training. *J Rehabil Res Dev.* 1991;28:19−26.

93. Shields RK, Dudley-Javoroski S. Musculoskeletal adaptations in chronic spinal cord injury: effects of long-term

soleus electrical stimulation training. *Neurorehabil Neural Repair.* 2007;21:169−179.

94. Dudley-Javoroski S, Littmann AE, Iguchi M, Shields RK. Doublet stimulation protocol to minimize musculoskeletal stress during paralyzed quadriceps muscle testing. *J Appl Physiol.* 2008;104:1574−1582.

95. Dudley-Javoroski S, Shields RK. Dose estimation and surveillance of mechanical loading interventions for bone loss after spinal cord injury. *Phys Ther.* 2008;88:387−396.

96. Dudley-Javoroski S, Shields RK. Muscle and bone plasticity after spinal cord injury: review of adaptations to disuse and to electrical muscle stimulation. *J Rehabil Res Dev.* 2008;45:283−296.

97. Yang YS, Koontz AM, Triolo RJ, Cooper RA, Boninger ML. Biomechanical analysis of functional electrical stimulation on trunk musculature during wheelchair propulsion. *Neurorehabil Neural Repair.* 2009;23: 717−725.

98. BeDell KK, Scremin AM, Perell KL, Kunkel CF. Effects of functional electrical stimulation-induced lower extremity cycling on bone density of spinal cord-injured patients. *Am J Phys Med Rehabil.* 1996;75:29−34.

99. Belanger M, Stein RB, Wheeler GD, Gordon T, Leduc B. Electrical stimulation: can it increase muscle strength and reverse osteopenia in spinal cord injured individuals? *Arch Phys Med Rehabil.* 2000;81:1090−1098.

100. Mohr T. Electric stimulation in muscle training of the lower extremities in persons with spinal cord injuries. *Ugeskr Laeger.* 2000;162:2190−2194.

101. Allen MR, Hogan HA, Bloomfield SA. Differential bone and muscle recovery following hindlimb unloading in skeletally mature male rats. *J Musculoskelet Neuronal Interact.* 2006;6:217−225.

102. Swift JM, Nilsson MI, Hogan HA, Sumner LR, Bloomfield SA. Simulated resistance training during hindlimb unloading abolishes disuse bone loss and maintains muscle strength. *J Bone Miner Res.* 2010;25(3): 564−574.

103. Qin YX, Lam H. Intramedullary pressure and matrix strain induced by oscillatory skeletal muscle stimulation and its potential in adaptation. *J Biomech.* 2009;42:140−145.

104. Lam H, Qin YX. The effects of frequency-dependent dynamic muscle stimulation on inhibition of trabecular bone loss in a disuse model. *Bone.* 2008;43:1093−1100.

105. Morey-Holton ER, Globus RK. Hindlimb unloading of growing rats: a model for predicting skeletal changes during space flight. *Bone.* 1998;22:83S−88S.

106. Morey-Holton ER, Globus RK. Hindlimb unloading rodent model: technical aspects. *J Appl Physiol.* 2002;92: 1367−1377.

107. Raggatt LJ, Partridge NC. Cellular and molecular mechanisms of bone remodeling. *J Biol Chem.* 2010;285: 25103−25108.

108. Rubin CT, Capilla E, Luu YK, et al. Adipogenesis is inhibited by brief, daily exposure to high-frequency, extremely low-magnitude mechanical signals. *Proc Natl Acad Sci USA.* 2007;104:17879−17884.

109. Luu YK, Capilla E, Rosen CJ, et al. Mechanical stimulation of mesenchymal stem cell proliferation and differentiation promotes osteogenesis while preventing dietary-induced obesity. *J Bone Miner Res.* 2009;24:50–61.

110. Luu YK, Pessin JE, Judex S, Rubin J, Rubin CT. Mechanical signals as a non-invasive means to influence mesenchymal stem cell fate, promoting bone and suppressing the fat phenotype. *BoneKEy Osteovision.* 2009;6:132–149.

111. Kawaguchi H, Akune T, Yamaguchi M, et al. Distinct effects of PPARgamma insufficiency on bone marrow cells, osteoblasts, and osteoclastic cells. *J Bone Miner Metab.* 2005;23:275–279.

112. Liu J, Wang Y, Pan Q, et al. Wnt/beta-catenin pathway forms a negative feedback loop during TGF-beta1 induced human normal skin fibroblast-to-myofibroblast transition. *J Dermatol Sci.* 2012;65:38–49.

113. Taes YE, Lapauw B, Vanbillemont G, et al. Fat mass is negatively associated with cortical bone size in young healthy male siblings. *J Clin Endocrinol Metab.* 2009;94: 2325–2331.

114. Bonewald LF. Osteocytes: a proposed multifunctional bone cell. *J Musculoskelet Neuronal Interact.* 2002;2: 239–241.

115. Bonewald LF. Osteocytes as dynamic multifunctional cells. *Ann N Y Acad Sci.* 2007;1116:281–290.

116. Bonewald LF, Johnson ML. Osteocytes, mechanosensing and Wnt signaling. *Bone.* 2008;42:606–615.

117. Armstrong VJ, Muzylak M, Sunters A, et al. Wnt/beta-catenin signaling is a component of osteoblastic bone cell early responses to load-bearing and requires estrogen receptor alpha. *J Biol Chem.* 7-13-2007;282: 20715–20727.

118. Robinson JA, Chatterjee-Kishore M, Yaworsky PJ, et al. Wnt/beta-catenin signaling is a normal physiological response to mechanical loading in bone. *J Biol Chem.* 2006;281:31720–31728.

119. Ke HZ, Richards WG, Li X, Ominsky MS. Sclerostin and Dickkopf-1 as therapeutic targets in bone diseases. *Endocr Rev.* 2012;33:747–783.

120. Li X, Liu P, Liu W, et al. Dkk2 has a role in terminal osteoblast differentiation and mineralized matrix formation. *Nat Genet.* 2005;37:945–952.

121. Li X, Zhang Y, Kang H, et al. Sclerostin binds to LRP5/6 and antagonizes canonical Wnt signaling. *J Biol Chem.* 2005;280:19883–19887.

122. Gong Y, Slee RB, Fukai N, et al. LDL receptor-related protein 5 (LRP5) affects bone accrual and eye development. *Cell.* 2001;107:513–523.

123. Robling AG, Turner CH. Mechanotransduction in bone: genetic effects on mechanosensitivity in mice. *Bone.* 2002;31:562–569.

124. Robling AG, Turner CH. Mechanical signaling for bone modeling and remodeling. *Crit Rev Eukaryot Gene Expr.* 2009;19:319–338.

125. Sawakami K, Robling AG, Ai M, et al. The Wnt co-receptor LRP5 is essential for skeletal mechanotransduction but not for the anabolic bone response to parathyroid hormone treatment. *J Biol Chem.* 2006;281:23698–23711.

126. Kiel DP, Hannan MT, Barton BA, et al. Insights from the conduct of a device trial in older persons: low magnitude mechanical stimulation for musculoskeletal health. *Clin Trials.* 2010;7:354–367.

127. Little RD, Carulli JP, Del Mastro RG, et al. A mutation in the LDL receptor-related protein 5 gene results in the autosomal dominant high-bone-mass trait. *Am J Hum Genet.* 2002;70:11–19.

128. Niziolek PJ, Warman ML, Robling AG. Mechanotransduction in bone tissue: the A214V and G171V mutations in Lrp5 enhance load-induced osteogenesis in a surface-selective manner. *Bone.* 2012;51:459–465.

129. Zhong Z, Zeng XL, Ni JH, Huang XF. Comparison of the biological response of osteoblasts after tension and compression. *Eur J Orthod.* 2013;35:59–65.

130. Lisse TS, Chun RF, Rieger S, Adams JS, Hewison M. Vitamin D activation of functionally distinct regulatory miRNAs in primary human osteoblasts. *J Bone Miner Res.* 2013;28:1478–1488.

131. Sengul A, Santisuk R, Xing W, Kesavan C. Systemic administration of an antagomir designed to inhibit miR-92, a regulator of angiogenesis, failed to modulate skeletal anabolic response to mechanical loading. *Physiol Res.* 2013;62:221–226.

132. Lian JB, Stein GS, van Wijnen AJ, et al. MicroRNA control of bone formation and homeostasis. *Nat Rev Endocrinol.* 2012;8:212–227.

133. Luzi E, Marini F, Sala SC, Tognarini I, Galli G, Brandi ML. Osteogenic differentiation of human adipose tissue-derived stem cells is modulated by the miR-26a targeting of the SMAD1 transcription factor. *J Bone Miner Res.* 2008;23:287–295.

134. Kapinas K, Kessler CB, Delany AM. miR-29 suppression of osteonectin in osteoblasts: regulation during differentiation and by canonical Wnt signaling. *J Cell Biochem.* 2009;108:216–224.

135. Mizuno Y, Yagi K, Tokuzawa Y, et al. miR-125b inhibits osteoblastic differentiation by down-regulation of cell proliferation. *Biochem Biophys Res Commun.* 2008;368: 267–272.

136. Zhang JF, Fu WM, He ML, et al. MiRNA-20a promotes osteogenic differentiation of human mesenchymal stem cells by co-regulating BMP signaling. *RNA Biol.* 2011;8: 829–838.

137. Ferreri SL, Talish R, Trandafir T, Qin YX. Mitigation of bone loss with ultrasound induced dynamic mechanical signals in an OVX induced rat model of osteopenia. *Bone.* 2011;48:1095–1102.

138. Hu M, Cheng J, Qin YX. Dynamic hydraulic flow stimulation on mitigation of trabecular bone loss in a rat functional disuse model. *Bone.* 2012;51(4):819–825.

139. Klein-Nulend J, Bacabac RG, Mullender MG. Mechanobiology of bone tissue. *Pathol Biol.* 2005;53:576–580.

140. Sikavitsas VI, Temenoff JS, Mikos AG. Biomaterials and bone mechanotransduction. *Biomaterials.* 2001;22:2581–2593.

141. Uddin SMZ, Cheng J, Lin W, Qin YX. Low-intensity amplitude modulated ultrasound increases osteoblastic mineralization. *Cell Mol Bioeng.* 2011;4:81–90.

142. Qin YX, Lam H, Ferreri S, Rubin C. Dynamic skeletal muscle stimulation and its potential in bone adaptation. *J Musculoskelet Neuronal Interact.* 2010;10:12−24.

143. Hu M, Tian GW, Gibbons DE, Jiao J, Qin YX. Dynamic fluid flow induced mechanobiological modulation of in situ osteocyte calcium oscillations. *Arch Biochem Biophys.* 2015;579:55−61.

144. Datta N, Pham QP, Sharma U, Sikavitsas VI, Jansen JA, Mikos AG. In vitro generated extracellular matrix and fluid shear stress synergistically enhance 3D osteoblastic differentiation. *Proc Natl Acad Sci USA.* 2006;103:2488−2493.

145. Mauney JR, Sjostorm S, Blumberg J, et al. Mechanical stimulation promotes osteogenic differentiation of human bone marrow stromal cells on 3-D partially demineralized bone scaffolds in vitro. *Calcif Tissue Int.* 2004;74:458−468.

146. Milan JL, Planell JA, Lacroix D. Computational modelling of the mechanical environment of osteogenesis within a polylactic acid-calcium phosphate glass scaffold. *Biomaterials.* 2009;30:4219−4226.

147. Pioletti DP. Biomechanics and tissue engineering. *Osteoporos Int.* 2011;22:2027−2031.

148. Caulkins C, Ebramzadeh E, Winet H. Skeletal muscle contractions uncoupled from gravitational loading directly increase cortical bone blood flow rates in vivo. *J Orthop Res.* 2007;25(6):732−740.

149. Garman R, Gaudette G, Donahue LR, Rubin C, Judex S. Low-level accelerations applied in the absence of weight bearing can enhance trabecular bone formation. *J Orthop Res.* 2007;25:732−740.

150. Qin YX, Kaplan T, Saldanha A, Rubin C. Fluid pressure gradients, arising from oscillations in intramedullary pressure, are correlated with the formation of bone and inhibition of intracortical porosity. *J Biomech.* 2003;36:1427−1437.

151. Zerath E, Canon F, Guezennec CY, Holy X, Renault S, Andre C. Electrical stimulation of leg muscles increases tibial trabecular bone formation in unloaded rats. *J Appl Physiol.* 1995;79:1889−1894.

152. Qin YX, Lin W, Rubin CT. Load-induced bone fluid flow pathway as defined by in-vivo intramedullary pressure and streaming potentials measurements. *Ann Biomed Eng.* 2002;30:693−702.

153. Xie L, Jacobson JM, Choi ES, et al. Low-level mechanical vibrations can influence bone resorption and bone formation in the growing skeleton. *Bone.* 2006;39:1059−1066.

154. Wang L, Ciani C, Doty SB, Fritton SP. Delineating bone's interstitial fluid pathway in vivo. *Bone.* 2004;34:499−509.

155. Wang L, Fritton SP, Cowin SC, Weinbaum S. Fluid pressure relaxation depends upon osteonal microstructure: modeling an oscillatory bending experiment. *J Biomech.* 1999;32:663−672.

156. Wang L, Fritton SP, Weinbaum S, Cowin SC. On bone adaptation due to venous stasis. *J Biomech.* 2003;36:1439−1451.

157. Stevens HY, Meays DR, Frangos JA. Pressure gradients and transport in the murine femur upon hindlimb suspension. *Bone.* 2006;39:565−572.

158. Valic Z, Buckwalter JB, Clifford PS. Muscle blood flow response to contraction: influence of venous pressure. *J Appl Physiol.* 2005;98:72−76.

159. Laughlin MH, Joyner M. Closer to the edge? Contractions, pressures, waterfalls and blood flow to contracting skeletal muscle. *J Appl Physiol.* 2003;94:3−5.

160. Kim SJ, Roy RR, Zhong H, et al. Electromechanical stimulation ameliorates inactivity-induced adaptations in the medial gastrocnemius of adult rats. *J Appl Physiol.* 2007;103:195−205.

161. Qin L, Appell HJ, Chan KM, Maffulli N. Electrical stimulation prevents immobilization atrophy in skeletal muscle of rabbits. *Arch Phys Med Rehabil.* 1997;78:512−517.

162. O'Gara T, Urban W, Polishchuk D, Pierre-Louis A, Stewart M. Continuous stimulation of transected distal nerves fails to prolong action potential propagation. *Clin Orthop Relat Res.* 2006;447:209−213.

CHAPTER 3.1

# Mechanobiology in Soft Tissue Engineering

MICHAEL T.K. BRAMSON • SARAH K. VAN HOUTEN • DAVID T. CORR

## 1. INTRODUCTION

Tissue engineering (TE) is a form of regenerative medicine that utilizes cells, engineering materials, and suitable biochemical factors to repair, improve, or replace biological tissues.[1] Beyond engineering healthy or developing tissue, some TE approaches seek to create models of tissue disease, dysfunction, or injury to enable mechanistic in vitro exploration.[2] Creating scalable tissue-engineered constructs that replicate the structure and function of musculoskeletal soft tissues (e.g., cartilage, ligament, tendon, skeletal muscle) is a challenging task; however, improvements in engineering approaches, specifically the addition of mechanical cues, have led to the development of more biomimetic constructs.[3-5] The central role of mechanobiology in the development, repair, and maintenance of human soft tissues has inspired the integration of various types of mechanical stimulation (e.g., compression, tension, shear) into current soft TE strategies. Indeed, tissue engineers frequently utilize custom bioreactors to deliver tissue-specific mechanical cues in vitro to elicit an appropriate mechanobiological response in their engineered constructs.[6] Administration of mechanical loading, in a controlled and reproducible manner, allows exploration of fundamental aspects of mechanobiology and its role in functional TE. Tissue engineers can leverage this mechanobiological response to accelerate construct maturation, tune the structure and biomechanical function of engineered soft tissues, or probe the role of mechanical cues in tissue development, disease, or dysfunction.

The vast majority of TE approaches utilize a scaffold, either natural or synthetic, to serve as a provisional niche for a population of cells.[7] Scaffold-based TE approaches have been widely used over the past several decades and remain the predominant TE approach for musculoskeletal soft tissues.[8] One of the great appeals of this approach is that the scaffold provides initial structure to the construct, wherein seeded cells can proliferate and deposit their own extracellular matrix (ECM). Owing to the many advances in bioprinting and biofabrication, a wide variety of scaffolds can be created, and their geometry and mechanical properties can be controlled during scaffold fabrication.[9-11] Scaffolds can be designed to be permanent or semipermanent, such that they will not degrade considerably in vivo and will provide a long-lasting structure to the construct, which may contribute to its mechanical function. Alternatively, biodegradable scaffolds can be created to provide an initial scaffold structure, which is remodeled or degraded with time as the cells replace it with their own ECM. The bulk material properties of scaffolds, such as elastic modulus and permeability/porosity, can be tuned through aspects of the scaffold network design, including pore size, as well as the size, topography, and alignment of the scaffold fibers.[12] This grants great flexibility to the properties and shapes of scaffolds that can be created for musculoskeletal soft tissues. However, great care must be taken in scaffold design, such that the scaffold is sufficiently strong to provide structure and stability to the developing construct yet is not so stiff that it stress-shields the cells, preventing them from receiving appropriate mechanical cues to produce matrix. This concern extends to biodegradable scaffolds, in which the rate of matrix deposition must balance the degradation rate of the scaffold. Thus, mechanical stimulation and mechanobiology feature centrally in permanent, semipermanent, and biodegradable scaffolds.

Despite the design flexibility and many other benefits offered by scaffolds, there are some concerns

Mechanobiology. https://doi.org/10.1016/B978-0-12-817931-4.00008-X

associated with their use, such as immune rejection[13] and scaffold degradation rate.[14] These limitations have led some to explore scaffold-free approaches for soft TE, which rely on cellular self-assembly and mechanobiological matrix production to form engineered constructs.[15,16] These scaffold-free approaches avoid the concerns associated with scaffold-based techniques and are sometimes described as more biomimetic because they rely on tissue (re)generation rather than matrix remodeling. However, many scaffold-free approaches are currently limited by technical challenges and costs associated with producing constructs large enough for clinical implementation.

Independent of the method, consideration of the native tissue's mechanical environment (e.g., the loads it experiences in vivo) is crucial to engineering soft tissues, as made evident by the numerous laboratories that incorporate static and dynamic mechanical stimulation in their TE strategies.[17] In this chapter, we will discuss how tissue engineers leverage the current understanding of mechanobiology to engineer musculoskeletal soft tissues, specifically cartilage, tendon, ligament, and skeletal muscle. For each soft tissue, we first discuss the mechanobiology of the native tissue and then explore current TE techniques utilizing mechanobiology to engineer said tissue, including applications of both scaffold-based and scaffold-free approaches. The progression of this chapter follows the evolution of the soft TE field, with increasing magnitudes and/or complexity of loading experienced by these tissues in vivo. We first explore the native mechanobiology and its incorporation into TE approaches of cartilage, a passive tissue that experiences relatively low compressive strains but high stresses, which can lead to high hydrostatic pressures and fluid flow. Then we explore these topics in tendons and ligaments, which are also passive tissues but experience tensile strains of much higher magnitude. Finally, we discuss the mechanobiology and TE approaches of skeletal muscle, a tissue that experiences both external loadings and self-generated contractile forces with activation, making it more challenging to study and to replicate in vitro. Having explored the role of mechanobiology in musculoskeletal soft TE, we then summarize our findings and discuss future directions that may enable increased biomimicry and clinical translation of engineered soft tissue constructs.

## 2. CARTILAGE

### 2.1. Cartilage Mechanobiology Review

Cartilage is a firm, tough, collagen-based connective tissue that is softer and more compliant than bone but is significantly stiffer than other soft tissues (e.g., tendon,

ligament, skin, muscle). There are three different types of cartilage found in the human body: fibro, elastic, and hyaline cartilage. In this review, we will focus on hyaline cartilage, located in the articular joints, as it is the most susceptible to loading and, thus, is highly influenced by mechanobiology. Joints in the human body are exposed to complex loading patterns in many activities of daily life, such as the static and dynamic compressive loading experienced during locomotion. Chondrocytes, the cells native to cartilage tissue, have cilia that serve as mechanical sensors (see Chapter 2.2), which play a central role in cartilage mechanobiology; in response to mechanical loading, chondrocytes deposit and remodel the ECM.[18] Dynamic compressive loading has been shown to activate and differentiate chondrocytes to aid in the maintenance, regeneration, and repair of cartilage in the joint,[18] whereas static compressive loading suppressed metabolic activity in cartilage explants.[19] Additionally, moderate exercise in young animals showed great benefits, leading to an increase in cartilage angiogenesis and matrix volume, as well as improved biomechanical properties.[19]

Much of cartilage development, maintenance, and repair relies on the chondrogenic differentiation of mesenchymal stem cells (MSCs). The mechanical forces that primarily affect MSC chondrogenic differentiation are hydrostatic pressure and cyclic compression.[20] Hydrostatic pressure can be further divided into compressive stress maintaining hyaline cartilage and tensile stress accelerating growth and ossification.[21] Hyaline cartilage's ability to withstand compressive forces depends on its organization of collagen fibers and the makeup of the ECM, specifically the high percentage of type II collagen ($\sim$95% of the collagen in the ECM) and tissue fluid ($\sim$80% of the tissue weight).[18] Hyaline cartilage is divided into three main regions: the pericellular, territorial, and interterritorial zones (Fig. 3.1.1). Although all three regions contribute to cartilage's strength, the alignment of the interterritorial zone is primarily responsible for the mechanical properties of the tissue.[18] Collagen fiber and ECM alignment are impacted by forces applied to the cells, in a manner similar to bone under Wolff's law. In other words, regions of cartilage adapt to the loads they experience, resulting in a complex, multilayered tissue that allows humans to perform essential activities of daily life. However, these distinct regions also make cartilage difficult to engineer because they exhibit notable structural differences and each region contributes differently to the mechanical properties of the overall tissue.[18] As hydrostatic pressure and cyclic compression are the main

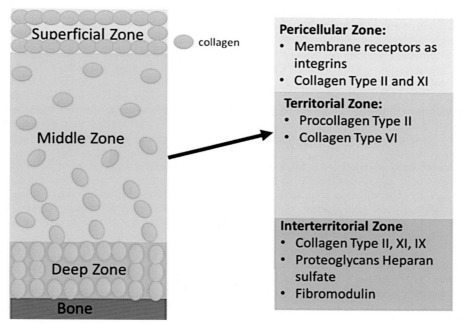

FIGURE 3.1.1 Hyaline cartilage is a complex collagenous tissue composed of three distinct zones that adapt to the loads they experience. This multilayered structure allows cartilage to withstand the complex loading and wear and tear of activities of daily life (e.g., joint articulation). The interterritorial region is primarily responsible for the mechanical properties of the tissue. (Credit: Author original.)

mechanical loadings experienced by hyaline cartilage, many cartilage TE approaches incorporate biomimetic exogenous mechanical stimuli to influence construct formation in vitro.

## 2.2. Engineering Cartilage

Articular cartilage has poor intrinsic healing and regenerative properties because it is avascular, aneural, and lymphatic.[22] As a result, there is demand for new ways to either replace cartilage tissue or improve its intrinsic healing. To accomplish this, many TE laboratories are seeking to leverage hyaline cartilage's native mechanobiology to create cartilage tissue constructs or replacements, while others are engineering cartilage constructs to serve as developmental or disease-state models. Developmental models can inform future cartilage TE strategies by providing insight into how tissues form and how construct maturation can be enhanced or accelerated. Additionally, these models can be used to study dysfunctional development or disease states to gain insight into the source(s) of the impairment/injury and its structural and functional consequences.

One of the most widely studied diseases affecting articular cartilage, with a significant clinical and economic burden, is osteoarthritis (OA). Currently, there is a paucity of in vitro models to study the progression of OA owing, in part, to the challenges in creating constructs with cartilage's three distinct zones. Replicating these zones is critical for a cartilage model in order to study the structural, compositional, and functional progression of OA. Because OA has been associated with changes to joint mechanical loading,[19] it is clear that mechanobiology plays an important role not only in cartilage development and maintenance but also in the progression of this disease. Thus many cartilage tissue engineers incorporate different modes (i.e., shear, compression, combined shear and compression), frequencies, durations, and amplitudes of mechanical loading into their cartilage models. Whether the overall goal is an in vitro model or a replacement tissue, knowledge of cartilage mechanobiology can be used in engineering constructs with desired tissue properties.

Experimentally, the combined application of compressive forces and low levels of shear stress has shown significant benefits, with a sixfold increase in the equilibrium modulus and an increase in oligomeric matrix protein expression in engineered cartilage constructs.[20] However, higher levels of shear stresses and compressive forces increased the production of proteoglycans, indicating a proinflammatory response.[20]

Thus, great care must be taken when incorporating mechanical stimulation into cartilage TE, as different magnitudes of loading may have contrasting effects. To avoid potential detrimental effects associated with excessive loading, many researchers in cartilage TE utilize rotating-wall bioreactors to apply low levels of shear stress to either chondrocytes or cell-loaded TE constructs through the flow of media. In this way, tissue engineers can reap the mechanobiological benefits while avoiding the proinflammatory response seen at higher magnitudes of loading.

In addition to shear stress, compressive loading has been studied extensively for use in cartilage TE, as it is the predominant loading experienced in vivo. Early efforts in this area utilized static compressive loading for mechanical stimulation; however, this had a number of negative effects, which prompted researchers to explore dynamic loading regimens. Dynamic compression increases solute transport and mechanically stimulates the cells. There have been numerous studies that aim to optimize the compressive waveform in terms of frequency, duty cycle, force magnitude, and duration in order to achieve the desired mechanobiological effects.[18−23] Although such studies provided invaluable mechanobiological insight, many were performed using whole tissue explants, and great care must be taken when extrapolating their findings to TE approaches because the developmental environment can be quite different than that of mature tissue. Moving to more developmental approaches, hydrostatic pressure has shown a positive effect on chondrogenic differentiation of MSCs in vivo. Specifically, cyclic hydrostatic pressure decreased collagen I messenger RNA (mRNA) levels, whereas cyclic tension led to an increase in collagen X (associated with mineralizing cartilage), which supports the hypothesis that hydrostatic pressure has a protective effect for the developing cartilage, whereas tension (or shear) encourages growth and ossification.[21] As seen with shear stress, in vitro studies showed that the biological response is sensitive to the magnitude of the applied hydrostatic pressure. Changes in hydrostatic pressure above or below physiologic levels were detrimental to the development of constructs, reducing cell viability and ECM secretion.[21] These discrepancies between in vivo and in vitro studies have not deterred tissue engineers from mechanically stimulating their constructs. Rather, due to the observed sensitivity to loading magnitude, these findings have fueled the development of new methods to deliver mechanical stimulation, of a controlled magnitude, to engineered cartilage constructs. For example, dynamic compression has shown beneficial results in developing tissue-like

constructs, when applied with 10% or lower strain magnitude at a loading frequency $\leq 1$ Hz.[20,23,24]

Utilizing either a scaffold-based or scaffold-free approach, tissue engineers seek to harness the benefits of mechanical loading in cartilage development and repair. As cartilage was one of the first tissues in which the role of mechanobiology in TE was explored, there is a vast body of work in that area. There are a number of detailed review articles that discuss the various methods in which mechanobiology is applied to cartilage TE.[19−22] In the sections that follow, we present some of the general results from using mechanobiology to engineer cartilage, as well as several novel ways of applying mechanical stimulation to cartilage tissue constructs. For a more comprehensive overview of the field, we direct the reader to the aforementioned review articles.

### 2.2.1. Scaffold-based

Cartilage tissue engineers often utilize various biomaterials, such as poly(lactic-co-glycolic acid), polyglycolic acid, collagen gel, polylactic acid, polycaprolactone, or poly(ethylene glycol), to create a scaffold. These materials may be fabricated into scaffolds with distinct, tunable mechanical properties using various common techniques (e.g., electrospinning, casting, bioprinting). Once a scaffold is seeded with cells, mechanical loading is typically applied using some form of bioreactor to stimulate the cells and elicit a mechanobiological response. Various methods can be utilized to apply these mechanical forces in a controlled and reproducible manner (Table 3.1.1).[18] For example, compressive forces may be applied directly to the seeded scaffold, using a piston-like system, to simulate the direct compressive loading experienced by articular joints, and hydrodynamic and shear stresses can be delivered through the fluid (i.e., culture media) in rotating bioreactors.

In the body, articular cartilage experiences both low-frequency and high-frequency compressive forces from voluntary muscle contraction that are required for joint movement.[21−23] Thus, TE approaches have been developed to apply a combination of low- and high-frequency compressive forces in order to better mimic the complex loads experienced by cartilage.[19,21,22] Similarly, constant perfusion of cell culture media can be used to mimic the shear stress experienced by chondrocytes in vivo when synovial fluid is released from cartilage with joint loading. The benefits of such mechanical loading have been realized in numerous prior studies. For example, constant perfusion of developing

**TABLE 3.1.1**
Bioreactors in Cartilage Engineering.

| Type of Bioreactor | Applied Load | Method of Application |
|---|---|---|
| Rotating-wall | Shear stress | Media rotate in a cylindrical bioreactor, applying shear forces to the cells/construct. |
| Direct compression apparatus | Compressive | Piston presses down on the cell culture to apply either static or dynamic loading. |
| Hydrodynamic focusing | Hydrodynamic | Inner and outer walls of cone-shaped bioreactor rotate, producing a wider range of shear forces. |

From Darling EM, Athanasiou KA. Review Articular Cartilage Bioreactors and Bioprocesses. 2003;9(1):9–26.

constructs increased cell proliferation, whereas hydrostatic pressure increased moduli, collagen expression, and proteoglycan production,[21,22] and intermittent compressive forces have been shown to increase sulfated glycosaminoglycan (sGAG) aggregation and matrix formation.[24]

To reap the benefits of mechanical loading, Mauck et al.[25] subjected cell-seeded agarose hydrogels to dynamic unconfined compression loading at physiologic strain levels. The bioreactor design used in their study is similar to the direct compression apparatus mentioned in Table 3.1.1, but it allows for sinusoidal compressive loading. They subjected the constructs to cyclic unconfined compression (10% peak-to-peak strain at 1 Hz loading frequency) delivered in 3 × 1-h on/off bouts, 5 days per week, and found that the dynamically loaded constructs exhibited an increase in peak stress, compression modulus, and sGAG content, as well as increased matrix formation.[25] Despite the improvements in these properties, the constructs' values did not achieve the range of normal cartilage; however, the increases in equilibrium moduli and peak stress show promise for monitoring tissue growth.[25] Bian et al. followed a similar approach to engineer functional cartilage tissue constructs. However, they loaded their agarose hydrogel constructs with a dynamic ±5% compressive strain superposed on a 10% unconfined compression tare strain, using impermeable loading

platens, and following the same 3 hours per day, 5 days per week loading protocol.[26] They also observed an increase in equilibrium modulus, dynamic modulus, glycosaminoglycan (GAG) content, and collagen production, creating constructs that had comparable Young modulus to that seen in native canine knee cartilage.[26] In human-MSC-laden hyaluronic acid gels, Bian and colleagues[27] were able to show that dynamic loading increased collagen and proteoglycan production while suppressing hypertrophy.

As growth factors and other soluble factors are also important for cartilage development and repair, some groups have explored their combination with mechanical loading to determine if there are any mechanical-soluble factor synergistic benefits. It was observed that, when given in combination, growth factors and hydrostatic pressure further improved the mechanical properties of cartilage constructs.[20] Additionally, Hung et al.[28] showed that chondrocyte-seeded constructs, when supplemented with transforming growth factor beta (TGF-β) and subjected to dynamic mechanical loading, displayed increased type II collagen production (Fig. 3.1.2). These findings suggest a cooperative role, such that mechanical loading may help the cells better utilize supplemented growth factors, particularly if the growth factor is part of a mechanosensitive signaling pathway. All the above mentioned methods rely on a scaffold to maintain the initial construct structure and support the cells as the construct is subjected to mechanical loading. Next, we will look at some scaffold-free approaches that utilize the cells' ability to self-assemble and self-organize into tissuelike constructs.

### 2.2.2. Scaffold-free

Scaffold-free methods to engineer cartilage typically utilize a monolayer or pellet culture, which allows the cells to aggregate into different shapes, based on cell-cell interactions. Because the shape of the pellet can be controlled by the shape of the culture boundary, tissue engineers are able to create cartilage constructs in desired, customizable shapes (Fig. 3.1.3).[29] Mechanical loading is typically introduced to these scaffold-free constructs using fluid-based means, such as shear stresses from rotational cultures or perfusion bioreactors. Like their scaffold-based counterparts, these scaffold-free constructs also exhibit pronounced mechanobiological benefits from mechanical stimulation. Cartilage constructs exposed to shear stress through rotation of the culture flask had greater structural stability than those grown in static conditions.[29] Additionally, constructs grown in rotational culture were thicker, with a larger diameter and more highly aligned

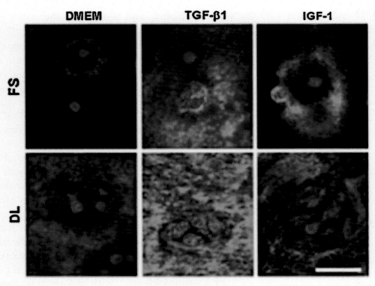

FIG. 3.1.2 Cartilage constructs, stained for type II collagen (green) and cell nuclei (red), showed increased levels of type II collagen when cultured with dynamic loading (DL) over free swelling (FS) culture. DL supplemented with transforming growth factor beta 1 yielded the highest levels of type II collagen, suggesting synergies for mechanical and biochemical stimulation in cartilage tissue engineering. *DMEM*, Dulbecco's modified Eagle's medium; *IGF-1*, insulin-like growth factor 1;*TGF-β1*, transforming growth factor beta 1. (Adapted from Hung CT, Mauck RL, C-b Wang C, Lima EG, Ateshian GA. A paradigm for functional tissue engineering of articular cartilage via applied physiologic deformational loading. *Ann Biomed Eng* 2004;34(1) 35–49, with permission.)

fibers, than those grown in static cultures. Moreover, rotational culture increased proteoglycan secretion, overall protein formation, and type II collagen production. In addition to cartilage-specific type II collagen production, rotational cultures showed no significant type I collagen formation, an isoform typically found in other soft tissues, suggesting this bioreactor specifically promotes cartilage formation. These structural and geometric changes were also reflected in improved mechanical properties; Young's modulus and stiffness of engineered cartilage were significantly higher when exposed to shear stress during culture than when cultured under static conditions. While the rotational flask is the main way that cartilage tissue engineers apply mechanical stimulation to scaffold-free constructs, Gilbert et al.[30] created a novel through-thickness perfusion bioreactor, which they used to increase the thickness and Young's modulus of their scaffold-free engineered cartilage.

## 2.3. Future Work

Having realized the benefits of mechanical stimulation, an essential future effort is to progress toward next-generation bioreactors that are able to provide physiologic mechanical loading in a controlled and multimodal fashion. By replicating the modes of loading (e.g., compression, shear), as well as their magnitudes, rates, and frequencies, tissue engineers can create in vitro environments that more closely recapitulate the mechanical stimuli that cartilage experiences in vivo. Such advances will enable expanded investigations into cartilage mechanobiology in healthy tissue, during development, and following injury or disease. Advances in scaffold design will benefit from the rapidly evolving bioprinting and biofabrication fields, as well as the exploration of new biomaterials, to create scaffolds with desired mechanical, physical, and chemical properties. In addition, future work will likely build on the observed synergistic benefits of combining growth factors with mechanical stimulation and explore the mechanobiological benefits of additional growth factors,[31] and combinations of growth factors,[32] when delivered with a variety of mechanical loading regimens. Such investigations can provide critical insight to inform future cartilage TE efforts, to create constructs that are mechanically similar or superior to healthy cartilage, and to scale-up construct fabrication to create implantable engineered cartilage replacements.

FIG. 3.1.3 Advances in biofabrication of scaffold-free cartilage constructs have allowed for improved shape control (e.g., spade) in cartilage tissue engineering, promoting the development of more biomimetic and clinically translatable constructs. (Adapted from Furukawa KS, Imura K, Tateishi T, Ushida T. Scaffold-free cartilage by rotational culture for tissue engineering. *J Biotechnol.* 2008;133(1): 134–145, with permission.)

## 2.4. Conclusion

Cartilage TE continues to advance toward developing constructs to replace injured, diseased, or degraded cartilage, particularly with the incorporation of mechanobiological cues. The design of biomaterial scaffolds not only affects the overall mechanical properties of the construct but also significantly influences the biological response (e.g., cell proliferation, matrix production). Different forms of mechanical stimulation, such as fluid-induced shear and direct static/dynamic compression, are applied to seeded scaffolds to reap the in vivo benefits of mechanical loading. While some of the mechanical forces experienced in vivo surprisingly had negative effects when delivered to developing scaffold constructs in vitro, others have shown tremendous mechanobiological promise for cartilage TE. Biomimetic mechanical stimulation has been shown to increase Young's moduli, protein production, fiber alignment, and type II collagen production in engineered cartilage. Tissue engineers are also developing new bioreactors to deliver mechanical stimuli in a variety of loading modes, which can be combined with growth factor delivery to capture their apparent synergistic benefits.

## 3. TENDON AND LIGAMENT

### 3.1. Tendon and Ligament Mechanobiology Review

Tendons and ligaments are highly collagenous hierarchical structures that withstand a wide range of forces and loading patterns in the human body. Generally, tendons connect muscle to bone, whereas ligaments connect bone to bone, and both are essential to normal biomechanical function and activities of daily life. Tendons transduce forces generated by skeletal muscle to bone to enable locomotion, while both tendons and ligaments provide passive joint stability.[33] Collagen, primarily types I and III, is the primary contributor to biomechanical function, making up more than 70% of the dry weight of both tissues.[34] Structurally, collagen fibrils run longitudinally along the tissue, grouped together into collagen fibers that similarly align to the axis of loading (i.e., the long axis of the tissue). At the next structural level, fibers are bundled together with sparse cells, tenocytes (tendon) or ligamentocytes (ligament), to form fascicles. Finally, a number of individually wrapped fascicles are bundled together and wrapped in a connective tissue sheath to form the whole tissue (Fig. 3.1.4).[35]

Despite the rather low number of cells in tendon and ligament, mechanobiology is an important consideration in TE replacements because of the prominent role of mechanical loading in matrix production and remodeling during tissue maturation and maintenance.[36] These tissues exhibit viscoelastic behavior, namely, strain-rate sensitivity and a strain-stiffening response at low loads. This distinct biomechanical profile is due largely to the highly aligned collagen fibers in these tissues.[37] However, ECM would not be present without the cells themselves. Tenocytes and ligamentocytes are the resident cells in tendon and ligament, respectively, which produce collagen and other components of the ECM (e.g., elastin, proteoglycans). These tissue-specific fibroblasts increase matrix production and remodel the existing matrix in response to mechanical loading.[38] In addition to mechanobiological benefits to cells, mechanical loading affects the deposited ECM; collagen molecules in low tension are preferentially cleaved, contributing to rapid production of aligned ECM that can withstand the loads experienced by tendon or ligament.[39] Thus, tendons and ligaments are dynamic tissues that adapt to the loads they experience, which is an important consideration in TE.[40] Experimentally, immobilization resulted in severe detrimental effects on tendon function and biomechanical recovery in a rat model,[41] and similar results have been found in computational models of human tendon and ligament.[40]

Mechanical loading also plays an important role in development and repair.[42] Cyclic loading, seen as early as embryonic development, promotes matrix production by fibroblasts and maturation of mechanical

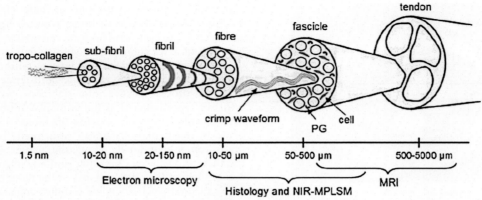

**FIG. 3.1.4** The structure of tendon, an aligned hierarchic collagen matrix with sparse cellularity, allows it to withstand the repeated high loads required for activities of daily life. Ligaments (not shown) typically display a similar highly organized collagen structure with limited cellularity. *MRI*, magnetic resonance imaging; *NIR-MPLSM*, near-infrared multiphoton laser-scanning microscope; *PG*, proteoglycan. (Adapted from Harvey AK, Thompson MS, Cochlin LE, Raju PA, Cui Z. Functional imaging of tendon. *Ann BMVA*. 2009;2009(8):1–11, with permission.)

properties.[43] In developing chick tendons, loading was shown to activate fibroblast growth factor and TGF-β signaling pathways, thereby upregulating tendon-specific gene expression (e.g., tenomodulin, scleraxis) and increasing type I collagen production, whereas immobilization decreased expression of these genes, resulting in smaller limbs and reduced type I collagen production.[44] Similar benefits have been observed in the healing response in adult tissues, where chicken tendons that were allowed constrained motion following surgical repair showed improved strength over those that were fully immobilized.[45] Although mechanical stimulation is crucial in tendon/ligament development, maintenance, and repair, overuse or overstimulation can contribute to inflammation and eventual failure of the tissue.[46] Thus, a healthy functional tendon or ligament requires ample, but not excessive, mechanical loading.

The strains and loading in tendon and ligament can also vary according to the location in the body and the primary function of the tissue. While these tissues are typically loaded in uniaxial tension, some tendons and ligaments are subjected to combined tension and torsion, and others experience significant compressive loading.[47] Loading is associated with unique gene expression profiles in tendon and ligament,[48] which may result in proinflammatory phenotypes with overuse or promote biomechanical development with moderate loading protocols.[49] The introduction of mechanical stimulation emulating in vivo loading patterns to TE approaches has significantly advanced

tendon and ligament engineering, with promise for future clinical translation.

## 3.2. Engineering Tendon and Ligament

Intrinsic healing of tendon and ligament is limited by poor vascularity and sparse cellularity. Although these traits are important for the tissues' high tensile strength, they prevent regeneration of the tissue following injury and instead allow for formation of biomechanically inferior scar tissue.[50] Severe tendon and ligament injuries are currently treated with autografts or allografts, which are limited by donor site morbidity and immune rejection, respectively. In addition, these grafts typically fail to recapitulate the biomechanical properties of the native tissue and are prone to rerupture.[33] As pioneered in articular cartilage, many researchers have turned to mechanobiology in tendon and ligament engineering to capture the important role of loading seen in the development and healthy maintenance of these mechanically essential tissues. The ultimate clinical goal of tendon and ligament TE is generating patient-specific constructs that can replicate the biomechanical properties of native tissues and integrate well with the injured host. Tendon and ligament typically experience periods of low-amplitude cyclic strain in normal activity, alternating with periods of rest. Although they are predominantly loaded in tension, tendons and ligaments may experience transverse and rotational forces, as well as direct compression.[34] TE methods often try to mimic these activities to recapitulate physiologic loading that the cells experience and elicit the desired mechanobiological response (e.g., matrix production and remodeling). For example, isolated human hamstring tenocytes subjected to periodic in vitro mechanical stimulation showed higher levels of collagen type I mRNA and procollagen proteins, when compared with continuously stretched cells.[51] In addition, compressive loads play a role in developing and maintaining fibrocartilaginous transition zones joining tendon/ligament and bone,[52] adding to the already complex mechanobiological profile of tendon and ligament and driving more complex TE strategies.

These tissues also exhibit region-specific properties; structural and functional heterogeneity must be considered when engineering their replacements.[53] For example, the tendon-bone junction, or enthesis, consists of graded transitions from tendon to bone with distinct biomechanical and morphologic properties.[54] These biomechanical heterogeneities at disparate tissue interfaces (e.g., ligament-bone entheses, tendon-bone entheses, and myotendinous junctions) play an important role in establishing the distinct mechanical strains that cells will experience in different regions of the tissue, which, in turn, helps establish the appropriate mechanobiological cues throughout the tissue for a given loading protocol or activity. As the importance of mechanical loading in the development, maintenance, and repair of tendon and ligament has gained notice, much like in the cartilage field, more research groups have incorporated mechanical stimulation in their TE approaches. Although the vast majority of tendon and ligament engineering is performed using scaffold-based platforms, there are a small, but growing, number of scaffold-free approaches being explored that leverage the ability of cells to self-assemble into functional constructs.

### 3.2.1. Scaffold-based

Existing TE strategies for tendon and ligament are largely scaffold-based, relying on decellularized structures, polymers, and/or gels to provide environmental cues and support for the developing constructs. Scaffolds also provide an initial supportive structure for the developing constructs to which mechanical loads can be applied, providing mechanical stimulation to the seeded cells. A suitable scaffold for tendon or ligament engineering must be biocompatible, and possibly biodegradable, such that it can provide mechanical support and promote tissue formation while also withstanding loading.[55] Some early work in these tissues studied the formation of constructs under static tension. Tenocytes isolated from embryonic chick metatarsal tendon (ECMT) and cultured in fixed-length fibrin gels formed constructs that generated significant static tension. The resulting engineered constructs exhibited viscoelastic behavior characteristic of tendon and formed de novo collagen matrix that, along with cells, aligned to the axis of static tension and contributed to enhanced biomechanical properties, when compared with immature constructs (Fig. 3.1.5).[56] This experiment illustrates the mechanobiological benefits of simple static strain on scaffold-based constructs. Similar results were seen with mature human tendon fibroblasts seeded in fixed-length fibrin gels, where cell-generated tissue tension was observed, and improvements in mechanical properties and matrix alignment were seen with construct maturation.[57] Moreover, disruption of tensile strain (i.e., tension release or unloading) resulted in matrix remodeling and an inflammatory phenotype in this engineered human tendon.[58] Also of note, quasi-static loading (slow, constant stretching) of ECMT-derived constructs produced larger increases in ultimate tensile strength (UTS) and modulus than unloaded constructs,[59] showcasing the importance of the multiple types of mechanical stimulation that tendons and ligaments experience in vivo. Although these studies have established the roles of static tension and slow (quasi-static) stretching in tendon (and ligament) engineering, they do not incorporate the complex dynamic loading these tissues experience in vivo.

In one of the pioneering studies of dynamic loading in tendon and ligament engineering, Altman et al.[60] investigated the benefits of applying combined tensile-compressive and torsional loading to cell-seeded collagen gels. Application of fairly slow cycles ($\sim$1 cycle/min) of 10% tensile strain and 25% rotational strain to bone-marrow-derived stem cells (BMSCs) seeded on collagen type I gels resulted in the formation of ligament-like tissues with aligned cells and collagenous matrix.[60] No exogenous factors were used in their cultures, highlighting the critical role of complex mechanical stimulation in engineering both tendon and ligament. As part of the functional TE paradigm,[1] Butler et al. explored a more physiologically mimetic mechanical loading protocol, where cell-gel-sponge constructs (i.e., cell-seeded collagen gels applied to collagen sponges) were subjected to dynamic loading after a 2-day acclimation in an incubator. Their loading protocol utilized tensile strain pulses (1-second duration, 2.4% strain amplitude) applied every 5 minutes for 8 hours a day, mimicking periods of activity. This dynamic loading significantly improved both the material and structural properties of the engineered constructs,[4] and the combination of cells, gel, sponge,

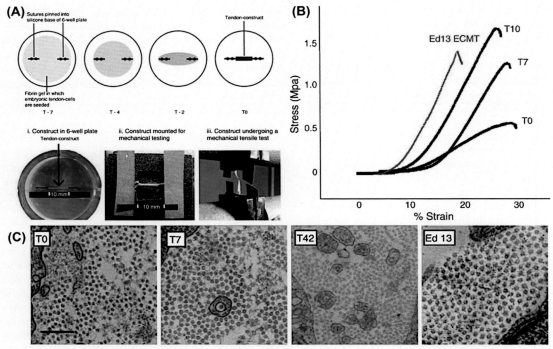

FIG. 3.1.5 A fibrin gel seeded with isolated tenocytes contracts around anchor points to form a construct under static tension. **(A)** Construct tensile mechanical characterization (i–iii) revealed maturation of mechanical properties with time in culture (T0–T10, T42) **(B)**, as well as the formation of an aligned collagenous matrix, similar to what is seen in vivo (Ed13) **(C)**. *ECMT*, embryonic chick metatarsal tendon. (Adapted from Chen JL, Yin Z, Shen WL, et al. Efficacy of hESC-MSCs in knitted silk-collagen scaffold for tendon tissue engineering and their roles. *Biomaterials*. 2010;31(36):9438–9451, with permission.)

and dynamic loading yielded their largest significant improvement in construct biomechanical properties. Although all approaches mentioned thus far have utilized fibrin or collagen gels, other scaffold materials have shown promise for promoting tendon and ligament development.

New complex polymeric scaffolds have been produced, at increasing rates, thanks to advances in fabrication techniques such as knitting/braiding, three-dimensional (3D) bioprinting, and electrospinning.[11] These fabrication techniques grant a great deal of flexibility in scaffold design and allow for a wide range of new materials to be explored for tendon and ligament scaffolds. For example, Chen et al.[55] produced bioengineered tendons by seeding human embryonic stem-cell-derived MSCs into knitted silk-collagen scaffolds and subjecting these constructs to dynamic mechanical strain. After mechanical loading (10% strain, at either 1 or 0.1 Hz for 2 hours a day), the constructs were implanted in mice. Histologically, the implanted

constructs showed similar ultrastructure to native tendon, which was also reflected in improved biomechanical properties, as compared to unloaded controls.

In addition to synthetic materials, some TE approaches have utilized decellularized tissues for their scaffolds, such as porcine small intestinal submucosa (SIS) or decellularized tendon. For example, SIS scaffolds seeded with 3T3 mouse fibroblasts and subjected to various loading regimens showed distinct gene expression profiles that suggested transition to a tendon-like or ligament-like phenotype.[61] In another application, tenocytes isolated from canines were seeded on SIS scaffolds, which were rolled into cellular tubes and mounted into custom loading frames in an incubator, to study the influence of static and dynamic mechanical stimulation.[62] After 16 days of static (1% –2% strain) or dynamic (30 minutes of 9% cyclic strain, alternating with 8-hour rest periods) stimulation, biomechanical evaluation of the resulting constructs revealed a crucial role for dynamic loading—significant improvements in biomechanical properties were seen when cells were incorporated in the scaffold and strain was transitioned from static to dynamic loading. In a similar study, decellularized rabbit tendons seeded with rabbit tenocytes showed a significantly increased modulus and UTS in response to 5 days of cyclic mechanical strain (1.25 N at 1 cycle/minute), as compared to unloaded controls.[63] These studies, and many similar ones, reinforce the importance of cyclic mechanical stimulation in scaffold-based tendon and ligament engineering.

The Flexcell tensile bioreactor system marked a significant advancement in mechanical conditioning of cells and TE constructs and has been explored for use in various TE approaches.[64] Application of a vacuum to deformable silicone membranes, in a six-well plate, enables delivery of controlled, reproducible mechanical stimulation as constructs develop. Flexcell systems have been used conventionally for two-dimensional monolayers, wherein cells typically align to the axis of loading. The use of a "Tissue Train" mold allows for generation of a 3D gel with embedded cells. In an early application of this technology, Al Banes' group engineered bioartificial tendons (BATs) by applying cyclic uniaxial strain to avian tendon fibroblasts embedded in collagen gels. The developing constructs were mechanically stimulated for 1 hour a day (1% strain at 1 Hz), and the BATs that received cyclic loading exhibited a cellular epitenon-like layer, an upregulation of tendon-specific collagen gene expression, and a nearly threefold increase in UTS,[65] illustrating that cyclic strain provides mechanobiological benefits that can be reflected in improved construct composition, structure, and biomechanical function. In a similar study, BMSCs were suspended in a collagen gel in the Flexcell "Tissue Train" mold and either statically or cyclically loaded. Static tension was self-generated by the construct, and cyclic strain (1% amplitude at 1 Hz) was delivered for 30 minutes a day for 7 days. While static conditions resulted in increased matrix production and upregulation of tenogenic markers, significant increases, particularly in collagen production and scleraxis expression, were also observed in the cyclically loaded group.[66] Matrix metalloproteinase profiling also suggested active remodeling of the matrix. Together, these studies show the Flexcell tensile bioreactor's ability to deliver controlled mechanical stimulation to the developing constructs and suggest its utility for future studies assessing mechanobiological effects in tendon, ligament, and other soft tissues.

The scaffold-based approaches reviewed here reveal the importance of mechanobiology in matrix synthesis and organization, gene expression, and construct biomechanical properties, which shows great promise for functional tendon and ligament TE.

### 3.2.2. Scaffold-free

Scaffold-free approaches to engineer tendon and ligament are fairly new and somewhat limited, and approaches incorporating mechanical stimulation are even fewer, due in part to the technical difficulty in applying relevant tensile loading to a construct without a provisional structure in place. These approaches rely on the ability of cells to self-assemble and produce their own matrix to create a functional construct, with no significant initial matrix present. The dominant means of scaffold-free tendon and ligament engineering utilizes self-assembled cellular sheets, which are then rolled into a tubular construct. Utilizing this method, tendon-derived stem cells isolated from rats have been cultured as sheets and rolled into tubes, which formed neotendon when implanted in a nude mouse model.[67] Similarly, periodontal ligament cell sheets have been shown to generate ligament-like tissue that has been histologically and biochemically characterized, showing promising results for periodontal ligament regeneration.[68] Histologic and gene expression results were quite compelling; however, there was no functional (biomechanical) component to these early scaffold-free studies. The Larkin group has used this cell sheet approach to grow self-organized tendons[69] and ligaments[70] that form around two metal loading pins in a Petri dish coated with laminin. The developing constructs experience self-generated tension during their formation and compaction phases of construct development. Engineered constructs were then rigorously assessed for structure (histology) and function (biomechanics), revealing the formation of enthesis-like structures at the tendon-bone and ligament-bone interfaces, and tissue-like biomechanical properties after implantation. While these engineered tissues self-assembled under a state of self-generated tension, dynamic loading was not applied during construct development.[69]

Our group has developed a platform for engineering scaffold-free tendon constructs, at the single-fiber scale, relying on self-assembly of cells in a fibronectin-coated agarose gel growth channel.[71] In this approach, cells self-assemble to form fibers that integrate into collagen-sponge disk anchors, at either end of the growth channel, such that the single fiber spans the full length of the growth channel (disk-to-disk) with no lateral adhesion. To enable dynamic loading, the entire cell-seeded growth channel assembly can be coupled to a modified Flexcell system, via vertical loading pins, through the collagen disk anchors

(Fig. 3.1.6).[72] In this way, precise, controlled mechanical stimulation can be delivered to the scaffold-free tendon fiber as it develops, in a controlled and reproducible manner. This ability to dynamically load engineered tendon/ligament fibers during development enables new lines of inquiry into the mechanobiology of fiber formation, development, and maturation. Using this approach, low-amplitude sinusoidal strain (0%—0.7%) was delivered in 8-hour intervals, with alternating 8-hour blocks of rest, to mimic periods of activity seen in embryonic chick tendons in ovo.[73] Loaded fibers exhibited extended longevity, improved biomechanical properties, upregulated tendon-specific gene expression, and highly aligned, compacted ECM with altered nuclear morphology, when compared with unloaded (self-generated static load) controls (Fig. 3.1.7).[72] These findings suggest the activation of mechanotransduction pathways, leading to increased matrix production and remodeling. Further exploration of loading patterns using this platform can inform other tendon (and ligament) TE strategies.

Whether using a scaffold or not, engineered tendon and ligament significantly benefit from mechanical stimulation, particularly bouts of low-amplitude cyclic strain, alternating with periods of rest. Both approaches enable complex loading of tendon and ligament constructs to elicit the desired mechanobiological response and bring us closer to a viable means of engineering tendon and ligament replacements.

## 3.3. Future Work

The stark results seen here with mechanical loading will continue to be supplemented with additional chemical and environmental stimuli to engineer a truly biomimetic or biomechanically superior tendon/ligament. Specifically, exploration of microenvironmental factors experienced in vivo (e.g., hypoxia and molecular crowding), and incorporation of multiple loading regimens and cell types, will advance TE approaches in tendon/ligament applications. Furthermore, advances in imaging modalities and genetic sequencing technologies will allow for a more in-depth evaluation of engineered

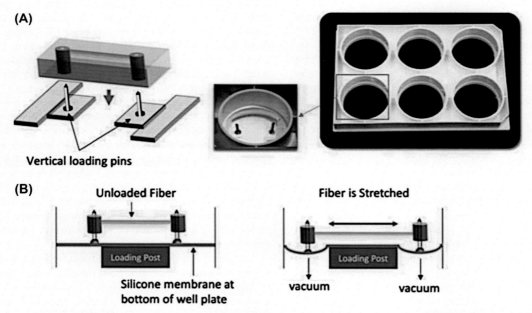

**(A)**

**Vertical loading pins**

**(B)**

**Unloaded Fiber**

Loading Post

**Silicone membrane at bottom of well plate**

**Fiber is Stretched**

Loading Post

**vacuum**          **vacuum**

FIG. 3.1.6 The Flexcell system allows for delivery of precise, controlled, repeatable mechanical stimulation using vacuum pressure and has been applied in both scaffold-based and scaffold-free TE approaches. For scaffold-free tendon fiber engineering, Flexcell plates were modified with vertical loading pins so that agarose growth channels could be coupled to the six individual wells in the system **(A)**. Computer-controlled application of vacuum pressure uniaxially strains the membranes, driving the loading pins apart to stretch the growth channel, and applies controlled dynamic strain to developing scaffold-free tendon fibers, emulating tensile strains seen in tendon development in vivo **(B)**. (Adapted from Mubyana K, Corr DT. Cyclic uniaxial tensile strain enhances the mechanical properties of engineered, scaffold-free tendon fibers. *Tissue Eng Part A.* 2018; 24(23—24):1808—1817, with permission.)

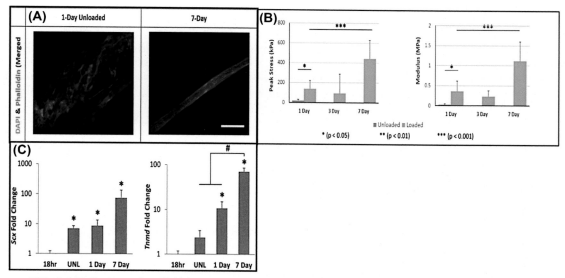

FIG. 3.1.7 Engineered scaffold-free tendon fibers subjected to low-amplitude sinusoidal cyclic strain during development show more tendon-like morphology **(A)** with increasing collagen alignment and compaction, which contributes to improved mechanical properties **(B)** and is correlated with upregulation of tendon-specific genes **(C)**. *DAPI*, 4′,6-diamidino-2-phenylindole. (Adapted from Mubyana K, Corr DT. Cyclic uniaxial tensile strain enhances the mechanical properties of engineered, scaffold-free tendon fibers. *Tissue Eng Part A*. 2018; 24(23—24):1808—1817, with permission.)

constructs. As the field continues to develop, we look forward to the production of a scalable, simplified model for engineering tendon and ligament replacements.

## 3.4. Conclusion

Tissue engineers have come a long way in understanding and utilizing the loading patterns seen in tendon and ligament development, maintenance, and repair. Early static cultures have been replaced with dynamic loading protocols, largely enabled by the development of numerous bioreactors. Although tendon and ligament constructs have formed in static gels, the field is moving toward dynamic loading protocols, mimicking the native environment (e.g., alternating low-amplitude cyclic sinusoidal strain and rest) and emulating observed periods of activity. While scaffold-based approaches are continuing to develop, in part through advancing fabrication techniques, scaffold-free approaches are gaining popularity as a means to study tendon and ligament development and inform future TE.

## 4. SKELETAL MUSCLE

### 4.1. Skeletal Muscle Mechanobiology Review

Skeletal muscles are complex soft tissues because of their ability to actively generate force when stimulated.

Similar to the hierarchical structure of tendons and ligaments (Fig. 3.1.4), skeletal muscles have a structural hierarchy in which the sarcomere serves as the basic repeating unit of structure. Each sarcomere is composed of two major myofilaments: thick myosin filaments within a lattice of thin actin filaments. These filaments are arranged in a highly organized structure, with alternating light and dark bands, giving the skeletal muscle its characteristic striated appearance. When activated, the interaction of actin and myosin crossbridges allows muscle to actively shorten and generate force. This force production is affected by a number of factors, including the myosin heavy chain (MHC) isoforms, which help define the contraction dynamics of the muscle and differentiate fast-twitch from slow-twitch skeletal muscles.

While the mechanisms of contraction are conserved among all skeletal muscles, the contractile and passive properties are heavily influenced by the composition of the ECM.[42] In general, the ECM contains different types of collagen, large glycoproteins, and proteoglycans with large GAG side chains.[42,74] Changes in ECM structure or mechanics will alter the ECM's ability to resist traction forces exerted by cells on integrin molecules, which in turn produces changes in the cytoskeleton and modulates the activity of transduction pathways. However, it is the variety of collagens and connective tissue types that can change muscle function (e.g., contractile properties and force transmission). This is because collagen is the primary organizational component of muscle, and mutations of different collagen levels can lead to muscle dystrophies and other musculoskeletal defects.[42] Thus, it is clear that mechanobiology plays a key role in normal muscle development, and many muscle diseases or dysfunctions can be associated with aberrant mechanical signaling.

During development, skeletal muscle is typically subjected to a combination of passive and active tension, addressing different developmental challenges.[74] Both passive and active tensions accumulate throughout muscle development, especially during elongation of the tissue. These tensions also play a role in integrin adhesion complex turnover and stabilization of clustered integrin complexes at the myotendinous junction.[42,75] Passive tension is primarily thought to maintain myofibrils at an optimal contractile length.[75] The different types of mechanical tensions experienced throughout development (i.e., passive and active tension) influence the passive mechanical properties of skeletal muscles, and these passive properties greatly influence muscle's active properties.

In addition to mechanical stimulation, the passive and active properties of skeletal muscles are influenced by neural stimulation. Neural stimulation influences muscle in a variety of ways, it: (1) controls cell phenotype, myosin expression, and contractile sarcomere assembly, (2) modulates the switching of fiber type (fiber type plasticity), and (3) induces contractility in differentiated myotubes. During myogenesis, neural impulses initiate Myf5 expression,[75] which plays a role in regulating skeletal muscle differentiation. In myotubes (multinucleated fibers formed by myoblast fusion), electric pulse stimulation (EPS) has been shown to induce metabolic reactions similar to those observed in vivo during exercise.[76] Electric impulses have also been used clinically to increase muscle tone and reduce recovery time for athletes[77]; however, EPS therapy by itself has not demonstrated the ability to increase biomechanical function in whole muscle. Owing to these observed benefits of native (in vivo) neural impulses or in vitro electrical stimulation on myogenesis, many tissue engineers have begun incorporating electrical stimulation into TE strategies to create functional skeletal muscle constructs.

Mechanical and electrical stimulation synergistically drive the elaboration of the contractile and passive properties of skeletal muscles. As myogenesis relies on both these modes of stimulation to ensure proper ECM composition, future artificial ECMs for TE applications should consider the critical role that electromechanical stimulations play in tissue development.[75] Moreover, because the passive mechanical properties of ECM also affect the tissue's contractile properties, effectively utilizing combined electrical and mechanical stimulation in vitro holds great mechanobiological potential for engineering functional skeletal muscle.

## 4.2. Engineering Skeletal Muscle

In the area of skeletal muscle TE, many are trying to either replicate or enhance aspects of tissue development to create functional skeletal muscle replacements or in vitro platforms to study dysfunctions and musculoskeletal diseases. One of the largest demands for skeletal muscle replacements is for the treatment of large muscle defects, such as volumetric muscle loss (VML), in which a significant amount of muscle tissue is lost, typically due to disease or major acute trauma. Current VML treatments use muscle grafts, either autografts or allografts; however, these have proven to be surgically challenging, with low success rates and modest functional recovery. As a result, there is great demand for a truly functional engineered skeletal muscle that can actively generate force, for improved VML treatments, and for effective surgical interventions for myopathies and congenital defects. Tissue engineers seek to create such functional tissue-like constructs by harnessing the mechanobiological advantages of mechanical stimulation and/or electrical stimulation.

Mechanical stimulation promotes myoblast alignment, elongation, proliferation, and fusion into myotubes, resulting in enhanced contractility and tetanic force production in mature skeletal muscles.[42,75] Electrical stimulation has been shown to induce muscle contraction and increase cell proliferation and differentiation, and it can be leveraged to direct fiber type specification, to either a fast-twitch or a slow-twitch fiber. Despite these significant benefits, care must be taken when applying electrical stimulation, as tetanic stimulations applied to a muscle can cause it to exhaust quickly and possibly cause permanent damage.[78]

In addition to stimuli, skeletal muscle TE approaches must consider cell type, as well as the scaffold material (if applying a scaffold-based technique). The choice of cell type is typically driven by the application and/or TE method, as some cell types are more conducive to certain TE approaches. Most researchers utilize some type of muscle progenitor cell (e.g., C2C12), whereas some use a heterogenic culture (e.g., myoblast/fibroblast coculture or minced muscle), and a few others use a more primitive cell source (e.g., stem cells, satellite cells). Because the mechanical properties of the ECM have such a profound influence on skeletal muscle's passive and active (contractile) properties, it is important to consider these properties when designing an appropriate scaffold. Ultimately, by incorporating various modes of stimulation and exploring the impact of cell type and scaffolds, tissue engineers aim to produce functional skeletal muscle constructs that can be utilized either as implants to treat injuries or muscular dystrophies or as idealized in vitro models to study development or disease progression.

### 4.2.1. Scaffold-based methods

As with cartilage, tendon, and ligament, the majority of skeletal muscle TE methods use either a natural or a synthetic scaffold to serve as a provisional structure, and cells are seeded into the scaffold to form tissue-like constructs. As scaffolds are responsible for providing initial structure and support to constructs in scaffold-based methods, candidate scaffold designs are frequently subjected to robust mechanical characterization to assess their potential utility. Skeletal muscle scaffolds are often engineered to mimic the stiffness and key compositional aspects of skeletal muscle ECM. Mechanical properties are typically tuned through the selection of scaffold material(s), as well as the design of the fiber network architecture. Surface patterning can also be used to alter the mechanical compliance of the ECM and to mimic or amplify mechanical cues, to influence cell shape or cytoskeletal tension.[79–81] Beyond scaffold design, mechanical stimulation can be directly provided to the construct using a bioreactor. To accomplish this, cell-seeded scaffolds are often cultured in bioreactors to allow for controlled mechanical stimulation, such as continuous (static) or cyclic (dynamic) tensile strains. Continuous stimulation has been shown to increase ECM formation, whereas cyclic stimulation increases fiber diameter and elasticity.[79,82]

Typical muscle bioreactor designs utilize a linear motor or motors to pull the scaffold in either uniaxial or biaxial tension, delivering mechanical forces and strains similar to what is experienced in vivo. For example, in one type of tensile bioreactor, the motor utilizes pins embedded in a gel to apply tensile strain to the construct (Fig. 3.1.8A), and the frequency, amplitude, duty cycle, and duration of the applied mechanical strain is controlled via a computer. Changing the mechanical strain applied to the scaffold causes a change in the mechanical stimuli transmitted to the cells. Therefore, scaffold-level strains can be prescribed to elicit the desired mechanobiological response. A simpler bioreactor approach does not actively strain

the construct, but it provides static fixation and stability such that the construct is subjected to its self-generated tensile strain. In this approach, an empty scaffold is held in place at each end (Fig. 3.1.8B), and when seeded with muscle cells, the contraction of the muscle cells causes the construct to bow, providing constant static strain. Additionally, some methods utilize two mobile pins to apply mechanical stimulation to cells seeded in a gel (Fig. 3.1.8C). Moving toward transplantation and clinical applications, another approach is to use the body as a bioreactor. In this strategy, seeded scaffolds are implanted, typically after a short period of in vitro culture, and the mechanical stimulation is provided by the body.[79,83] One study showed significant improvements when muscle constructs were also subjected to in vitro mechanical stimulation for 1 week before implantation. Constructs that were mechanically stimulated before implantation showed notable structural and functional improvements at the injury site, as well as the formation of vasculature and neural structures in the construct.[79,83]

In addition to mechanical stimulation, tissue engineers have begun applying electrical stimulation to the developing skeletal muscle constructs, due in part to the significant benefits observed with electrical stimulation in muscle development. In most scaffold-based approaches, cells are embedded into hydrogels (e.g., Matrigel, collagen, gelatin), and electrical stimulation is applied to the developing constructs using platinum electrodes. Alternatively, some researchers have provided electrical stimulation to cells seeded onto electrospun fibers, in order to harness both topological and bioelectric cues.[80,84] Some early work utilized field stimulation, whereas more recent approaches have shifted focus to tuning the stimulation waveform (e.g., amplitude, pulse width, duty cycle) to more closely mimic dynamic neural impulses experienced in vivo. More specifically, alternating, biphasic stimuli have become attractive waveform options because they avoid the buildup of ions at the stimulating electrodes, which can cause a local pH change and cytotoxicity. These electrical stimuli, inspired by impulses from the central nervous system, drive muscle differentiation and contraction. The aforementioned approaches, and most others for skeletal muscle engineering, employ a direct contact method, in which the electrical stimulation is applied directly to the cells in culture or to the scaffold (Fig. 3.1.9). A contactless method of applying electrical stimulation was developed to decrease the chances of electrically damaging constructs during stimulation.[85] Together, these results provide striking evidence supporting the value of incorporating electrical

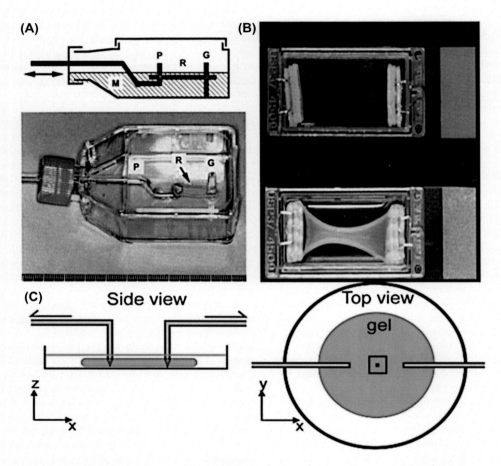

FIG. 3.1.8 Some of the various methods created to apply mechanical stimulation to developing skeletal muscle constructs. In one approach, **(A)** the construct (R) is cultured in media (M) between an immobilized ground pin (G) and straining pin (P) that can be moved by a motor to apply controlled strains to the developing construct. Alternately, in **(B)** a chamber slide model, floatation bars at either end of the slide hold a collagen gel seeded with isolated muscle cells, and muscle cell contraction produces self-generated static strain, resulting in the bowed construct shown in the lower panel. Additionally, some methods pull directly on the construct, such as **(C)** the pin-based system in which two mobile pins are used to pull the gel in opposite directions to apply tensile strains to the developing construct (cell-seeded gel). (Adapted from Qazi TH, Mooney DJ, Pumberger M, Geißler S, Duda GN. Biomaterials based strategies for skeletal muscle tissue engineering: existing technologies and future trends. *Biomaterials*. 2015;53:502–521; Vader D, Kabla A, Weitz D, Mahadevan L. Strain-induced alignment in collagen gels. Langowski J, ed. *PLoS One*. 2009;4(6):e5902, with permission.)

stimulation, in addition to mechanical stimulation, in TE strategies for skeletal muscle.

With the benefits of both mechanical strain and electrical stimulation realized for skeletal muscle TE, some recent efforts have applied both mechanical and electrical stimulation concurrently, to determine if they offer synergistic benefits. For example, Cole et al.[86] developed a bioreactor in which cell-seeded scaffolds could receive combined cyclic strain and single phasic neural cues to simultaneously deliver electrical and mechanical stimulation. Such systems offer great flexibility because key features of both the mechanical and electrical

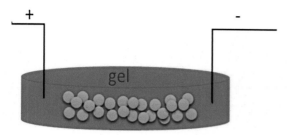

FIG. 3.1.9 Schematic of direct electrical stimulation of cells (orange spheres) in a hydrogel (pink) disk scaffold, in which stimulating electrodes (+ and −) are directly applied to the developing construct. (Credit: Author original.)

stimuli can be controlled (e.g., duty cycle, frequency, voltage, amplitude, duration) to deliver the intended impulses for the appropriate mechanobiological response. Many skeletal muscle tissue engineers anticipate that application of both mechanical and electrical stimuli to scaffold-based constructs will not only promote improved formation and maturation but also control the fiber type composition within constructs, which may prove important because the body is composed of different skeletal muscle fiber types.

### 4.2.2. Scaffold-free methods
Similar to tendon and ligament engineering, scaffold-free skeletal muscle engineering typically utilizes either static or dynamic uniaxial tensile strain to provide mechanical stimulation to developing constructs. Applied strain may incorporate alternating periods of high-frequency and low-frequency strain, as well as periods of rest, to mimic movement patterns seen in animal and human models. While a bioreactor is typically used to apply the mechanical strain, applying mechanical stimulation to a construct without the provisional structure of a scaffold poses unique challenges. The Larkin laboratory addressed this challenge by allowing scaffold-free skeletal muscle constructs to develop in vitro, then implanting the developed constructs in the hind limbs of rats and relying on the animal to provide further in vivo mechanical stimulation.[87] After implantation, the muscle constructs exhibited significant increases in cellularity and MHCs, and a decrease in collagen type I (Fig. 3.1.10)—impressive mechanobiological findings in a preclinical rodent model, which hints toward potential clinical applications.

Mechanical loading can also be applied to the constructs as they develop, in order to mimic key aspects of myogenesis. Our group uses an approach, adapted from our scaffold-free tendon and ligament fiber engineering, to provide controlled cyclic mechanical strain

to developing muscle fibers. As previously discussed, we utilize a modified Flexcell Tissue Train plate to apply sinusoidal tensile strain, at a variety of frequencies and amplitudes (Fig. 3.1.6).[72] In this approach, the entire growth channel assembly is strained, such that the tensile stimuli are delivered to the cells without the need to directly grip the fiber. Thus controlled mechanical stimulation can be applied to the developing fiber, even before any significant structure is present. In this way, mechanical stimuli can prompt cells to self-assemble and deposit their own (native) ECM.

In scaffold-free approaches, electrical stimulation is typically applied directly to the cells utilizing a stimulating electrode and a generated pulse waveform, designed to simulate neural impulses.[72] Our laboratory utilizes a biphasic waveform delivered through the vertical loading pins in our modified Flexcell plates.[88] In this system the vertical loading pins that enable mechanical stimulation of the growth channels also double as stimulating electrodes, and biphasic electrical stimulation directs myoblast polarization and promotes fiber formation. The Larkin group, similar to their mechanical stimulation methods, implants constructs that are predeveloped in vitro to utilize the native tissue's neural stimulation and reap the benefits of electrically stimulating cells.[87] Electrical stimulation has proven beneficial for engineering skeletal muscles, utilizing both scaffold-based and scaffold-free methods.

Similar to scaffold-based methods, some researchers utilize a combination of mechanical and electrical stimulation in their scaffold-free approaches to engineer skeletal muscle-like constructs. Early experiments from the Corr group showed that mechanical strain and biphasic electrical stimulation were each beneficial to fiber formation and development. However, when delivered in combination, their benefit was more than additive, revealing potential electromechanical stimulation synergies.[88] Combined stimulation produced skeletal muscle fibers that notably stiffened when activated with $Ca^{2+}$, likely a result of actin-myosin crossbridging in response to calcium activation. Our ongoing research aims to build from this to provide the appropriate mechanobiological stimuli to engineer fibers that are able to produce a physiologically relevant specific force.

### 4.2.3. Measuring force production
The primary goal of soft TE is to create functional tissue or tissue-like constructs. Consequently, for skeletal muscle, it is critical to assess the active, contractile properties, in addition to the passive properties of the construct. Force production in engineered muscle is

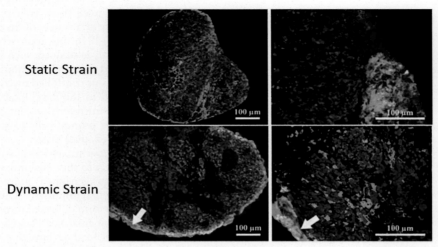

Static Strain

Dynamic Strain

FIG. 3.1.10 Immunostaining of skeletal muscle constructs in vitro (static strain), and the same constructs explanted following 1 week of implantation in a rat model (dynamic strain). Staining for sarcomeric myosin heavy chains (red), collagen I (green), and cell nuclei (blue) revealed that sarcomeric myosin was present at higher levels in explanted constructs, and formation of an epimysium-like outer layer of collagen was also observed in dynamically strained explants (indicated by white arrows). (Adapted from Williams ML, Kostrominova TY, Arruda EM, Larkin LM. Effect of implantation on engineered skeletal muscle constructs. *J Tissue Eng Regen Med.* 2013;7(6):434–442, with permission.)

often measured in response to electrical stimulation[85,87,89] and can be robustly characterized by determining the construct's force-length and force-velocity properties, using common muscle physiology experimental approaches. As an alternative to electrical stimulation, calcium activation can be utilized to test force production.[88] The active and passive mechanical properties can be determined by traditional soft tissue tensile characterization methods performed in an active (stimulated) and passive (nonstimulated) state, respectively. This is typically done by attaching the muscle construct to a force transducer, then stretching the construct to failure to evaluate tensile properties such as Young's modulus, failure strain, toe-in region, and peak stress.[72] While some research groups have been able to engineer skeletal muscle constructs that produce force or calcium-activated changes in stiffness,[85,87–89] the specific force production is lower than that of native skeletal muscle tissues. These findings show great promise and highlight the need for continued improvements in skeletal muscle engineering to achieve truly functional muscle fibers or muscle-like constructs.

### 4.3. Future Work

Although the benefits of mechanical and electrical stimulation for skeletal muscle engineering have been widely appreciated, it is still fairly uncommon for researchers to apply both mechanical and electrical stimuli in combination. However, an increasing number of research groups, having realized their potential synergies, have started to incorporate both mechanical and electrical stimulations into their TE strategies. Some scaffold-based methods have been able to produce functional engineered skeletal muscle constructs[85,87,89]; however, the force production properties of these constructs do not match physiologic force production. Future studies can explore the modes of stimulation necessary to achieve greater contractile function, and the next generation of bioreactors will greatly aid in these efforts.

Additionally, some scaffold-based constructs have been implanted into immunocompromised animal models to study how they will integrate with the native tissue. This is an important step toward clinical translation. However, these constructs may provoke an adverse immune response when implanted in an immunocompetent model, and future studies will need to validate the immune tolerance for any engineered muscle constructs.

Owing to their formation via self-assembly, scaffold-free approaches grant unique insight into the development of muscle, in a manner that mimics key aspects of embryonic myogenesis. When combined with precise delivery of electrical and mechanical stimuli during development, made possible by novel scaffold-free bioreactors, scaffold-free studies can probe the specific contributions of electromechanical stimulation to the evolution of structural hierarchy in muscle construct formation and how it affects the contractile and passive properties. Such exploration can provide unprecedented insight into the role of mechanobiology in the structural and functional properties of engineered skeletal muscles.

## 4.4. Conclusion

Engineering functional skeletal muscle poses unique challenges because, in addition to resisting external forces, the construct must be able to generate contractile forces when activated. Mechanical loading, neural impulses, and ECM composition play a key role in the contractile and passive properties of a skeletal muscle. Mechanical loading directs organization, proliferation, and elongation of cells and promotes matrix deposition and remodeling. Neural impulses control aspects such as cell phenotype, myosin expression, and contractile sarcomere assembly. Tissue engineers seek to develop functional skeletal muscle constructs that can be used as models of disease or development or as functional implants to treat large injuries or myopathies. The majority of skeletal muscle constructs are created using a scaffold as a provisional structure; however, there are also scaffold-free approaches that utilize cellular self-assembly to engineer skeletal muscles. In general, mechanical and/or electrical stimulation results in structural, compositional, and functional benefits to these constructs, highlighting the important role of mechanobiology in skeletal muscle engineering. Gaining a more complete understanding of the mechanobiology behind skeletal muscle development and functional maturation can inform future TE approaches and ultimately lead to truly functional skeletal muscle constructs and engineered muscle replacements.

## 5. CONCLUSION

Musculoskeletal soft TE is a complex discipline seeking to create soft tissue constructs (e.g., cartilage, tendon, ligament, skeletal muscle) as functional tissue replacements for injury or as in vitro models of disease or development. Tissue engineers attempt to recapitulate key features of in vivo tissue development, repair, or maintenance using precise combinations of cells, signals, and scaffolds in vitro. The application of mechanical stimulation to the developing constructs has gained traction over the past few decades owing to the

important role of mechanobiology observed in vivo and the promising in vitro results. Native musculoskeletal soft tissues typically adapt to the loads they experience by depositing and remodeling their ECM, contributing to improved function and biomechanical properties, as compared with unloaded tissues. These observations have direct implications in TE. By incorporating dynamic loading protocols in TE, similar to what the target tissue(s) of interest experience in vivo, researchers can mimic the native mechanobiological environment to promote matrix production and tissue organization. This is evidenced in a variety of engineered constructs that have shown increased collagen production and alignment, and more biomimetic mechanical properties, when subjected to tissue-specific dynamic loading. However, engineers must be judicious in their application of mechanical stimulation, as overstimulation of engineered constructs can lead to the development of pathologic conditions or inflammatory phenotypes.

Studies assessing the role of mechanobiology in soft TE can generally be split into two groups: scaffold-based approaches and nascent scaffold-free approaches. The majority of TE strategies are scaffold-based, relying on incorporation of seeded cells into a scaffold that provides initial structure and support to a developing construct, and may be remodeled or degraded as the construct develops. Scaffold design varies with applications, as tissue engineers aim to replicate the structure of functional tissues. Rapid advances in new biomaterials and fabrication techniques have granted great flexibility to scaffold design, which can enable more complex scaffold-based approaches to engineer functional soft tissues. In addition, scaffold-free approaches for TE have garnered interest, as research groups seek to incorporate developmental approaches to engineer constructs, relying on the intrinsic ability of cells to self-assemble and deposit their own ECM. While these approaches each have their unique advantages and disadvantages, both are moving toward integration of dynamic loading protocols to engineer biomimetic biomechanically superior constructs for further clinical applications.

Cartilage is the first soft tissue in which the role of mechanobiology was extensively studied. Hyaline cartilage, which lines the joints and is largely exposed to compressive forces in activities of daily life, is particularly mechanosensitive. It has three distinct layers, each organized and aligned to allow for the tissue to support complex movements of joints required for locomotion. Compression, shear stress, and hydrostatic pressure all play a role in tissue organization and development and the maintenance of healthy tissue. Thus, tissue engineers have incorporated similar stimuli using bioreactors in their cartilage TE approaches. In both scaffold-based and scaffold-free approaches, dynamic compressive loading and shear stress have been applied using piston-based bioreactors and rotating wall bioreactors, respectively. Results have been promising, with increased type II collagen production and improved biomechanical properties observed in constructs subjected to dynamic loading.

Tendon and ligament are passive tissues that experience largely tensile strains of higher magnitudes. This is reflected in a highly aligned hierarchical collagenous matrix that is responsible for the resilience of these tissues. Alternating periods of cyclic strain and rest, mimicking periods of activity seen in vivo, have contributed to stronger and more organized constructs in both scaffold-based and scaffold-free approaches. Skeletal muscle presents a more complex TE challenge, as its functionality relies on the ability to actively generate force, as well as withstand external loading. To accomplish this, both scaffold-based and scaffold-free approaches have sought to integrate both mechanical and electrical stimuli in engineering skeletal muscles. Although functional muscle constructs have been created, their specific force production has not yet achieved values similar to native tissue.

Together, these applications show the importance of considering native tissue mechanobiology in musculoskeletal soft TE. Independent of the TE method, incorporating the appropriate mechanobiology is crucial to engineering more biomimetic biomechanically advanced tissues that may serve as dysfunction or disease state models and eventually move toward clinical applications.

# REFERENCES

1. Butler DL, Goldstein SA, Guilak F. Functional tissue engineering: the role of biomechanics. *J Biomech Eng*. 2000; 122(6):570–575.
2. Benam KH, Dauth S, Hassell B, et al. Engineered in vitro disease models. *Annu Rev Pathol Mech Dis*. 2015;10: 195–262.
3. Salinas EY, Hu JC, Athanasiou K. A guide for using mechanical stimulation to enhance tissue-engineered articular cartilage properties. *Tissue Eng Part B Rev*. 2018; 24(5):345–358.
4. Butler DL, Juncosa-Melvin N, Boivin GP, et al. Functional tissue engineering for tendon repair: a multidisciplinary strategy using mesenchymal stem cells, bioscaffolds, and mechanical stimulation. *J Orthop Res*. 2008;26(1):1–9.
5. Cittadella Vigodarzere G, Mantero S. Skeletal muscle tissue engineering: strategies for volumetric constructs. *Front Physiol*. 2014;5:362.
6. Martin I, Wendt D, Heberer M. The role of bioreactors in tissue engineering. *Trends Biotechnol*. 2004;22(2):80–86.
7. O'Brien FJ. Biomaterials & scaffolds for tissue engineering. *Mater Today*. 2011;14(3):88–95.
8. Henkel J, Hutmacher DW. Design and fabrication of scaffold-based tissue engineering. *BioNanoMaterials*. 2013;14(3–4):171–193.
9. Hutmacher DW, Sittinger M, Risbud MV. Scaffold-based tissue engineering: rationale for computer-aided design and solid free-form fabrication systems. *Trends Biotechnol*. 2004;22(7):354–362.
10. Hutmacher DW, Cool S. Concepts of scaffold-based tissue engineering—the rationale to use solid free-form fabrication techniques. *J Cell Mol Med*. 2007;11(4):654–669.
11. Chen G, Ushida T, Tateishi T. Scaffold design for tissue engineering. *Macromol Biosci*. 2002;2:67–77.
12. Lin S, Hapach LA, Reinhart-King C, Gu L. Towards tuning the mechanical properties of three-dimensional collagen scaffolds using a coupled fiber-matrix model. *Mater (Basel, Switzerland)*. 2015;8(8):5376–5384.
13. Crupi A, Costa A, Tarnok A, Melzer S, Teodori L. Inflammation in tissue engineering: the Janus between engraftment and rejection. *Eur J Immunol*. 2015;45(12): 3222–3236.
14. Zhang H, Zhou L, Zhang W. Control of scaffold degradation in tissue engineering: a review. *Tissue Eng Part B Rev*. 2014;20(5):492–502.
15. Verissimo AR, Nakayama K. Scaffold-free biofabrication. In: *3D Printing and Biofabrication*. Cham: Springer International Publishing; 2017:1–20.
16. DuRaine GD, Brown WE, Hu JC, Athanasiou KA. Emergence of scaffold-free approaches for tissue engineering musculoskeletal cartilages. *Ann Biomed Eng*. 2015;43(3): 543–554.
17. Guilak F, Butler DL, Goldstein SA, Baaijens FPT. Biomechanics and mechanobiology in functional tissue engineering. *J Biomech*. 2014;47(9):1933–1940.
18. Darling EM, Athanasiou KA. Articular cartilage bioreactors and bioprocesses. *Tissue Eng*. 2003;9(1):9–26.
19. Griffin TM, Guilak F. The role of mechanical loading in the onset and progression of osteoarthritis. *Exerc Sport Sci Rev*. 2005;33(4):195–200.
20. Zhang L, Hu J, Athanasiou KA. The role of tissue engineering in articular cartilage repair and regeneration. *Crit Rev Biomed Eng*. 2009;37(1–2):1–57.
21. Carter DR, Beaupré GS, Wong M, et al. The mechanobiology of articular cartilage development and degeneration. *Clin Orthop Relat Res*. 2004:69–77.
22. Temenoff JS, Mikos AG. Review: tissue engineering for regeneration of articular cartilage. *Biomaterials*. 2000; 21(5):431–440.
23. Anderson DE, Johnstone B. Dynamic mechanical compression of chondrocytes for tissue engineering: a critical review. *Front Bioeng Biotechnol*. 2017;5:76.
24. Jin M, Frank EH, Quinn TM, Hunziker EB, Grodzinsky AJ. Tissue shear deformation stimulates proteoglycan and protein biosynthesis in bovine cartilage explants. *Arch Biochem Biophys*. 2001;395(1):41–48.

25. Mauck RL, Soltz MA, Wang CCB, et al. Functional tissue engineering of articular cartilage through dynamic loading of chondrocyte-seeded agarose gels. *J Biomech Eng.* 2000; 122(3):252−260.

26. Bian L, Fong JV, Lima EG, et al. Dynamic mechanical loading enhances functional properties of tissue-engineered cartilage using mature canine chondrocytes. *Tissue Eng Part A.* 2010;16(5):1781−1790.

27. Bian L, Zhai DY, Zhang EC, Mauck RL, Burdick JA. Dynamic compressive loading enhances cartilage matrix synthesis and distribution and suppresses hypertrophy in hMSC-laden hyaluronic acid hydrogels. *Tissue Eng Part A.* 2012;18(7−8):715−724.

28. Hung CT, Mauck RL, C-b Wang C, Lima EG, Ateshian GA. A paradigm for functional tissue engineering of articular cartilage via applied physiologic deformational loading. *Ann Biomed Eng.* 2004;34(1):35−49.

29. Furukawa KS, Imura K, Tateishi T, Ushida T. Scaffold-free cartilage by rotational culture for tissue engineering. *J Biotechnol.* 2008;133(1):134−145.

30. Gilbert E, Mosher M, Gottipati A, et al. A novel through-thickness perfusion bioreactor for the generation of scaffold-free tissue engineered cartilage. *Processes.* 2014; 2(3):658−674.

31. Elder BD, Athanasiou KA. Synergistic and additive effects of hydrostatic pressure and growth factors on tissue formation. *PLoS One.* 2008;3(6). e2341.

32. Jin M, Emkey GR, Siparsky P, Trippel SB, Grodzinsky AJ. Combined effects of dynamic tissue shear deformation and insulin-like growth factor I on chondrocyte biosynthesis in cartilage explants. *Arch Biochem Biophys.* 2003; 414(2):223−231.

33. Jung H-J, Fisher MB, Woo SL-Y. Role of biomechanics in the understanding of normal, injured, and healing ligaments and tendons. *Sports Med Arthrosc Rehabil Ther Technol.* 2009;1(1), 9-9.

34. Kannus P. Structure of the tendon connective tissue. *Scand J Med Sci Sport.* 2000;10(6):312−320.

35. Harvey AK, Thompson MS, Cochlin LE, Raju PA, Cui Z. Functional imaging of tendon. *Ann BMVA.* 2009; 2009(8):1−11.

36. Lavagnino M, Wall ME, Little D, Banes AJ, Guilak F, Arnoczky SP. Tendon mechanobiology: current knowledge and future research opportunities. *J Orthop Res.* 2015; 33(6):813−822.

37. Zitnay JL, Weiss JA. Load transfer, damage, and failure in ligaments and tendons. *J Orthop Res.* 2018;36(12): 3093−3104.

38. Galloway MT, Lalley AL, Shearn JT. The role of mechanical loading in tendon development, maintenance, injury, and repair. *J Bone Joint Surg Am.* 2013;95(17): 1620−1628.

39. Ruberti JW, Hallab NJ. Strain-controlled enzymatic cleavage of collagen in loaded matrix. *Biochem Biophys Res Commun.* 2005;336(2):483−489.

40. Wren TA, Beaupré GS, Carter DR. Tendon and ligament adaptation to exercise, immobilization, and remobilization. *J Rehabil Res Dev.* 2000;37(2):217−224.

41. Murrell GAC, Lilly EG, Goldner RD, Seaber AV, Best TM. Effects of immobilization on achilles tendon healing in a rat model. *J Orthop Res.* 1994;12(4):582−591.

42. Kjaer M. Role of extracellular matrix in adaptation of tendon and skeletal muscle to mechanical loading. *Physiol Rev.* 2004;84(2):649−698.

43. Kuo CK, Pan XS, Li J, Brown EB. Embryo movements regulate tendon mechanical property development. *Philos Trans R Soc Lond B Biol Sci.* 2018;373(1759):20170325.

44. Havis E, Bonnin M-A, Esteves de Lima J, Charvet B, Milet C, Duprez D. TGFβ and FGF promote tendon progenitor fate and act downstream of muscle contraction to regulate tendon differentiation during chick limb development. *Development.* 2016;143(20):3839−3851.

45. Hitchcock TF, Light TR, Bunch WH, et al. The effect of immediate constrained digital motion on the strength of flexor tendon repairs in chickens. *J Hand Surg Am.* 1987; 12(4):590−595.

46. Herod TW, Veres SP. Development of overuse tendinopathy: a new descriptive model for the initiation of tendon damage during cyclic loading. *J Orthop Res.* 2018;36(1): 467−476.

47. Fang F, Lake SP. Multiscale strain analysis of tendon subjected to shear and compression demonstrates strain attenuation, fiber sliding, and reorganization. *J Orthop Res.* 2015;33(11):1704−1712.

48. Banes AJ, Horesovsky G, Larson C, et al. Mechanical load stimulates expression of novel genes in vivo and in vitro in avian flexor tendon cells. *Osteoarthr Cartil.* 1999;7(1): 141−153.

49. Andarawis-Puri N, Flatow EL, Soslowsky LJ. Tendon basic science: development, repair, regeneration, and healing. *J Orthop Res.* 2015;33(6):780−784.

50. Hildebrand KA, Frank CB. Scar formation and ligament healing. *Can J Surg.* 1998;41(6):425−429.

51. Huisman E, Lu A, McCormack RG, Scott A. Enhanced collagen type I synthesis by human tenocytes subjected to periodic in vitro mechanical stimulation. *BMC Musculoskelet Disord.* 2014;15:386.

52. Docking S, Samiric T, Scase E, Purdam C, Cook J. Relationship between compressive loading and ECM changes in tendons. *Muscles Ligaments Tendons J.* 2013; 3(1):7−11.

53. Fang F, Lake SP. Experimental evaluation of multiscale tendon mechanics. *J Orthop Res.* 2017;35(7):1353−1365.

54. Thomopoulos S, Genin GM, Galatz LM. The development and morphogenesis of the tendon-to-bone insertion − what development can teach us about healing -. *J Musculoskelet Neuronal Interact.* 2010;10(1):35−45.

55. Chen JL, Yin Z, Shen WL, et al. Efficacy of hESC-MSCs in knitted silk-collagen scaffold for tendon tissue engineering and their roles. *Biomaterials.* 2010;31(36): 9438−9451.

56. Kalson NS, Holmes DF, Kapacee Z, et al. An experimental model for studying the biomechanics of embryonic tendon: evidence that the development of mechanical properties depends on the actinomyosin machinery. *Matrix Biol.* 2010;29(8):678−689.

57. Giannopoulos A, Svensson RB, Heinemeier KM, et al. Cellular homeostatic tension and force transmission measured in human engineered tendon. *J Biomech*. 2018; 78:161–165.

58. Bayer ML, Schjerling P, Herchenhan A, et al. Release of tensile strain on engineered human tendon tissue disturbs cell adhesions, changes matrix architecture, and induces an inflammatory phenotype. *PLoS One*. 2014; 9(1):e86078.

59. Kalson NS, Holmes DF, Herchenhan A, Lu Y, Starborg T, Kadler KE. Slow stretching that mimics embryonic growth rate stimulates structural and mechanical development of tendon-like tissue in vitro. *Dev Dyn*. 2011;240(11): 2520–2528.

60. Altman G, Horan R, Martin I, et al. Cell differentiation by mechanical stress. *FASEB J*. 2002;16(2):270–272.

61. Gilbert TW, Stewart-Akers AM, Sydeski J, Nguyen TD, Badylak SF, Woo SL-Y. Gene expression by fibroblasts seeded on small intestinal submucosa and subjected to cyclic stretching. *Tissue Eng*. 2007;13(6): 1313–1323.

62. Androjna C, Spragg RK, Derwin KA. Mechanical conditioning of cell-seeded small intestine submucosa: a potential tissue-engineering strategy for tendon repair. *Tissue Eng*. 2007;13(2):233–243.

63. Angelidis IK, Thorfinn J, Connolly ID, Lindsey D, Pham HM, Chang J. Tissue engineering of flexor tendons: the effect of a tissue bioreactor on adipoderived stem cell–seeded and fibroblast-seeded tendon constructs. *J Hand Surg Am*. 2010;35(9):1466–1472.

64. Banes AJ, Gilbert J, Taylor D, Monbureau O. A new vacuum-operated stress-providing instrument that applies static or variable duration cyclic tension or compression to cells in vitro. *J Cell Sci*. 1985;75:35–42.

65. Garvin J, Qi J, Maloney M, Banes AJ. Novel system for engineering bioartificial tendons and application of mechanical load. *Tissue Eng*. 2003;9(5):967–979.

66. Kuo CK, Tuan RS. Mechanoactive tenogenic differentiation of human mesenchymal stem cells. *Tissue Eng Part A*. 2008; 14(10):1615–1627.

67. Ni M, Rui YF, Tan Q, et al. Engineered scaffold-free tendon tissue produced by tendon-derived stem cells. *Biomaterials*. 2013;34(8):2024–2037.

68. Hasegawa M, Yamato M, Kikuchi A, Okano T, Ishikawa I. Human periodontal ligament cell sheets can regenerate periodontal ligament tissue in an athymic rat model. *Tissue Eng*. 2005;11(3–4):469–478.

69. Larkin LM, Calve S, Kostrominova TY, Arruda EM. Structure and functional evaluation of tendon-skeletal muscle constructs engineered in vitro. *Tissue Eng*. 2006;12(11): 3149–3158.

70. Ma J, Smietana MJ, Kostrominova TY, Wojtys EM, Larkin LM, Arruda EM. Three-dimensional engineered bone-ligament-bone constructs for anterior cruciate ligament replacement. *Tissue Eng Part A*. 2012;18(1–2): 103–116.

71. Schiele NR, Koppes RA, Chrisey DB, Corr DT. Engineering cellular fibers for musculoskeletal soft tissues using directed self-assembly. *Tissue Eng Part A*. 2013;19(9–10): 1223–1232.

72. Mubyana K, Corr DT. Cyclic uniaxial tensile strain enhances the mechanical properties of engineered, scaffold-free tendon fibers. *Tissue Eng Part A*. 2018;24(23–24): 1808–1817.

73. Schiele NR, Marturano JE, Kuo CK. Mechanical factors in embryonic tendon development: potential cues for stem cell tenogenesis. *Curr Opin Biotechnol*. 2013;24(5): 834–840.

74. Lemke SB, Schnorrer F. Mechanical forces during muscle development. *Mech Dev*. 2017;144(PtA):92–101.

75. Klumpp D, Horch RE, Beier JP. Skeletal muscle tissue engineering. In: *Tissue Engineering Using Ceramics and Polymers*. 2nd ed. 2014:524–540.

76. Gregory CM, Bickel S. Recruitment patterns in human skeletal muscle during electrical stimulation perspective. *Phys Ther*. 2005;85(4):358–364.

77. Holcomb WR. A practical guide to electrical therapy. *J Sport Rehabil*. 1997;6(3):272–282.

78. Nikolic N, Aas V. Electrical pulse stimulation of primary human skeletal muscle cells. *Myogenesis*. 2019;1889: 17–24.

79. Qazi TH, Mooney DJ, Pumberger M, Geißler S, Duda GN. Biomaterials based strategies for skeletal muscle tissue engineering: existing technologies and future trends. *Biomaterials*. 2015;53:502–521.

80. Yul Lim J, Donahue HJ. Cell sensing and response to micro-and nanostructured surfaces produced by chemical and topographic patterning. *Tissue Eng*. 2007;13(8): 1879–1891.

81. Auluck A, Mudera V, Hunt NP, Lewis MP. A three-dimensional in vitro model system to study the adaptation of craniofacial skeletal muscle following mechanostimulation. *Eur J Oral Sci*. 2005;113(3): 218–224.

82. Boldrin L, Elvassore N, Malerba A, et al. Satellite cells delivered by micro-patterned scaffolds: a new strategy for cell transplantation in muscle diseases. *Tissue Eng*. 2007; 13(2):253–262.

83. Vader D, Kabla A, Weitz D, Mahadevan L. Strain-induced alignment in collagen gels. *PLoS One*. 2009;4(6):e5902.

84. Donnelly K, Khodabukus A, Philp A, Deldicque L, Dennis RG, Baar K. A novel bioreactor for stimulating skeletal muscle in vitro. *Tissue Eng. Part C Methods*. 2010;16(4): 711–718.

85. Ahadian S, Ramón-Azcón J, Ostrovidov S, et al. A contactless electrical stimulator: application to fabricate functional skeletal muscle tissue. *Biomed Microdevices*. 2013;15(1):109–115.

86. Cole K, Henano N, Miller T, Pawelski K. *Mechanical and Electrical Stimulation Device for the Creation of a Functional Unit of Human Skeletal Muscle in Vitro*. 2015:1–107.

87. Williams ML, Kostrominova TY, Arruda EM, Larkin LM. Effect of implantation on engineered skeletal muscle constructs. *J Tissue Eng Regen Med*. 2013;7(6): 434–442.

88. Koppes RA. *Dynamic Skeletal Muscle Contraction and Tissue Engineering: Using Drosophila Melangastor as a Genetically Manipulable Experimental Model Species to Investigate the Role of Myosin in the Underlying Mechanisms of Force Depression and Force Enhancement, and the Development of a Electromechanical Bioreactor for Tissue Engineering of Single Fiber Mammalian Skeletal Muscle.* 2013 (Doctoral dissertation).

89. Rao L, Qian Y, Khodabukus A, Ribar T, Bursac N. Engineering human pluripotent stem cells into a functional skeletal muscle tissue. *Nat Commun.* 2018;9(1): 126.

# CHAPTER 3.2

# Intracellular Force Measurements in Live Cells With Förster Resonance Energy Transfer—Based Molecular Tension Sensors

JING LIU, PHD

## 1. PICONEWTON FORCES GOVERN CELLULAR BIOLOGICAL PROCESSES

Mechanical forces have emerged as an important player in various biological processes. First recognized in muscle cells, biomechanical activities and functions have been widely discovered and investigated in various cell types.[1] Forces dominate the development of cells and organs by shaping cells, tissues, and organs and regulating signaling pathways between and within cells.[2–4] Furthermore, various subcellular components have been identified to generate, sense, and transmit piconewton (pN) forces that comprehensively regulate the molecular or cellular functions (Fig. 3.2.1).[2,5–14] We know, for instance, that motor proteins are able to move along the cytoskeleton and play a major role in bidirectional transport in the cytoplasm, which is essential for cell physiology, plasticity, morphogenesis, and survival.[8] Forces generated by the movement of motor proteins are passed on/along the cytoskeleton to mechanosensors, which can perceive and translate the force, activating various downstream signaling pathways that play important roles in the regulation of cell morphogenesis and polarity. For example, cells probe tissue stiffness by mechanosensitive ion channels and focal adhesions (FAs),[2,6,11,15–18] whereas cell-cell junctions perceive mechanical tension between neighboring cells.[5,19] Therefore it is critical to read out the functional output induced by specific motor proteins or mechanosensors that are under load, and achieving mechanical force measurements at a molecular scale with molecular specificity in situ is needed.

Traditional techniques used to investigate biomechanics of cells include atomic force microscopy,[20,21] traction force microscopy,[22] optical/magnetic tweezers,[23,24] and DNA calibrated force sensors.[25] However, none of these approaches is capable of measuring intracellular or intramolecular forces in live cells with high specificity and spatiotemporal resolution. The recent emergence of molecular tension sensors, which are based on Förster resonance energy transfer (FRET), has enabled real-time monitoring of molecular pN-scale mechanics within living cells.[5,6,9,11,12,19,26] This technique has been extended to various motor proteins and sensing proteins and offers the possibilities to (1) better understand the mechanical architecture underlying the shapes of cells, tissues, or organs when combined with other force methods at different scales and (2) identify primary mechanotransducers and more generally relate molecular tensions to molecular, cellular, or tissue functions.

## 2. PRINCIPLE OF FÖRSTER RESONANCE ENERGY TRANSFER BASED MOLECULAR TENSION SENSORS

### 2.1. A Genetically Encoded Tension Sensor Unit

A genetically encoded newtonmeter, named "tension sensor unit," plays a critical role in reading out the pN force borne by individual molecules within cells (Fig. 3.2.2A). The newtonmeter functions as a molecular spring with a FRET pair conjugated on each end. As the spring is deformed, sensitive energy transfer occurs between donor and acceptor fluorophores (Fig. 3.2.2B). According to Hooke's law, the applied forces can be deduced based on the deformation of

Mechanobiology. https://doi.org/10.1016/B978-0-12-817931-4.00009-1

**161**

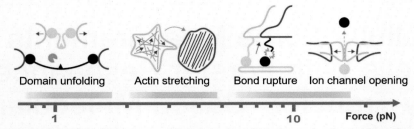

FIG. 3.2.1 Subcellular biological processes that are governed by piconewton forces. (From Roca-Cusachs P, Conte V, Trepat X Quantifying forces in cell biology. *Nat Cell Biol*. 2017;19:742–751.)

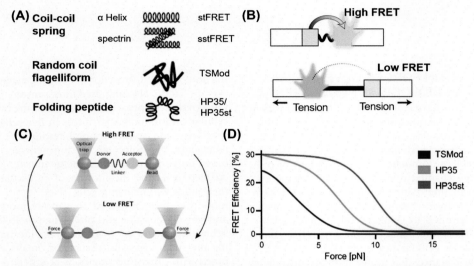

FIG. 3.2.2 Principle of Förster resonance energy transfer (FRET)-based molecular newtonmeters. **(A)** Three types of genetically encoded molecular springs: a coil-coil spring with α-helix or a spectrin repeat, a random coil flagelliform tension sensor module (TSMod), and a villin headpiece fast-folding peptide.[43] **(B)** Principle of the FRET change when force is applied on the molecular newtonmeter. **(C)** Optical trapping is used to calibrate and calculate the elastic constant of the tension sensor module.[12] **(D)** A calibration curve that is used to convert FRET efficiency to piconewton forces.[44]

the newtonmeter (details are introduced later in this chapter). By inserting this newtonmeter into a protein that bears mechanical forces, it is possible to trace the force dynamics at a single-molecule resolution. So far, there are several types of FRET-based tension sensor modules (TSMods) that are categorized by different genetically encoded newtonmeters (Fig. 3.2.2A), and details of them are summarized in Table 3.2.1.

The first reported molecular tension sensor, stretch-sensitive FRET (stFRET), was based on an α-helix that lengthens when tension is applied[27] (Fig. 3.2.2A) and was composed of a pair of green fluorescence proteins, Cerulean and Venus, linked on both ends of the α-helix. The length of amino acids in the helix is optimized in the stFRET tension sensor unit to ensure maximum FRET in

the relaxed state. In addition, a similar coiled-coil spring TSMod, spectrin repeat stretch-sensitive FRET (sstFRET) (Fig. 3.2.2A),[28] consisting of three folded α-helices of a spectrin repeat, was designed to increase the sensitivity of the tension sensor for ultralow force detection (1–4 pN). The second category of genetically encoded newtonmeter, TSMod, contains a 40-amino-acid sequence derived from the spider silk protein flagelliform, consisting of 5 amino acid repeats presumably acting as an entropic spring (Fig. 3.2.2A).[6] Both types of sensor springs are assumed to elongate in response to the tension applied along the axis, where the FRET between a pair of flanking fluorescent proteins varies monotonically along with the applied force. However, the elongation of amino acids can be enabled by a small

**TABLE 3.2.1**
**Summary of Representative FRET-Based Genetically Coded Molecular Tension Sensors.**

| Category | Name | Description | Force Range | Proteins |
|---|---|---|---|---|
| Coil-coil spring | stFRET[27] | Stable α-helix | Low piconewton range (~1–7 pN) | Collagen, spectrin, ffilamin a, α-actinin[27–29] |
| | sstFRET[28] | Spectrin repeat | Ultralow piconewton range (~1–4 pN) | α-Actinin[28] |
| Random coil flagelliform | TSMod[6] | (GPGGA)8 | 1–6 pN | Vinculin,[6,30–34] E-cadherin,[5,35,36] VE-cadherin,[37,38] β-spectrin,[39] MUC1,[40] PECAM-1,[41] integrin,[16] talin,[18] myosin IIb[42] |
| Folding peptide | HP35[15] HP35st[15] | Villin headpiece peptide Stabilized villin headpiece peptide | 6–8 pN 9–11 pN | Talin[15] Talin[15] |

FRET, Förster resonance energy transfer; stFRET, stretch-sensitive Förster resonance energy transfer; sstFRET, spectrin repeat stretch-sensitive Förster resonance energy transfer; TSMod, tension sensor module.

range of forces (typically < 6 pN), which limits the sensitivity and dynamic range of the force measurement possible using stFRET and TSMod sensors. Therefore a third type of molecular tension sensing module based on a folding peptide was proposed to extend the range of force detection beyond 10 pN. This TSMod uses the 35-amino-acid-long villin headpiece peptide (HP35), which is an ultrafast-folding peptide that undergoes an unfolding/folding transition. Although mechanical forces around 7 pN are sufficient to trigger the unfolding/folding transition, a stabilized mutant (HP35st) responds to mechanical forces around 10 pN (Fig. 3.2.2D).

## 2.2. Principle of Förster Resonance Energy Transfer to Force

Similar to a real mechanical spring, the Hooke's law or force-displacement relationship of the genetically encoded newtonmeter must be calibrated using known forces and displacements. As the FRET efficiency is proportional to the inverse sixth power of the distance between fluorophores, obviously a linear Hooke's law is not applicable for the tension sensor. Therefore experimental measurements and calibration are needed to measure the force sensitivity of these stFRET, sstFRET, HP35, and TSMods. To this end, single-molecule force spectroscopy is currently the only method that can provide sufficiently resolved information on the pN sensitivity, reversibility, loading rate sensitivity, and hysteresis of a TSMod (Fig. 3.2.2C).[6]

To perform the calibration of a full-length TSMod, including donor and acceptor fluorophores, the protein is purified from the cytoplasm of cells and attached via

terminal cysteines and DNA strands to two microbeads. Focused laser beams allow the application of pN forces at several different loading rates, allowing highly sensitive force/extension-relaxation measurements to be made. The observed protein extensions can be correlated to the expected FRET efficiency changes. Reversibility, loading rate sensitivity, and the hysteresis of individual molecules can be readily measured. The final calibration curve of FRET force is deduced from the measurements, as shown in Fig. 3.2.2D. Obviously, for the TSMod, the curve indicates a nonlinear response of FRET to the force, with a linear slope in the range between 1 and 6 pN. Single-molecule force spectroscopy can also identify the upper limit of the TSMod in detecting molecular forces.[6] To further extend the detection range, additional genetically encoded TSMods have been developed to cover higher force regimes. Two versions of the 35-amino-acid-long villin headpiece peptide (HP35 and HP35st) have been reported to be capable of responding to forces of approximately 10 pN (Fig. 3.2.2D).[27,28]

## 2.3. Förster Resonance Energy Transfer Measurements

FRET refers to the dipole-dipole interactions of two fluorescent molecules with overlapping excitation and emission wavelengths.[45] When the donor fluorophore is in its electronic excited state, it may transfer its excitation energy to another nearby fluorophore, i.e., the acceptor. Such a dipole-dipole interaction leads to the fact that the FRET interaction depends on the inverse sixth power of the distance between the donor and acceptor fluorophores. The phenomenon typically functions over a distance less than 10 nm. cyan fluorescent protein (CFP)/ yellow fluorescent protein (YFP)-like donor/acceptor pairs have been used for FRET readout in most of the genetically encoded newtonmeters. The nonradiative energy transfer from the excited donor fluorophore to the acceptor reduces the fluorescence quantum yield of the donor; while in the absence of the acceptor molecule, the donor only decays through fluorescence. Such a reduction of quantum yield leads to reduced fluorescence intensity and lifetime of the donor molecule. FRET may therefore be measured from differences in fluorescence lifetimes, steady-state intensities, or anisotropy. Lifetime-based FRET and intensity-based FRET are discussed later. These are the main FRET measurement techniques that have been applied to molecular tension sensors, but other techniques also exist.

### 2.3.1. Lifetime-based Förster resonance energy transfer

In lifetime-based FRET, the FRET efficiency $E$ is calculated based on the lifetime of the donor molecule: $E = 1 - \frac{\tau_{AD}}{\tau_D}$, where $\tau_{AD}$ is the fluorescence lifetime of the donor molecule in the presence of the acceptor and $\tau_D$ is the fluorescence lifetime of the donor molecule without the acceptor. The fluorescent lifetime, which typically lasts for only a couple of nanoseconds, is usually obtained from fluorescence lifetime imaging microscopy (FLIM).[46–48] FLIM typically uses a time-domain approach, i.e., time-correlated single photon counting, or a frequency-domain method to acquire the fluorescence lifetime. From the FLIM analysis, the decay behavior of the donor molecule can be characterized by exponential decay curves (Fig. 3.2.3A). In the presence of the acceptor molecule, the donor molecule shows faster decay behavior, which suggests smaller lifetimes. In FLIM-FRET, only two samples (donor only and donor-acceptor) are necessary to perform effective and reliable FRET measurements, although microenvironmental characterization of the donor molecule is needed to exclude effects from other factors, such as pH, ions, and refractive index. Therefore in molecular tension sensing, a sensor-tagged protein of interest lacking the acceptor is probably one of the best controls to use as a donor-only construct. FLIM-FRET has been performed to evaluate the TSMod in vinculin, VE-cadherin, PECAM-1, β-spectrin, MUC1, and E-cadherin; in cells; and *Caenorhabditis elegans*.[5,6,18,19,30]

### 2.3.2. Intensity-based Förster resonance energy transfer

Theoretically the FRET efficiency can also be simply calculated from the intensity of donor molecules, $E = 1 - \frac{I_{AD}}{I_D}$, where $I_{AD}$ is the fluorescence intensity of the donor molecule in the presence of the acceptor and $I_D$ is the fluorescence intensity of the donor molecule without the acceptor. This may be achieved sequentially before and after photobleaching of the acceptor within the same sample to ensure no change in donor concentration (Fig. 3.2.3C), as intensity measurements are directly dependent on fluorophore concentration. However, this approach assumes that all acceptor molecules were totally photobleached while the donor molecules were not affected. Incompletion of the acceptor photobleaching or partial donor bleaching will underestimate the FRET efficiency.

A more reliable and widely used intensity-based FRET measurement is sensitized emission (Fig. 3.2.3B). However, multiple measurements and control constructs are required for this method. Upon continuous donor excitation, FRET efficiency can be determined as $E = \frac{I_{AD}}{I_{AD} + \gamma I_{DA}}$, where $I_{AD}$ and $I_{DA}$ are the fluorescence intensities of TSMods (donor + acceptor) in the acceptor channel and the donor channel, respectively, while $\gamma$ accounts for differences in fluorescence quantum yields and spectral bleed-through. To this end, both donor-only and acceptor-only constructs are required as controls. This approach is as fast as standard fluorescence intensity imaging and is therefore well suited for fast biological processes. Most studies using molecular tension microscopy have used such sensitized emission and FRET index approaches.

## 3. INTRACELLULAR FORCE SENSING WITH MOLECULAR TENSION SENSORS

### 3.1. Protein of Interest and Related Tension Sensors

In principle, the FRET sensor can be inserted into any host protein, in any site suspected to bear mechanical forces (Fig. 3.2.4). However, the insertion of a genetically encoded newtonmeter with a total size that may exceed 600 amino acids is not an innocuous modification. Therefore knowledge of the protein sequence and structure is essential to avoid insertion sites that are likely to disrupt protein functions. In addition, it is wise to avoid known binding domains, posttranslationally modified sites such as phosphorylation sites, localization sequences, and highly structured domains.

In addition to regular function or expression validation analysis, such as Western blot and immunofluorescence, it is important to check the mechanical function of the new sensor and to test whether the protein of interest is under tension. Therefore a comparison with a tensionless control construct is required. Fig. 3.2.4 shows a list of control constructs for the molecular tension sensor. This

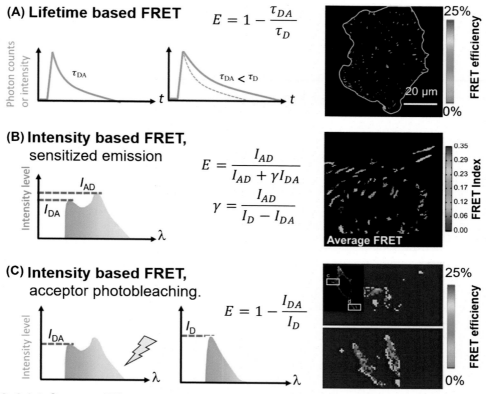

**(A) Lifetime based FRET** $$E = 1 - \frac{\tau_{DA}}{\tau_D}$$

**(B) Intensity based FRET,** sensitized emission $$E = \frac{I_{AD}}{I_{AD} + \gamma I_{DA}}$$ $$\gamma = \frac{I_{AD}}{I_D - I_{DA}}$$

**(C) Intensity based FRET,** acceptor photobleaching. $$E = 1 - \frac{I_{DA}}{I_D}$$

FIG. 3.2.3 Summary of Förster resonance energy transfer (FRET) measurement for tension sensing. (Left) Schematic drawing of different measurements. (Right) Representative FRET images reporting the force in live cells. **(A)** Fluorescence lifetime of the donor molecule is used to calculate the FRET efficiency, which is directly reflected by the reduction of lifetime of the donor molecules in the presence of acceptor molecules. **(B)** Intensity-based FRET measurements through sensitized acceptor emission. Acceptor intensity corrected from spectral bleed-through and direct excitation requires additional measurements with acceptor excitation and on control constructs. **(C)** Intensity-based FRET measurements through acceptor photobleaching. FRET efficiency is calculated by the intensity increase when acceptor molecules are photobleached. (Schematic drawings are from Gayrard, C., Borghi, N. FRET-based molecular tension microscopy. *Methods*. 2016;94: 33–42, and FRET images on the right are from Grashoff C, Hoffman BD, Brenner MD, et al. Measuring mechanical tension across vinculin reveals regulation of focal adhesion dynamics. *Nature*. 2010;466:263–266; Fangjia Li AC, Liu S, Reeser A, et al. Vinculin force sensor detects tumor-osteocyte interactions. *Sci Rep*. 2019; 9:5615 and Meng F, Saxena S, Liu Y, et al. The phospho-caveolin-1 scaffolding domain dampens force fluctuations in focal adhesions and promotes cancer cell migration. *Mol Biol Cell*. 2017;28:2190–2201.)

includes the sensor module appended at the C or N terminus of the protein, or the sensor-tagged protein truncated or mutated at one of its binding ends (Fig. 3.2.4C and D). As all the abovementioned constructs lack a binding site on at least one side of the sensor, the sensor is presumably under no tension. Therefore theoretic maximum FRET efficiency is expected than the efficiency of their full-length internally tagged counterparts. It is worth noting that not all tension sensors exhibit higher FRET for these tensionless constructs. For instance, spectrin exhibits insignificant

tension in HEK cells,[29] terminally tagged spectrins exhibited lower FRET than that of corresponding cytosolic sensors,[39] and the mutated myosin tension sensor is still capable of detecting uniaxial propagating forces.[42]

## 3.2. Mechanics of Focal Adhesion: Force Sensing and Cell Migration

Migration of the cell is mainly mediated by the FA to the extracellular matrix (ECM). The FAs usually grow when forces are high and disassemble when forces are

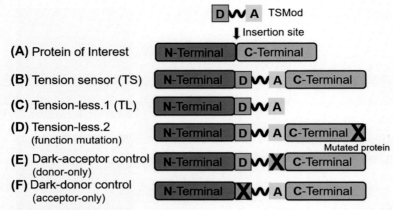

FIG. 3.2.4 Overview of protein-specific tension sensor constructs. **(A)** Insertion site of the tension sensor module (TSMod) into the protein of interest. **(B)** Molecular tension sensors (TS) with the TSMod inserted between the C and N terminal of the protein of interest. **(C,D)** Functional tensionless (TL) control constructs, where the sensor is in a state of relaxation. **(E,F)** Donor-/acceptor-only control constructs where either the acceptor or donor fluorescence molecule in TSMod is mutated with nonfluorescent counterparts.

low.[2,51] In FAs, integrin receptors, which are at the core of these adhesion complexes, do not directly bind to the actomyosin network but instead recruit numerous cytoplasmic molecules that mediate the binding to filamentous (f)-actin stress fibers.[52,53] Recent advances in super-resolution microscopy have successfully disclosed the nanoscale architecture of these molecules in adhesion complexes (Fig. 3.2.5A).[50] However, it was unclear which adapter molecules bear the cell adhesion forces in such highly dynamic and constant turnover sites. Therefore to determine the mechanisms underlying their mechanosensitivity, it is necessary to individually detect the force response of each receptor in FAs.

The first single-molecule calibrated tension sensor peptide, TSMod, was used to investigate the biomechanical behavior of the FA protein vinculin (Fig. 3.2.5B), which connects integrins to the actin filaments.[1,49] Vinculin is sensitive to mechanical stimulation and induces signaling for cell migration through ECMs.[2] This seminal work not only verified the prediction that the vinculin tension is dependent on actomyosin activity but also further discovered that the vinculin force is highly dynamic and heterogeneous during FA turnover.

Two other FA molecules, talin and integrin, have recently been investigated by FRET-based tension microscopy. Both molecules are central to FA function. Talin has a long rod domain connected to the f-actin cytoskeleton and activates the integrin receptor, which directly connects with the ECM. Given that the talin and integrin forces were expected to exceed 1–6 pN, HP35 and HP35st tension sensor molecules, which

were developed to extend the mechanical force sensing regime to 6–11 pN, were used.[27–29] It was found that both talin and integrins are crucial for FA mechanosensing because not only integrin activity is interactively regulated but also talin mediates an indispensable mechanical linkage between integrins and the f-actin cytoskeleton.

## 3.3. Force Propagation in the Cytoskeleton

Mechanical forces generated along the actin filaments through type II myosin are central to embryogenesis, organ development, muscle and heart contraction, and many other crucial cellular functions. In a recent work, Hart et al. inserted the TSMod into nonmuscle myosin II (NMII) molecules (Fig. 3.2.6A). NMII is a hexameric motor protein that can cross-link actin filaments and cause filament contraction through ATP hydrolysis. They found that the NMIIB tension sensors are capable of detecting dynamic changes in FRET efficiency along the actomyosin filaments (Fig. 3.2.6B). Moreover, they showed that the inhibition of NMII motor activity significantly increases the FRET efficiency of the NMIIB tension sensor. The FRET efficiency percentages of NMIIB TSMod increased to $26.76 \pm 1\%$ in blebbistatin-treated cells (Fig. 3.2.6C), which is a myosin inhibitor, compared with only $18.47 \pm 2.7\%$ recorded before blebbistatin treatment.[42] In addition, the NMIIB TSMod detected mechanical heterogeneity in actomyosin-associated force during cell migration.

In another work, TSMod was inserted into the β-spectrin ortholog of *Caenorhabditis elegans C. elegans* to detect force generation and dynamics in the highly

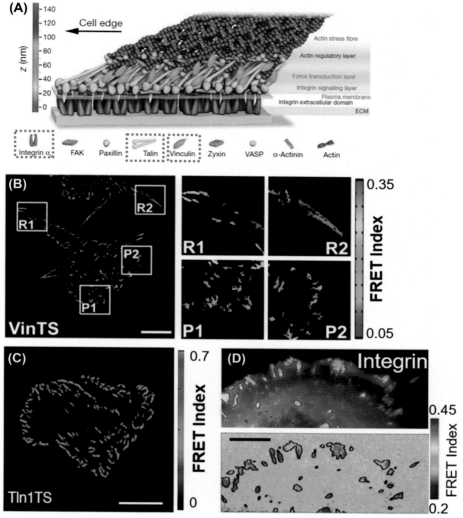

FIG. 3.2.5 Tension sensors report the force architecture in focal adhesions of cells on extracellular matrix. **(A)** The molecular architecture of the focal adhesion site.[50] The **(B)** vinculin tension sensor (VinTS),[6] **(C)** talin tension sensor,[15] and **(D)** integrin tension sensors[16] suggest a heterogeneous mechanical landscape of focal adhesions in a single cell. *ECM*, extracellular matrix; *FRET*, Förster resonance energy transfer.

ordered networks of the cortical spectrin-based system.[39,54,55] Because the insertion of TSMod did not affect the molecular function of β-spectrin, the β-spectrin tension sensor was able to establish that a constitutive spectrin tension of about 1.5 pN, on average, is present in live neuron cells. These spectrin tension sensors have also been used to detect the force in embryos so as to investigate the effect of mechanobiology in developmental biology.

## 3.4. Force Transmission in Cell-Cell Junctions

The cohesion of epithelial and endothelial tissues relies on the mechanical integrity of cadherin-based cell-cell junctions. Cadherins are transmembrane receptors that mediate hemophilic adhesion between neighboring cells and connect to intracellular cytoskeletal networks by recruiting adaptor proteins. Preliminary work with the vinculin tension sensor has found

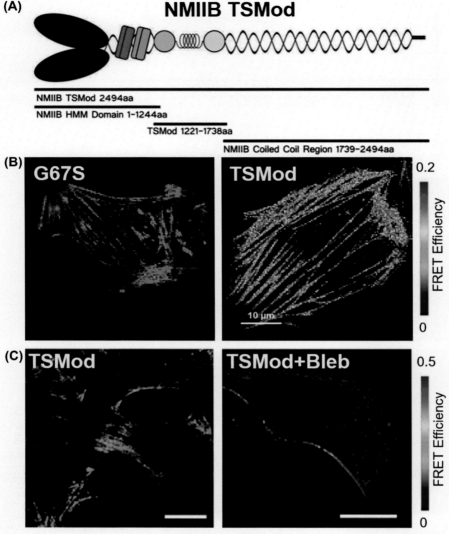

**FIG. 3.2.6** A myosin tension sensor enabled the direct visualization of force distribution and dynamics on the actin cytoskeleton. **(A)** The nonmuscle myosin IIB tension sensor (NMIIB TSMod). **(B)** Förster resonance energy transfer (FRET) image of individual U2OS cells expressing donor-only control construct (G67S) and NMIIB tension sensors (TSMod). **(C)** The myosin inhibitor blebbistatin (Bleb) significantly reduced the force on actomyosin. (From Hart RG, Kota D, Li F, et al. Myosin II tension sensors visualize force generation within the actin cytoskeleton in living cells. *bioRxiv*. 2019, 623249.)

that it also experiences mechanical tension in adherens junctions.[6] To study E-cadherin tension, the TSMod was inserted between the juxtamembrane region and the catenin-binding domain of the cytoplasmic tail of E-cadherin (Fig. 3.2.7A).[5,19] It has been found that E-cadherin supports mechanical loads of a few pN, which depends on its association with catenins and

actomyosin activity. It was also found that the tension in E-cadherin increases in intercellular junctions, which could potentially indicate that E-cadherin is a mechanotransducer of intercellular and extracellular stresses.

A desmoglein-2 (DSG-2) force sensor was developed based on the TSMod to directly measure mechanical

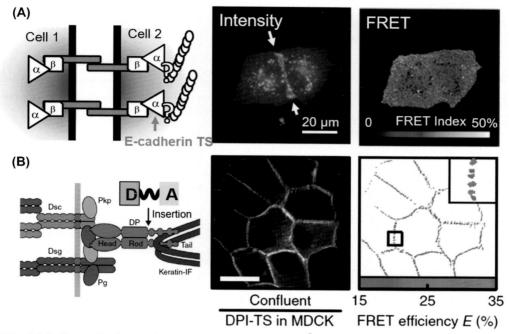

FIG. 3.2.7 The molecular tension sensors (TS) **(A)** E-cadherin[5] and **(B)** desmosomes report the force architecture on junctions between adjacent cells.[56,57] *FRET*, Förster resonance energy transfer; *MDCK*, Madin-Darby canine kidney. DPI-TS, desmoplakin I tension sensor.

forces across desmosome (Fig. 3.2.7B).[56,57] It was found that when expressed in human cardiomyocytes, the force sensor reported high tensile loading of DSG-2 during contraction. Additionally, when expressed in Madin-Darby canine kidney (MDCK) epithelial or epidermal (A431) monolayers, the sensor also reported tensile loading. Finally, higher DSG-2 forces were observed in three-dimensional MDCK acini than in two-dimensional monolayers.

## 4. PERSPECTIVES

The FRET-based molecular tension sensors have enabled the investigation of molecular mechanisms that govern the mechanobiology of human cells for the first time. We have learned a lot from these sensors and related experiments. They demonstrate that many biomechanical processes in cells are more complex than was previously thought, and there is no doubt that new methods and approaches are needed to establish a better and comprehensive understanding of intracellular force generation and transduction. For

instance, new TSMods with different stiffness are needed to detect the low range or high range forces or to enable more sensitive force detection of single molecules.[58] New FRET fluorophore pairs with better dipole-dipole interactions and better photostability are necessary to improve the sensitivity and range of the force detection. This will also enable the possibility of multiplex force detection in single cells, which will enable the direct visualization of force transduction between different subcellular structures.[59] In addition, emerging super-resolution imaging methods have disclosed that individual tension sensing proteins/molecules in subcellular structures are significantly heterogeneous (Fig. 3.2.5A), and new imaging techniques are in a high demand to visualize and analyze the force generation and transduction in these complex at the nanoscale. Finally, further investigation of mechanotransduction in cells calls for external loading or disturbance, and the recent progress in mechanical manipulation with optical tweezers or trapping in single cells should allow molecular mechanical loading in situ.

## ACKNOWLEDGMENT

The author gratefully acknowledge funding from the Burroughs Wellcome Fund.

## REFERENCES

1. Mitchison TJ, Salmon ED. Mitosis: a history of division. *Nat Cell Biol.* 2001;3:E17–E21.
2. Geiger B, Spatz JP, Bershadsky AD. Environmental sensing through focal adhesions. *Nat Rev Mol Cell Biol.* 2009;10:21.
3. Mammoto T, Ingber DE. Mechanical control of tissue and organ development. *Development.* 2010;137:1407.
4. Keller R. Physical biology returns to morphogenesis. *Science.* 2012;338:201.
5. Borghi N, Sorokina M, Shcherbakova OG, et al. E-cadherin is under constitutive actomyosin-generated tension that is increased at cell-cell contacts upon externally applied stretch. *Proc Natl Acad Sci USA.* 2012;109:12568–12573.
6. Grashoff C, Hoffman BD, Brenner MD, et al. Measuring mechanical tension across vinculin reveals regulation of focal adhesion dynamics. *Nature.* 2010;466:263–266.
7. Yanagida T, Iwaki M, Ishii Y. Single molecule measurements and molecular motors. *Philos Trans R Soc Lond B Biol Sci.* 2008;363:2123–2134.
8. Hirokawa N, Niwa S, Tanaka Y. Molecular motors in neurons: transport mechanisms and roles in brain function, development, and disease. *Neuron.* 2010;68:610–638.
9. Jurchenko C, Salaita KS. Lighting up the force: investigating mechanisms of mechanotransduction using fluorescent tension probes. *Mol Cell Biol.* 2015;35:2570–2582.
10. LaCroix AS, Rothenberg KE, Berginski ME, Urs AN, Hoffman BD. Construction, imaging, and analysis of FRET-based tension sensors in living cells. *Methods Cell Biol.* 2015;125:161–186.
11. Liu J, Wang Y, Gog WI, et al. Talin determines the nanoscale architecture of focal adhesions. *Proc Natl Acad Sci USA.* 2015;112:E4864–E4873.
12. Freikamp A, Cost AL, Grashoff C. The piconewton force awakens: quantifying mechanics in cells. *Trends Cell Biol.* 2016;26:838–847.
13. Polacheck WJ, Chen CS. Measuring cell-generated forces: a guide to the available tools. *Nat Methods.* 2016;13:415–423.
14. Gates EM, LaCroix AS, Rothenberg KE, Hoffman BD. Improving quality, reproducibility, and usability of FRET-based tension sensors. *Cytometry.* 2019;95:201–213.
15. Austen K, Ringer P, Mehlich A, et al. Extracellular rigidity sensing by talin isoform-specific mechanical linkages. *Nat Cell Biol.* 2015;17:1597–1606.
16. Morimatsu M, Mekhdjian AH, Adhikari AS, Dunn AR. Molecular tension sensors report forces generated by single integrin molecules in living cells. *Nano Lett.* 2013;13:3985–3989.
17. Buxboim A, Swift J, Irianto J, et al. Matrix elasticity regulates lamin-A,C phosphorylation and turnover with feedback to actomyosin. *Curr Biol.* 2014;24:1909–1917.
18. Kumar A, Ouyang M, Van Den Dries K, et al. Talin tension sensor reveals novel features of focal adhesion force transmission and mechanosensitivity. *J Cell Biol.* 2016;213:371–383.
19. Ng MR, Besser A, Brugge JS, Danuser G. Mapping the dynamics of force transduction at cell-cell junctions of epithelial clusters. *Elife.* 2014;3:e03282.
20. Kirmizis D, Logothetidis S. Atomic force microscopy probing in the measurement of cell mechanics. *Int J Nanomed.* 2010;5:137–145.
21. Besch S, Snyder KV, Zhang PC, Sachs F. Adapting the Quesant© Nomad™ atomic force microscope for biology and patch-clamp atomic force microscopy. *Cell Biochem Biophys.* 2003;39:195–210.
22. Munevar S, Wang Y-l, Dembo M. Traction force microscopy of migrating normal and H-ras transformed 3T3 fibroblasts. *Biophys J.* 2001;80:1744–1757.
23. Jiang G, Giannone G, Critchley DR, Fukumoto E, Sheetz MP. Two-piconewton slip bond between fibronectin and the cytoskeleton depends on talin. *Nature.* 2003;424:334–337.
24. Smith SB, Finzi L, Bustamante C. Direct mechanical measurements of the elasticity of single DNA molecules by using magnetic beads. *Science.* 1992;258:1122–1126.
25. Shroff H, Reinhard BM, Siu M, et al. Biocompatible force sensor with optical readout and dimensions of 6 nm3. *Nano Lett.* 2005;5:1509–1514.
26. Roca-Cusachs P, Conte V, Trepat X. Quantifying forces in cell biology. *Nat Cell Biol.* 2017;19:742–751.
27. Gayrard C, Borghi N. FRET-based molecular tension microscopy. *Methods.* 2016;94:33–42.
28. Freikamp A, Mehlich A, Klingner C, Grashoff C. Investigating piconewton forces in cells by FRET-based molecular force microscopy. *J Struct Biol.* 2017;197:37–42.
29. Meng F, Suchyna TM, Sachs F. A fluorescence energy transfer-based mechanical stress sensor for specific proteins in situ. *FEBS J.* 2008;275:3072–3087.
30. Meng F, Sachs F. Visualizing dynamic cytoplasmic forces with a compliance-matched FRET sensor. *J Cell Sci.* 2011;124:261.
31. Meng F, Sachs F. Orientation-based FRET sensor for real-time imaging of cellular forces. *J Cell Sci.* 2012;125:743.
32. Fangjia Li AC, Liu S, Reeser A, et al. Vinculin force sensor detects tumor-osteocyte interactions. *Sci Rep.* 2019;9:5615.
33. van den Dries K, Meddens MB, de Keijzer S, et al. Interplay between myosin IIA-mediated contractility and actin network integrity orchestrates podosome composition and oscillations. *Nat Commun.* 2013;4:1412.
34. Chang C-W, Kumar S. Vinculin tension distributions of individual stress fibers within cell–matrix adhesions. *J Cell Sci.* 2013;126:3021.
35. Gautrot JE, Malmström J, Sundh M, et al. The nanoscale geometrical maturation of focal adhesions controls stem cell differentiation and mechanotransduction. *Nano Lett.* 2014;14:3945–3952.
36. Rubashkin MG, Cassereau L, Bainer R, et al. Force engages vinculin and promotes tumor progression by enhancing

PI3K activation of phosphatidylinositol (3,4,5)-triphosphate. *Cancer Res.* 2014;74:4597.

37. Sim JY, Moeller J, Hart KC, et al. Spatial distribution of cell-cell and cell-ECM adhesions regulates force balance while maintaining E-cadherin molecular tension in cell pairs. *Mol Biol Cell.* 2015;26:2456—2465.

38. Rolland Y, Marighetti P, Malinverno C, et al. The CDC42-interacting protein 4 controls epithelial cell cohesion and tumor dissemination. *Dev Cell.* 2014;30:553—568.

39. Tornavaca O, Chia M, Dufton N, et al. ZO-1 controls endothelial adherens junctions, cell—cell tension, angiogenesis, and barrier formation. *J Cell Biol.* 2015;208:821.

40. Daneshjou N, Sieracki N, van Nieuw Amerongen GP, et al. Rac1 functions as a reversible tension modulator to stabilize VE-cadherin trans-interaction. *J Cell Biol.* 2015;208:23.

41. Krieg M, Dunn AR, Goodman MB. Mechanical control of the sense of touch by β-spectrin. *Nat Cell Biol.* 2014;16:224.

42. Paszek MJ, DuFort CC, Rossier O, et al. The cancer glycocalyx mechanically primes integrin-mediated growth and survival. *Nature.* 2014;511:319.

43. Conway DE, Breckenridge MT, Hinde E, et al. Fluid shear stress on endothelial cells modulates mechanical tension across VE-cadherin and PECAM-1. *Curr Biol.* 2013;23:1024—1030.

44. Hart RG, Kota D, Li F, et al. Myosin II tension sensors visualize force generation within the actin cytoskeleton in living cells. *bioRxiv.* 2019, 623249.

45. Förster T. Zwischenmolekulare energiewanderung und fluo-reszenz., English translation. *Ann Phys.* 1948;6:54—75.

46. Liu J, Jing X, Iishi S, Shalaev V, Irudayaraj J. Quantifying the local density of optical states of nanorods by fluorescence lifetime imaging. *New J Phys.* 2014;16, 063069.

47. Vidi PA, Liu J, Salles D, et al. NuMA promotes homologous recombination repair by regulating the accumulation of the ISWI ATPase SNF2h at DNA breaks. *Nucleic Acids Res.* 2014;42:6365—6379.

48. Salvador Moreno N, Liu J, Haas KM, et al. The nuclear structural protein NuMA is a negative regulator of 53BP1 in DNA double-strand break repair. *Nucleic Acids Res.* 2019;47:2703—2715.

49. Case LB, Waterman CM. Integration of actin dynamics and cell adhesion by a three-dimensional, mechanosensitive molecular clutch. *Nat Cell Biol.* 2015;17:955.

50. Horton ER, Byron A, Askari JA, et al. Definition of a consensus integrin adhesome and its dynamics during adhesion complex assembly and disassembly. *Nat Cell Biol.* 2015;17:1577.

51. Calderwood DA, Campbell ID, Critchley DR. Talins and kindlins: partners in integrin-mediated adhesion. *Nat Rev Mol Cell Biol.* 2013;14:503.

52. Kanchanawong P, Shtenge G, Pasapera AM, et al. Nanoscale architecture of integrin-based cell adhesions. *Nature.* 2010;468:580—584.

53. Meng F, Saxena S, Liu Y, et al. The phospho-caveolin-1 scaffolding domain dampens force fluctuations in focal adhesions and promotes cancer cell migration. *Mol Biol Cell.* 2017;28:2190—2201.

54. Baines AJ. The spectrin—ankyrin—4.1—adducin membrane skeleton: adapting eukaryotic cells to the demands of animal life. *Protoplasma.* 2010;244:99—131.

55. Xu K, Zhong G, Zhuang X. Actin, spectrin, and associated proteins form a periodic cytoskeletal structure in axons. *Science.* 2013;339:452.

56. Baddam SR, Arsenovic PT, Narayanan V, et al. The desmosomal cadherin desmoglein-2 experiences mechanical tension as demonstrated by a FRET-based tension biosensor expressed in living cells. *Cells.* 2018;7.

57. Price AJ, Cost A-L, Ungewiß H, et al. Mechanical loading of desmosomes depends on the magnitude and orientation of external stress. *Nat Commun.* 2018;9:5284.

58. LaCroix AS, Lynch AD, Berginski ME, Hoffman BD. Tunable molecular tension sensors reveal extension-based control of vinculin loading. *Elife.* 2018;7.

59. Ringer P, Weißl A, Cost A-L, et al. Multiplexing molecular tension sensors reveals piconewton force gradient across Talin-1. *Nat Methods.* 2017;14:1090—1096.

CHAPTER 4.1

# Multiscale Models Coupling Chemical Signaling and Mechanical Properties for Studying Tissue Growth

VIJAY VELAGALA • WEITAO CHEN • MARK ALBER • JEREMIAH J. ZARTMAN

## 1. INTRODUCTION

Organ growth is tightly regulated during development. For example, our right hand grows to a size similar to our left hand despite being disconnected from each other. How do cells locally sense the global size of their collective community (the organ)? What directs them to stop growing? How do multiple organ systems coordinate growth? These questions of organ size control remain unanswered to a significant extent.[1] Extensive coupling between mechanical and chemical processes occurring in cells makes answering these fundamental questions challenging. Furthermore, deregulation of size control mechanisms can lead to congenital disabilities or promote tumor progression.[2,3] Therefore, a better understanding of the mechanics of size control is necessary for advancing the fields of tissue engineering, regenerative medicine, and cancer biology.

During morphogenesis, cells communicate with each other to facilitate coordinated movements, shape changes, and divisions. As one of the four major tissue types, epithelial cells cover the surface of organs and are important for defining organ function.[4] Epithelial tissues play a major role in shaping the organs during the morphogenesis at early developmental stages. They also provide robust support for the formation of the structure of an embryo. Epithelial tissues are key to multiple morphogenetic processes such as invagination,[5] bending,[6] and neural tube formation.[7] These morphogenetic processes are regulated by multiple biochemical factors (defined by gene expression patterns and signaling regulatory networks) and mechanical elements, including actomyosin contractility and extracellular matrix (ECM) stiffness.[8] A better understanding of the underlying mechanisms of how these factors regulate morphogenesis remains a significant challenge in developmental biology.[9] Moreover, developing systems-level mechanistic computational models for studying how cells respond to mechanical cues in regulating tissue size remains a daunting task. In this chapter, we discuss key chemical and mechanical cues regulating tissue growth as well as approaches used to model biochemical signaling and mechanical processes in developing epithelial tissues. Further, we discuss progress toward developing computational model simulations that couple biochemical and mechanical processes. A selected set of computational studies of the specific example of organogenesis—the developing *Drosophila* wing—is described to highlight how computational models can be instrumental in investigating mechanisms of organ growth. We conclude with a perspective of next steps required toward developing predictive multiscale models of tissue growth.

## 2. BIOCHEMICAL SIGNALS REGULATE TISSUE GROWTH

A broad range of factors determines the growth rate and final size of an organ.[10,11] These factors include hormone availability, growth factors, and an organ-intrinsic developmental program encoded by the genome. Experiments over several decades have begun to unravel how cells sense the global organ size (reviewed in Ref. 12). Early grafting experiments revealed that size control could be organ autonomous or nonautonomous. For example, Twitty et al. surgically removed embryonic tissue that would form the leg in a large *Salamander* species. The embryonic tissue was then transplanted into an embryo of a smaller species and

Mechanobiology. https://doi.org/10.1016/B978-0-12-817931-4.00010-8

allowed to grow.[13] Strikingly, the leg grew to the original size of the donor species, completely independent of the host species. This experiment suggested that the factors regulating the size of the leg in this species are intrinsic. Broadly, growth control regulators can be divided into organ-intrinsic and organ-extrinsic classes. Organ-extrinsic regulators provide information to the organ about the developmental status of the organism. This information includes nutrition, developmental stages, and availability of hormones such as insulin and steroids. In contrast, organ-intrinsic regulators act locally in individual cells within the tissue. In general, they provide information about the local cellular environment, positional cues, and local cell-cell adhesions.[14]

Size control mechanisms are better understood once we identify the characteristic factors that regulate organ growth. As a powerful model system, the *Drosophila* wing imaginal disc is one of the most intensively studied systems for size regulation[15,16] (Fig. 4.1.1). The wing disc grows from a cluster of 30–50 cells present inside a *Drosophila* embryo. During the larval stages, it expands in size to approximately 30,000–50,000 cells.[17,18] Growth of the wing disc is regulated by a combination of extrinsic and intrinsic regulators. One of the key extrinsic regulators known to regulate the growth of the wing disc is insulin signaling. Insulin signaling regulates cell growth and proliferation in response to nutrient availability. Genetic studies in *Drosophila* highlight the importance of insulin signaling as mutations in the components of this pathway affect the final size of the tissue (i.e., the wing).[19] While extrinsic regulators mediate growth in response to the external factors, intrinsic regulators control local cell-cell interactions to coordinate cellular proliferation.

One of the most well-established intrinsic regulators is the Hippo signaling pathway.[15] Hippo signaling regulates cell growth, proliferation, and apoptosis.[20–24] The Hippo pathway consists of a highly conserved core kinase cassette regulated by multiple upstream factors. Two kinases, Warts and Hippo, play a central role in the transduction of Hippo signaling.

Loss of function in either of the two kinases results in a substantial overgrowth of imaginal discs and of corresponding adult size.[25–29] Thus, Hippo signaling plays a key role in regulating organ size by restricting cell proliferation and promoting apoptosis.

Besides Hippo signaling, several intercellular signaling pathways such as the Notch, epidermal growth factor receptor (EGFR), bone morphogenetic protein (BMP)/Decapentaplegic (Dpp), and Wingless (Wg) signaling pathways regulate growth in *Drosophila*

wing disc.[30–37] Dpp and Wg belong to a group of molecules called morphogens that are involved in pattern formation and specifying positional information to cells. These morphogens pattern the wing disc by forming gradients along the anterior-posterior and dorsal-ventral axis of the wing disk (Fig. 4.1.1B and C). Genetic studies have revealed the importance of morphogens in regulating organ size (reviewed in Refs. 38–40). For example, Dpp plays a crucial role in regulating the final size of the *Drosophila* wing disc. Overexpression of Dpp in a wing disk increases the size of the wing, whereas insufficient Dpp leads to smaller wing size.[41–44] Although the concentration of the Dpp is not homogeneous, cell proliferation is uniform to first approximation throughout the wing disk during later stages of growth.[17,45] Additionally, many regulators of the cell cycle are also key in tuning overall organ growth, include dMyc and Cyclin D (reviewed in Ref. 46). Overall, the growth of a wing disk is regulated by multiple extrinsic and intrinsic biochemical signaling pathways with multiple feedback loops between pathways.

## 3. MECHANICAL SIGNALING REGULATING TISSUE GROWTH

As an additional layer to the complexity of biochemical signaling, living cells and tissues experience a dynamic and heterogeneous mechanical environment. How cells respond to the biomechanical environment is critical for tissue homeostasis. The processes by which cells convert these physical/mechanical cues into biochemical signals and integrate them into a specific cellular response is referred to as mechanotransduction.[47,48] Deregulation of mechanotransduction leads to many diseases.[49] For example, in the beating heart, the stretching of cardiac muscle cells releases chemical signals that regulate heart function. Genetic perturbation of these stretching mechanisms leads to heart disease in a mouse model.[50] In general, mechanical signals are transmitted from the ECM and the surrounding cells to the nucleus through different cytoskeletal structures. Broadly, the mechanical signal transmission machinery can be divided into four components (reviewed in Ref. 51, Fig. 4.1.2). The first component is the focal adhesion (FA) assembly, which plays a crucial role in transmitting forces from the ECM to the cytoskeleton.[55] The second component is the cytoskeleton machinery (actin, myosin, and microtubules), which transmits forces from FAs and neighboring cell junctions to the nucleus. The third component is the linker of nucleoskeleton and cytoskeleton (LINC) complex, which is present on the nuclear envelope and connects the

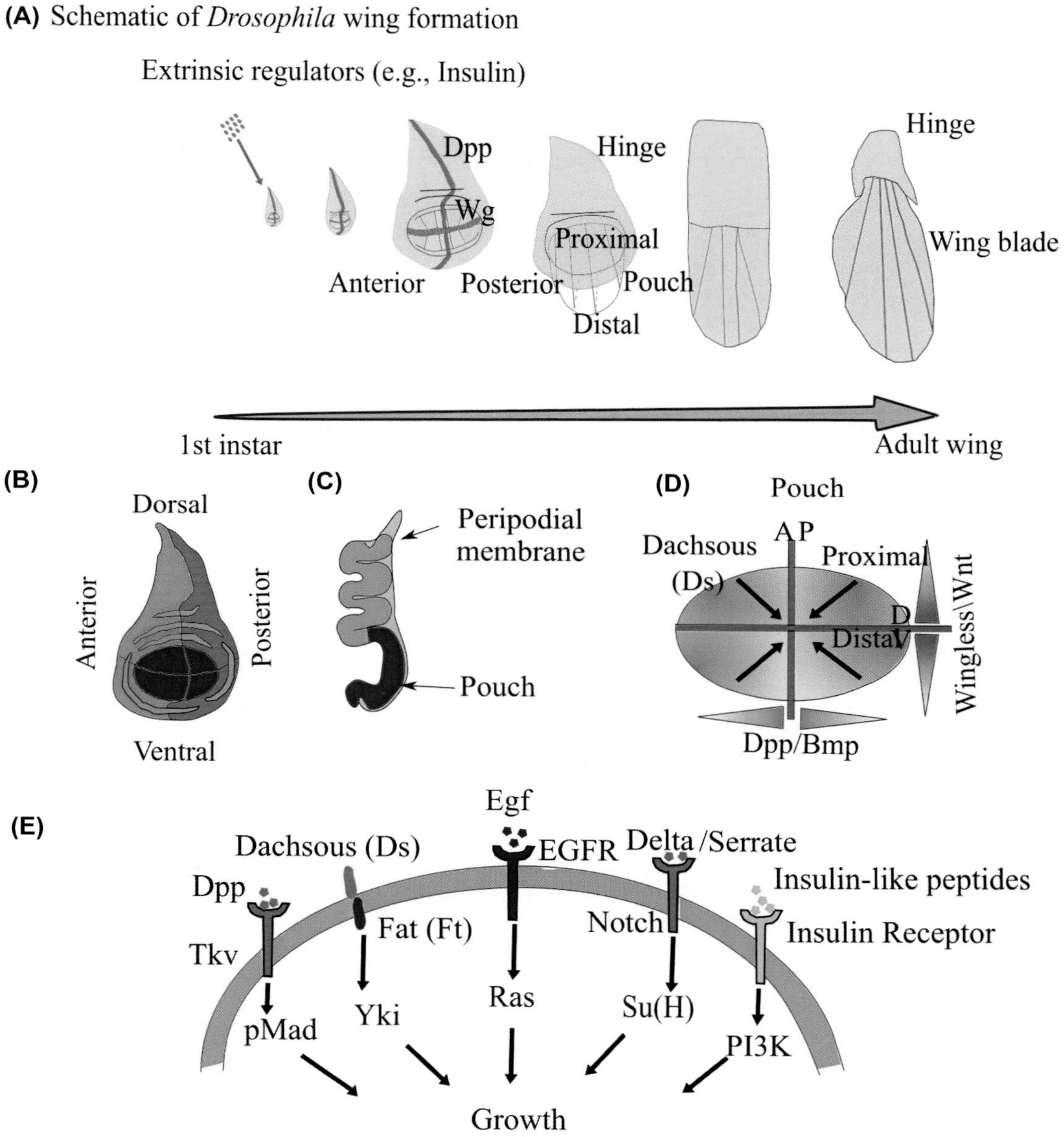

**(A)** Schematic of *Drosophila* wing formation

Extrinsic regulators (e.g., Insulin)

1st instar                                    Adult wing

**(B)**     **(C)**     **(D)**

**(E)**

FIG. 4.1.1 **Biochemical signals regulating the growth of wing disc in *Drosophila*.** **(A)** Schematic of a wing disc growth. **(B and C)** Top **(B)** and lateral **(C)** views of a wing disk with dorsal (D)-ventral (V) and anterior (A)-posterior (P) axes defined. **(D)** Formation of the bone morphogenetic protein homolog Decapentaplegic (Dpp) and WNT homolog Wingless (Wg) gradients in the wing disc pouch generates an orthogonal Cartesian coordinate system. Moreover, the Dachsous/Fat gradient defines the proximal-distal axis. **(E)** Summary of biochemical and genetic signals regulating growth in a wing disc.

cytoskeleton to the nuclear cytoskeleton such as lamin. The forces are directly transmitted from actin to the nucleus through LINC complexes. The final component is the nuclear pore complex, which controls the transport of RNA and proteins between the nucleus and cytoplasm. Forces are transmitted from ECM and adjacent cells to the cytoskeleton through FAs and adherens junctions. LINC complexes, which directly couple the

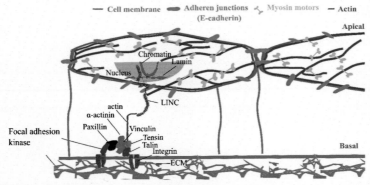

FIG. 4.1.2 **Key components defining the mechanical properties of the cell.**    Stiffness in the extracellular matrix (ECM) and other mechanical cues are transmitted through integrins. Integrins bind to ECM and sense mechanical cues. Once integrins bind to ECM, talin binds to them and undergoes conformational changes resulting in the opening of vinculin-binding sites. Vinculin binds to Talin and reinforces the actin filaments. Forces are transmitted to the nucleus through the actin and myosin contractile network. In addition to forces from the ECM, actomyosin filaments sense forces from adjacent cells through adherens junctions. These forces are transmitted to the nucleus through LINC complex, which connects the nucleoskeleton and cytoskeleton. These forces lead to a change in gene expression patterns in the nucleus. (Figure Adapted from Cell Migration lab — cell adhesion. http://www.reading.ac.uk/cellmigration/adhesion.html; Puckelwartz MJ, Kessler E, Zhang Y, et al. Disruption of nesprin-1 produces an Emery Dreifuss muscular dystrophy-like phenotype in mice. *Hum Mol Genet*. 2009;18(4):607–620. https://doi.org/10.1093/hmg/ddn386 and Vasquez CG, Martin AC. Force transmission in epithelial tissues. *Dev Dynam*. 2016;245(3):361–371. https://doi.org/10.1002/dvdy.24384.[52–54].)

cytoskeletal components to nucleoskeletal elements such as lamins, can lead to changes in nuclear stiffness and shape as the forces are transmitted. Subsequently, the change in stiffness can influence gene expression by detaching chromatin from the nuclear envelope.[56] The generated proteins and RNA are then shuttled between cytoplasm and nucleus through the nuclear pore complex.

The mechanisms by which cells convert mechanical forces into biochemical signals and molecular responses still remain elusive.[47] Unraveling these mechanisms is needed to formulate general design principles of mechanobiology.[57–59] Mechanical forces are key to the regulation of cell and tissue growth. For example, in mammalian systems, shearing forces generated by blood flow regulates the growth of blood vessels during embryogenesis.[60] Additionally, studies show that the application of mechanical tension increases the proliferation of mammalian cell cultures.[61,62] Conversely, compressive forces inhibit cell proliferation.[63] Besides mechanical tension, the stiffness of ECM regulates both growth and differentiation.[64] For example, Dupont et al. cultured mammalian epithelial cells on substrates with varying stiffness and found that mechanical signals are relayed to nuclear effectors, specifically the Yorkie-homologue YAP (Yes-associated

protein) and the transcriptional coactivator TAZ (transcriptional coactivator with PDZ-binding motif, also known as WWTR1) through tension in the cytoskeleton and Rho GTPase activity to control differentiation of mesenchymal stem cells.[64]

A key pathway involved in mechanotransduction is the Hippo pathway (Fig. 4.1.3). YAP and TAZ, homologues of the *Drosophila* transcriptional factor Yorkie, are the downstream effectors of the Hippo pathway. Interestingly, their study showed that YAP and TAZ levels increased significantly when the stiffness of the ECM is increased in mammalian epithelial cells. As YAP and TAZ are genetically linked to organ size control, this study demonstrates that the stiffness of ECM influences tissue growth. In addition to ECM stiffness, cell morphology also regulates the levels of YAP and TAZ. For example, Wada et al. showed that cell density, stress fibers, and cell morphology regulate YAP phosphorylation in fibroblast cells.[65]

In *Drosophila*, perturbation of the cytoskeletal tension in cells influences tissue growth. For example, Rauskolb et al. showed that increasing cytoskeletal tension increases the growth of the wing disc through regulation of Yorkie activity.[66] An increase in actomyosin contractility recruits the Ajuba Lim protein Jub at the adherens junctions. Binding of Jub to adherens junctions regulates Hippo signaling by inhibiting Warts. Ajuba recruits Warts to adherens junction as a function of cortical tension, preventing the phosphorylation of Yorkie, which prevents nuclear localization and transcriptional activation[67] (Fig. 4.1.3). Thus, actomyosin-mediated cytoskeletal tension indirectly acts as an upstream regulator of the growth controlling the Hippo signaling pathway. Similarly, Pan et al. reported that genetically altering the growth rates in *Drosophila* imaginal discs decrease cytoskeletal tension in the faster-growing clones. This reduced tension downregulates the Jub and Wts recruitment to the junctions resulting in downregulation of Yki activity.[68] Additionally, recent investigations show that reducing the levels of spectrin, a contractile protein at the cytoskeletal membrane interfaces, promotes tissue overgrowth.[69,70] For example, Deng et al. observed that the loss of a spectrin-based membrane skeleton protein (SBMS) in the wing disc leads to defects in Hippo signaling. Further, they observed elevated levels of phosphorylated myosin in the cell cortex. Additionally, studies show that cell density also acts as a regulator of Hippo signaling in mammalian cell cultures. For example, cells that are cultured at high density have low YAP activity. Cells that are cultured at low density have high YAP activity.[71–73] These experimental studies confirm that mechanical feedback plays an important role in maintaining uniform growth rates in developing tissues. Theoretical models explain the uniform growth throughout the wing disc by assuming compressive stresses inhibit growth (Fig. 4.1.4).[75] In summary, mechanical cues directly or indirectly regulate the tissue growth and size through regulation of Hippo signaling or other growth factors. A key gap in the field is to derive quantitative relationships between signaling activities, levels of forces in tissue, and cellular growth rates.

## 4. MODELING DYNAMICS OF BIOCHEMICAL SIGNALS IN TISSUES

Mathematical modeling is a critical methodology for testing hypothesized mechanisms of morphogenesis and growth control due to its capability to simulate complex biological systems *in silico* with high efficiency.[76] It has been widely used to understand pattern formation in a variety of biological systems, especially those experiencing complex morphological changes. As a foundational work, Alan Turing developed a simple mathematical model consisting of reaction-diffusion equations to generate complex spatial patterns.[77] Since then, Turing-based theory has been used to explain generalized mechanisms of morphogenesis in numerous systems.[78–85]

Another conceptual framework, positional information,[86] explains the formation of spatial patterns of gene expression in various tissues or organs through the localized secretion, transport, and downstream uptake of key biomolecules called morphogens.[30,38,87,88]

This general approach has been extended to produce mathematical models of biochemical signaling networks in growing tissues (Fig. 4.1.5). One modeling approach employs partial differential equations (PDEs) on a closed domain to model biochemical signaling network at the macroscale with a moving boundary that represents the growing tissue. Moving boundary problems associated with PDEs are usually challenging to solve numerically but can be tackled using a Lagrangian framework. Alternatively, an immersed boundary method[90] or level set method can be used.[91]

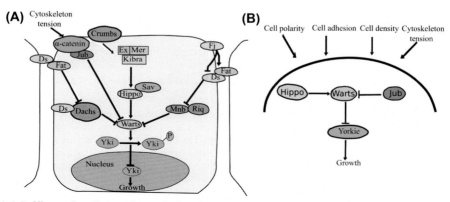

FIG. 4.1.3 **Hippo signaling pathway in *Drosophila*.** **(A)** Summary of Hippo signaling; **(B)** Mechanical factors regulating Hippo signaling. *Ajub*, Ajuba; *Crb*, Crumbs; *Ds*, Dachsous; *Ex*, Expanded; *Fj*, Four-jointed; *Jub*, *Drosophila* Ajuba; *Mer*, Merlin; *Minibrain (Mnb)*, a DYRK family kinase; *Sav*, Salvador; *Yki*, Yorkie; *Riquiqui (Riq)*, a scaffold for protein-protein interactions.

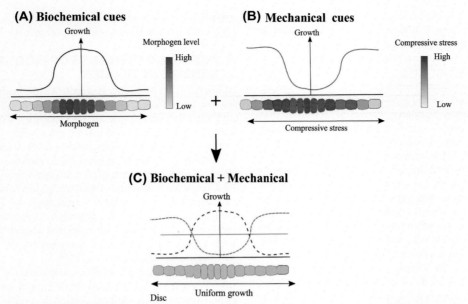

FIG. 4.1.4 **Biochemical and mechanical cues regulate organ growth.** **(A)** Formation of morphogen gradient in a tissue. Cells in the center of the tissue have higher concentration of the morphogen/growth, which leads to higher stimulation of growth. **(B)** Formation of a pressure gradient along the tissue axis. Cells become compressed in the center as they respond to higher levels of growth factors by growing faster. **(C)** In this conceptual model, combining both mechanical and chemical cues may explain the uniform growth of the tissue. (The figure is redrawn and adapted from Shraiman BI. Mechanical feedback as a possible regulator of tissue growth. *Proc Natl Acad Sci*. 2005;102(9):3318–3323. https://doi.org/10.1073/pnas.0404782102.[74])

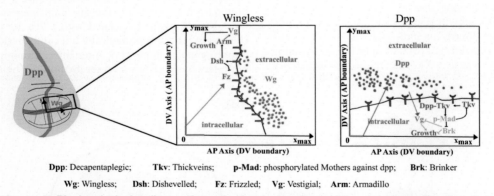

**Dpp**: Decapentaplegic;    **Tkv**: Thickveins;    **p-Mad**: phosphorylated Mothers against dpp;    **Brk**: Brinker

**Wg**: Wingless;    **Dsh**: Dishevelled;    **Fz**: Frizzled;    **Vg**: Vestigial;    **Arm**: Armadillo

FIG. 4.1.5 **Biochemical signaling pathway in developing *Drosophila* wing disc.** The Dpp ligand represents Dpp signal concentration along the AP axis, Wg forms a concentration gradient along the DV axis. Dfz and Tkv represent receptors. Arm and phosphorylated (p-MAD) are signal transducers that impact downstream modulators of growth and proliferation through transcriptional regulation, including Brinker (BRK) and Vestigial (VG). Not shown are additional growth regulators through Notch and Hippo signaling.[89] This example biochemical/gene regulatory network can be represented by a system of differential equations (Table 4.1.1). *Dpp*, Decapentaplegic; *Tkv*, Thickveins; *p-MAD*, phosphorylated Mothers against decapentaplegic; *Brk*, Brinker; *Wg*, Wingless; *Dsh*, dishevelled; *Fz*, frizzled; *Vg*, vestigial; *Arm*, armadillo.

**TABLE 4.1.1**
**The System of Partial Differential Equations Which Describes Dynamics of Dpp or Wg Signaling and Intracellular Regulatory Network in Two Dimensions (Related to Fig. 4.1.5).**

| Description | Equations |
|---|---|
| [L]: Ligand | $$\frac{\partial[L]}{\partial t} + \nabla\cdot(\vec{V}[L]) = D_L\Delta[L] - k_{on}[L][R] \\ + k_{off}[LR] + v_L(x,y) \\ - d_L(x,y)[L]$$ (4.1.1) |
| [R]: Receptor | $$\frac{\partial[R]}{\partial t} + \nabla\cdot(\vec{V}[R]) = -k_{on}[L][R] + k_{off}[LR] \\ + v_R(x,y) - d_R[R]$$ (4.1.2) |
| [LR]: Complex after binding | $$\frac{\partial[LR]}{\partial t} + \nabla\cdot(\vec{V}[LR]) = +k_{on}[L][R] \\ - k_{off}[LR] \\ - d_{LR}[LR]$$ (4.1.3) |
| [S$_1$]: Downstream signaling 1 (i.e., transducers) | $$\frac{\partial[S_1]}{\partial t} + \nabla\cdot(\vec{V}[S_1]) = v_{S_1}([LR])[S_1] \\ - d_{S_1}[S_1]$$ (4.1.4) |
| [S$_2$]: Downstream signaling 2 (i.e., transcriptional output) | $$\frac{\partial[S_2]}{\partial t} + \nabla\cdot(\vec{V}[S_2]) = v_{S_2}([S_1])[S_2] \\ - d_{S_2}[S_2]$$ (4.1.5) |

An alternative approach uses a particle-based structure to describe the population dynamics of chemical species involved in the biological system. This class of models is advantageous when the number of molecules is not large enough for applying the continuous PDE model and in incorporating the intrinsic noise into

consideration. In particular, stochastic simulation algorithms (SSAs) are developed for computing chemical master equations describing reaction kinetics. Both have been applied extensively to biological systems, including cell lineages in epithelia[92–95] and tumor growth.[96–98] Details about multiple modeling approaches are discussed in the following sections.

## 4.1. Continuum Models of Biochemical Signaling in Growing Tissue Domains

Continuum models describe the dynamics of concentrations of molecules through interactions that form the biochemical signaling network. In particular, ordinary differential equations are used to describe the temporal dynamics of gene regulatory circuits or cell populations to understand the interplay among different species without spatial information.[99,100] PDEs, usually reaction-diffusion equations, are applied to model the dynamics of diffusive molecules at cell or tissue level.[45,101,102] Each equation usually consists of a production term describing biosynthesis, a degradation term describing the destruction or conversion of the molecular species, and reaction terms corresponding to binding and unbinding activities, feedback regulations, and other terms related to diffusion or directed transport.[102] In the equations of diffusive molecules, a Laplace-Beltrami operator is used for describing diffusion on a curved surface (e.g., cell membrane).[101] Otherwise, the regular Laplace operator is applied for describing diffusion in multidimensional domains.[104,105]

Studying biochemical signals at either the cellular or tissue level can involve morphogenesis under similar time scales to the diffusion of growth factors in the tissue.[106] This results in using advection-reaction-diffusion systems, taking into account cell shape changes, cell movements, or proliferation.[106] Consequently, it requires solving the system of equations on a closed spatial domain with a moving boundary. Multiple methods have been developed to tackle such challenging problems numerically. In Peksin et al.,[90] an immersed boundary method was proposed to study the fluid-structure interactions where the fluid is described in an Eulerian frame and the structure is represented in a Lagrangian frame.

Similarly, an explicit representation of the moving tissue interface is also used in the front-tracking method.[107,108] Both the immersed boundary method and front-tracking method can be applied directly to study the biochemical signaling network on a moving domain by discretizing the moving boundary into meshes. Adaptive mesh refinement and interpolation/extrapolation techniques are usually required for achieving accuracy and stability under significant geometric changes when applied to biological systems.[109–111]

In another set of methods such as the level set method[91] or phase-field model,[112,113] the moving interface is represented by a function implicitly. Usually, a level set function is defined on a higher dimensional space with opposite signs associated with inside or outside the evolving domain, and the interface can be naturally extracted from the zero contours (level set method) or a narrow band close to the zero contours (phase-field method) (Fig. 4.1.6). Instead of explicitly tracking the moving interface, the level set function is evolved in time following another differential equation in addition to the biochemical system.

This type of method only requires fixed Cartesian grids to deal with dynamic interfaces.[101,116,117] However, the computation is usually more expensive due to the additional equations and higher dimension of the computational domain. The other method, Lagrangian framework, projects the moving domain into a fixed one and solves the resulting equation system before projecting the solution back.[118] This method can only be applied to systems with a simple shape change.[92,119,120] Otherwise, it is challenging to construct the projection and solve the resulting system.[121]

## 4.2. Discrete Stochastic Models of Biochemical Signaling in Tissues

Discrete stochastic models consider different biochemical molecules as a collection of particles and involve mesoscopic-scale, stochastic reaction-diffusion kinetics following a Markov process. This type of model is usually preferred over continuum models when the number of molecules for a chemical species is too small. It provides more details about the reaction-diffusion processes at the microscopic scale (Fig. 4.1.7). Various SSAs have been developed to solve stochastic biochemical models. In particular, the Gillespie algorithm has been developed for stochastic reactions to solve the chemical master equations using a kinetic Monte Carlo algorithm.[122] In this algorithm, populations for each reactant are considered separately, and stochasticity is introduced by randomly determining which reaction will occur next and when it will occur. For diffusion processes, Smoluchowski equations are used as the discretized version of stochastic diffusion equations and solved by updating positions of diffusing molecules.

In another approach, instead of tracking the behavior of individual molecules, a fixed number of compartments are considered to cover the entire domain. The Gillespie method is applied to simulate migration between neighboring compartments due to diffusion. Both approaches can be coupled with the SSA for simulating reactions to complete the numerical simulation of the reaction-diffusion system.[123] To deal with the domains with complex geometry, spatial SSA has recently been combined with a smoothed dissipative particle dynamics (SDPD) method to solve the advection-reaction-diffusion equation in the Lagrangian coordinate system.[124] The approach uses SSA or SDPD methods to simulate mass reaction-diffusion processes and SDPD to describe advection dynamics by treating each particle as a voxel in the spatial SSA.

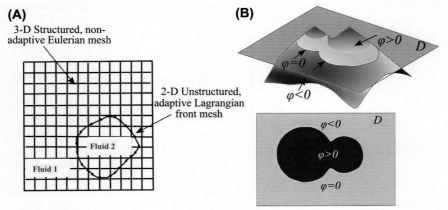

FIG. 4.1.6 **(A)** Three-dimensional Level Set function. **(B)** Design domain with the zero level set. (Figure adapted and redrawn from Wang Y, Luo Z, Zhang N, Kang Z. Topological shape optimization of microstructural metamaterials using a level set method. *Comput Mater Sci*, 2014;87:178–186. https://doi.org/10.1016/j.commatsci.2014.02.006 and CFD for the Design and Optimization of Slurry Bubble Column Reactors. https://doi.org/10.5772/intechopen.71361.[114,115])

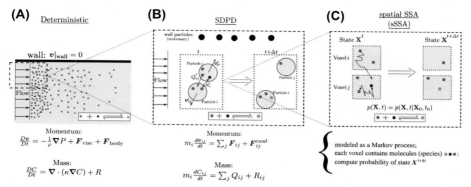

**(A)** Deterministic   **(B)** SDPD   **(C)** spatial SSA (sSSA)

Momentum:
$$\frac{Dv}{Dt} = -\frac{1}{\rho}\nabla P + F_{\text{visc}} + F_{\text{body}}$$

Mass:
$$\frac{DC}{Dt} = \nabla \cdot (\kappa\nabla C) + R$$

Momentum:
$$m_i \frac{dv_i}{dt} = \sum_j F_{ij} + F_{ij}^{\text{rand}}$$

Mass:
$$m_i \frac{dC_i}{dt} = \sum_j Q_{ij} + R_{ij}$$

$$p(\mathbf{X},t) = p(\mathbf{X},t|\mathbf{X}_0,t_0)$$

{ modeled as a Markov process;
each voxel contains molecules (species) ●●;
compute probability of state $X^{t+\Delta t}$

FIG. 4.1.7 Comparison of discrete models: **(A)** a typical deterministic approach, **(B)** smoothed dissipative particle dynamics (SDPD), **(C)** stochastic simulation algorithm (SSA), for the case of a channel flow with dilute species. (Figure from Drawert B, Jacob B, Li Z, Yi T-M, Petzold L. A hybrid smoothed dissipative particle dynamics (SDPD) spatial stochastic simulation algorithm (sSSA) for advection—diffusion—reaction problems. *J Comput Phys.* 2019;378:1—17. https://doi.org/10.1016/j.jcp.2018.10.043.[124])

## 5. MODELING BIOMECHANICAL PROPERTIES OF EPITHELIAL TISSUE

Multiple models have been developed to explain how a simplified description of cell biomechanics can explain epithelial morphogenesis (reviewed in Ref. 125, Fig. 4.1.8). Epithelial tissues are soft materials that can be treated as viscoelastic materials exhibiting both viscous and elastic properties. Elastic materials return to their original state after deformation when mechanical stress is released. In viscous materials, mechanical stress relaxes through remodeling of the internal structure, which may result in permanent deformation.

Rheological studies show that epithelial monolayers dissipate stress in strained tissues either by cell divisions or actomyosin turnover.[130] Investigation of the mechanical response of the cell-cell junctions shows that they exhibit elastic behavior on shorter time scales and viscous behavior on longer time scales.[131,132]

A large class of models treats epithelial tissues as a continuum or viscoelastic material by neglecting the cytoplasm and cell junctions. These models employ finite element or similar methods for discretizing the tissue for simulation purposes. In particular, the continuum approximation averages over length scales larger than a typical diameter of a cell. As a result, many cellular processes such as cell migration and adhesion, directed division, or cell polarity cannot be cast into a continuum formulation. On the other hand, discrete models or cell-based approaches treat cells as discrete entries. They are most suitable for studying different cell-level processes when compared with continuum models. Here, we discuss a selection of increasingly used tissue modeling approaches used to study various aspects of morphogenesis. Further, we discuss how these multiscale models account for viscoelastic behavior and mechanical heterogeneity as most multicellular systems are heterogeneous.

### 5.1. Continuum Models of Tissue Mechanics

Continuum models are used to model tissues at a larger scale with reduced complexity. In continuum models, tissue can be described as either an elastic material, a viscous fluid, or as a more complex viscoelastic material (Fig. 4.1.8A). They have been developed for soft elastic tissues to investigate various morphogenetic applications.[133,134] Usually, shape changes in a growing tissue occur due to two factors. First, addition or removal of material to tissue changes its local stress-free reference state, which makes the tissue change its initial configuration. Second, tissue elastically deforms to fit the change in tissue configuration due to growth. These elastic deformations make the total growth compatible by allowing no discontinuities in the tissue. Together, continuum models divide the total shape change tensor ($F = F^* \cdot G$) during the growth into two parts, the elastic deformation gradient tensor $F^*$ and growth tensor $G$. This theory of finite volumetric growth has been applied to many morphogenetic processes including ventral furrow formation, which involves cellular apical constriction in *Drosophila*.[135]

On longer time scales, tissues can be modeled as a viscous liquid. For example, Ranft et al. showed that embryonic tissues show liquidlike behavior with well-defined shear and bulk viscosities on long time scales.[136] This behavior is due to the dynamic reorganization of tissue induced by cell division and apoptosis. For example, Dillon et al. assumed the tissue in a growing limb as a viscoelastic liquid and used

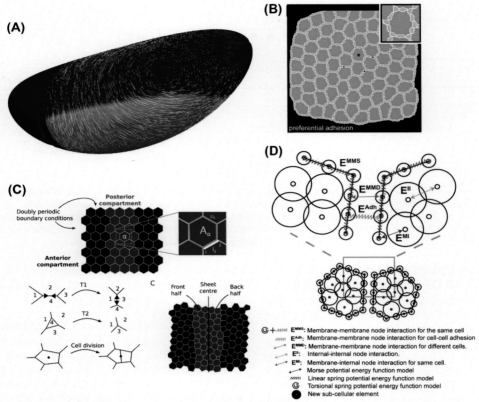

FIG. 4.1.8 Different computational models of epithelial tissues. **(A)** Continuum model simulation demonstrating the tissue flow in an embryo during germband extension. Green indicates the ventral region where the planar polarized activity of myosin is assumed. White arrows indicate tissue flow. **(B)** Cellular Potts model simulations of cell sorting for two hypothetical conditions. **(C)** Diagram of a vertex model implemented to test multiple scenarios of size control mechanisms for maintaining homeostasis of posterior compartments of *Drosophila* embryos. **(D)** Diagram of nodes of subcellular element model showing the interactions between the subcellular elements. ((A) Figure modified with permission from Dicko M, Saramito P, Blanchard G, Lye C, Sanson B, Étienne J. Geometry can provide long-range mechanical guidance for embryogenesis. PLoS Comput Biol, 2017;13(3):e1005443.[126] **(B)** Figure modified with permission from Larson DE, Johnson RI, Swat M, Cordero JB, Glazier JA, Cagan RL. Computer simulation of cellular patterning within the Drosophila pupal eye. *PLoS Comput Biol*, 2010;6(7):e1000841.[127] **(C)** Figure modified with permission from Kursawe J, Brodskiy PA, Zartman JJ, Baker RE, Fletcher AG. Capabilities and limitations of tissue size control through passive mechanical forces. *PLoS Comput Biol*, 2015;11(12):e1004679. https://doi.org/10.1371/journal.pcbi. 1004679.[128] **(D)** Figure modified with permission from Nematbakhsh A, Sun W, Brodskiy PA, et al. Multi-scale computational study of the mechanical regulation of cell mitotic rounding in epithelia. *PLoS Comput Biol*, 2017; 13(5):e1005533.[129])

Navier-Stokes equations to describe the fluid flow.[137] Similarly, Streichan et al. investigated the role of local stresses generated by molecular motors on global tissue flows during *Drosophila* gastrulation.[138] Here, the tissue flow is described in terms of the Stokes equation, and the stresses are governed by Maxwell viscoelasticity.[138] The Maxwell model of viscoelasticity can be denoted by a purely viscous damper and a purely elastic spring connected in series. Additionally, continuum models can be used to study the influence of local heterogeneities on tissue growth globally. Heterogeneity is defined as the variability of behavioral or mechanical properties of cells within a cell population (reviewed in Ref. 139).

Most multicellular systems are heterogeneous, and the heterogeneity within cell populations can affect their mechanical properties. For example, Tozluoğlu et al. have developed a continuum model to predict the positions of folds in a wing disc.[140] In their model, heterogeneity was incorporated into the tissue by considering differential spatial and temporal growth rates along the apical surface and also by varying the stiffness of the apical and basal surface. Using this model, they were able to generate patterns of folds like that of the experimental ones with the apical surface having the greater stiffness. When compared with other model methods, continuum modeling readily captures the impact of local forces and movements of the tissue globally. However, this approach does not model individual cell behavior, cell migration, or cell division.

## 5.2. Discrete Models of Tissue Mechanics

### 5.2.1. Cellular Potts models

Cellular Potts models are agent-based, on-lattice models[141] (Fig. 4.1.8B). In on-lattice models, cells are represented by a single or a group of pixels or voxels (nodes of the lattice), coupled with potentials that define the interaction between these nodes based on their states. In this case, the positions of cells or cell elements are governed by either stochastic (where stochasticity is adapted to capture random cell movement) or ordinary differential equations. In this approach, these objects extend spatially and stay on lattice sites.[142] These objects can be referred to as biological cells, compartments, or subcellular particles.

Cellular Potts simulations depend on the effective energy that describes cell behaviors and interactions. The energy of each cell is determined through its interactions such as cell-to-cell adhesion, and cellular properties such as target volume.

The equation for the effective energy can be written as:

$$H_{eff} = \sum_{i,j} J\big(\tau(\sigma(\bar{i})), \tau(\sigma(\bar{i}))\big)\big(1 - \delta(\sigma(\bar{i}), \sigma(\bar{j}))\big)$$
$$+ \sum_{\sigma} \lambda_{vol}(\sigma)(v(\sigma) - v_t(\sigma))^2 \quad (6)$$

The terms $\bar{i}$ denotes the vector of integers representing the coordinates of the pixel (nodes) on a lattice, $\sigma(\bar{i})$ is the cell number of specific type occupying the pixel $\bar{i}$, and $\tau(\sigma(\bar{i}))$ is the type of the cell. The term $J(\tau(\sigma(\bar{i})), \tau(\sigma(\bar{i})))$ represents the boundary energy per unit area between two cells. Further, $v(\sigma)$ represents the volume of the cell, $H_{eff}$ is the total effective energy of the system. The term $V_t(\sigma)$ represents the target volume of the cell. The first term in the equation represents the surface energy between two neighboring cells. The second term acts to minimize the volume variations of cells from their target volumes. A Monte Carlo simulation tries to minimize the effective energy by flipping the indices of pixels on the boundaries of cells using a Metropolis algorithm. Here, the attempt is accepted with a probability that is a function of the change in the effective energy. Eventually, the cells on the lattice evolve to minimize the total effective energy locally, leading to the motility and migration of cells.

Cellular Potts models have been used to study many morphogenetic contexts. These include blood vessel formation,[143] somitogenesis,[144] cell sorting,[145] limb formation,[146,147] and cyst formation.[148] For example, the cellular Potts model has been used to study the mechanisms behind the patterning of ommatidium cells in the *Drosophila* eye.[149] This model included two groups of cells—pigment cells and cone cells, and heterogeneity was specified by considering differential adhesion between the cell types. This study inferred that the topology and geometry of the ommatidium could be reproduced by making the interfacial tension dependent on the cell adhesion levels. Similarly, a cellular Potts model was used to find that programmed cell death and previously unappreciated changes in apical cell areas play critical roles in the large-scale organization of the emerging eye field.[127] Cellular Potts models are very flexible and usually efficient for simulating even a large number of cells. However, cellular Potts models are usually considered as phenomenological rather than mechanistic.[150]

### 5.2.2. Vertex-based models

Vertex models are agent-based, off-lattice models, where each cell is represented as a polygon with vertices and edges shared between adjacent cells[151,152] (Fig. 4.1.8C). The location of vertices and connection between them define the model geometry. Movement of the vertices occurs when they experience forces. The following equation gives the force acting on a vertex:

$$F_i = \frac{\partial U}{\partial x_i} \quad (7)$$

Here, $F_i$ denotes the force acting on vertex i. $U$ denotes the energy function with respect to x. Further, $x_i$ represents the junctional direction of the vertex. The energy function depends on the cell's deformability, cortical contractility, and cell-cell adhesion:

$$U = \sum_{\alpha=1}^{N} \left( \frac{K_\alpha}{2}\left(A_\alpha - A_\alpha^{(0)}\right)^2 + \frac{\Gamma_\alpha}{2}P_\alpha^2 \right) + \sum_{i,j} \Lambda_{ij} l_{ij} \quad (8)$$

The first term represents the area elasticity, with $K_\alpha$ denoting the area elasticity coefficient. The current area of the cell is indicated by $A_\alpha$, with $A_\alpha^{(0)}$ being the target area. The second term represents the perimeter contractility. $\Gamma_\alpha$ represents the perimeter contractility, and $P_\alpha$ represents the perimeter of the cell. The final term represents the energy contribution from line tension with $\Lambda_{ij}$ being the line tension coefficient and $l_{ij}$ the edge length. Cell division is introduced into the model by dividing the mother cell and introducing a new edge between the daughter cells. Cell rearrangements are modeled through the exchange of neighboring cell bonds, which result in topological changes in the cellular network.

Different approaches have been used to simulate the movement of vertices. One approach is to derive the forces from the energy equation and move the cells according to overdamped Newtonian mechanics.[153] Another approach is to assume that the entire tissue evolves quasistatically and relaxes to its local steady state at each time step. Here, the local minimum of the energy function is computed for every time step.

Vertex models have been employed to simulate many epithelial morphogenetic processes. As an early example, Odel et al. used a vertex model to show that the ventral furrow formation in the *Drosophila* embryo is due to spatial patterning of mechanical properties.[154] They modeled a ring of cells in the dorsal-ventral plane with the apical surface actively contracting. In addition to tissue deformation, vertex models have been used to test hypothesized mechanisms of organ size control,[155] including integrating biochemical gene regulatory networks into the vertex model.[105] Vertex models have also been used to identify physical mechanisms governing the de novo formation of the dorsoventral boundary in *Drosophila* wing discs.[156] Further, they have been extended to investigate 3D morphogenesis. For example, vertex models were used to study the formation of dorsal appendages from follicular epithelium through a combination of sheet bending and cell intercalation.[157] Recently, Barton et al. developed an active vertex model that can introduce dynamics into the vertex models.[158] This model combines the vertex model for describing epithelial tissue mechanics with the active matter dynamics from soft matter physics.

Vertex models can be further extended to explain the viscoelastic behavior of the epithelial junctions. For example, Staddon et al. modified the vertex model to account for the viscoelastic behavior of the epithelial cell junctions.[159] In this study, the authors used an optogenetic tool called TULIP (tunable, light-controlled interacting protein tags which dimerizes upon light activation)[160] to activate actomyosin contractility in epithelia for varying durations and observed the dynamics of junctional remodeling. For shorter durations, the epithelial junctions show elastic behavior. For longer durations of activation, the epithelial junctions underwent an irreversible deformation. Previous vertex models are not able to predict this

viscoelastic behavior of the epithelial junctions. So, the authors modified the existing vertex model by incorporating an irreversible junctional remodeling term into the energy equation. This term accounts for the change in the resting length of the junction when the strain exceeds a certain threshold. By incorporating this term, this model was able to predict the viscoelastic behavior of the epithelial junctions for varying durations of actomyosin activation.[159] Vertex models naturally simulate densely packed epithelia due to the polygonal shapes of the apical surfaces of cells. Further, they can include cell-neighbor rearrangements when compared with other agent-based models. However, these models neglect the cell-matrix adhesion and the influence of subcellular elements (SCEs) such as the cytoskeleton. Scaling up vertex models to multicellular systems with large numbers of cells and incorporation of a coupled signaling submodel is also computationally expensive.[161]

### 5.2.3. Subcellular element models

The cell-based off-lattice SCE modeling approach was developed to determine the impact of the mechanical properties of individual cells and their subcellular components on the emerging properties of growing multicellular tissue. SCE models include detailed representations of membrane and cytoplasm of each cell by using different sets of elements (nodes), and their mechanical properties are described using viscoelastic interactions between nodes (Fig. 4.1.8D). This yields a coarse-grained molecular dynamics type representation of the cytoskeleton and membrane.[162] Biomechanical and adhesive properties of cells are modeled through viscoelastic interactions between elements represented by phenomenological potential functions that represent close-range repulsion of neighboring SCEs to prevent their overlapping, and medium-range attraction between elements of the same or different cells to simulate the adhesive forces. Parameters of potential functions can be adjusted to calibrate model representations of biomechanical properties of a cell by directly using experimental data. Namely, an SCE model is used to perform in silico bulk rheology experiments on a single cell resulting in scaling the parameters.[163] This results in the SCE model capturing biomechanical properties of a biological system and remaining biologically relevant regardless of the number of elements used to represent each cell in the model.

SCE models were successfully used to study a variety of biological problems such as the role of cytosolic pressure, ECM stiffness in regulating the cellular movements and change in cell shapes.[164] Also, they have been extended to predict the redistribution of mechanical forces generated by cells in a tissue and for studying the rheology of tissue, the growth of tissue, and other cellular phenomena.[163] For example, Christey et al. used an SCE models to construct a 3D model of epidermal development, in which the cell behavior is coupled to neighboring cell-cell interactions.[165] Nemat-bakhsh et al. used an SEM to predict the quantitative contributions of cytoplasmic pressure, cell-cell adhesion, and cortical stiffness to mitotic cell rounding and expansion (Fig. 4.1.8D).[129] Most SCE models have been parallelized and implemented on powerful GPU computer clusters which substantially decrease computational time. Integration of inference tools, which estimates the subcellular distribution of forces using SCE simulation, is one of the recent developments in this area.[166,167] A major consideration is to identify the best level of coarse-graining of the model to avoid overfitting the model.

## 6. COUPLED CELL SIGNALING AND MECHANICAL MODELS TO INVESTIGATE EPITHELIAL MORPHOGENESIS

In the above sections, we have discussed various modeling approaches for studying morphogenesis from an either biochemical or mechanical perspective. However, morphogenesis is a process controlled by a combination of both mechanical and biochemical cues. Hence, it is essential for models to have a tight mutual coupling between mechanical and genetic cues for a better understanding of morphogenesis. For example, extracellular signals regulate cytoskeleton and surface adhesion dynamics downstream of the gene regulatory network. Regulation of cytoskeleton leads to changing dynamics of force generation. Further, multiscale models also need to consider the effect of forces on the signaling environment as cells experience deformation and mechanical stress. Integration of these feedback loops into a single framework provides insight into the interplay between molecular mechanisms and the mechanical properties of the tissue. Several multiscale models have been developed to study growing tissues or organs.[38,39,90–92] However, these models sacrifice either the detailed cell structure or the fine mesh to compute the chemical dynamics accurately throughout the tissue to reduce computational costs.[89,168] Models of cell biomechanics describe how physical forces and deformations exerted by the cells lead to the formation of complex shapes during morphogenesis. The biochemical module simulates how gene expression and different molecular mechanisms control cell behaviors, such as shape changes or cell growth and division.

Recent implementations of computational models on powerful high-performance computer clusters have enabled the ability to combine both submodels of dynamic gene regulatory circuits with more realistic and flexible tissue dynamics submodels. Below, we discuss a few examples where biochemical signals are coupled to mechanical models to investigate various aspects of morphogenesis. For example, Menshykau et al. used a continuum model to show that the ligand-receptor-based Turing mechanisms, in combination with tissue-specific gene expression, can control the branching of embryonic lungs.[169] In this model, a single 3D tissue layer is considered and the reaction-diffusion equations are solved on this 3D domain. Further, the tissue is allowed to grow at the sites of strongest signaling in the direction normal to the domain's surface. However, continuum models ignore individual cell mechanics.

Cell-based models (vertex, cellular Potts, and SCE models) are very suitable to study the interplay between signaling and cell mechanics as they provide detailed representations for individual cells, which can be easily extended to a description of the tissue. Cells can be described as extended objects with a volume whether represented by cellular Potts, vertex models, or SCE models, and they respond to inputs from the environment impacting their proliferation, adhesion, and migration (Fig. 4.1.9). Cells represented in cell-based models can also alter the environment through the regulation of gene expression. Hybrid models combine continuous and discrete (cell-based) submodels. A continuum submodel describes the dynamics of the environment, such as extracellular molecules and the reactions occurring in the cells that influence the decision-making outcomes of the cells. A discrete submodel describes the dynamics of the individual cells such as growth, movement and division. A hybrid model is called multiscale when the behavior of a discrete cell submodel is driven by a continuous (differential equations) submodel operating at multiple lengths and time scales (Fig. 4.1.10).

Vertex models can integrate biochemical signaling by discretizing the reaction-diffusion equations using different numerical techniques. Here, the polygons in the vertex model act as a mesh for discretizing these equations. For example, Schilling et al. coupled the differential equations governing explicit two-dimensional production of Hedgehog (Hh) production, diffusion, and transduction to a vertex model.[171] The Hh diffusion equation was discretized by Finite Volume Method by using the cells as local control volumes. After each growth step, the diffusion step was executed, and then the remaining reaction equations were solved numerically. Further, the model assumed that the localized

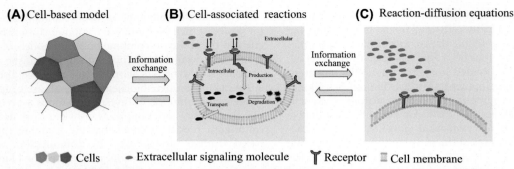

**(A)** Cell-based model    **(B)** Cell-associated reactions    **(C)** Reaction-diffusion equations

🛑🛑🛑 Cells    ● Extracellular signaling molecule    Υ Receptor    ▨ Cell membrane

FIG. 4.1.9 **Multiscale models integrate biochemical signaling and cell mechanics.** Models across different scales are integrated with the exchange of information. **(A)** A cell-based model represents tissue and cellular scales (e.g., cell activation states). **(B)** Cell-associated reactions are described by ordinary differential equations. **(C)** Reaction-diffusion equations describe the diffusion and binding of the extracellular ligands. Combining the models results in a hybrid multiscale model. (Figure inspired by Cilfone NA, Kirschner DE, Linderman JJ. Strategies for efficient numerical implementation of hybrid multi-scale Agent-based models to describe biological systems. *Cell Mol Bioeng*, 2015;8(1):119–136. https://doi.org/10.1007/s12195-014-0363-6.[170])

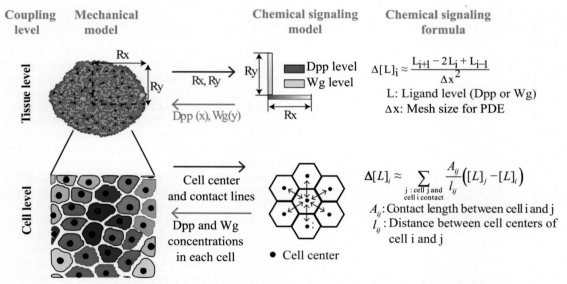

| Coupling level | Mechanical model | Chemical signaling model | Chemical signaling formula |
|---|---|---|---|

Tissue level

Rx, Ry →
← Dpp (x), Wg(y)

■ Dpp level
□ Wg level

$$\Delta[L]_i \approx \frac{L_{i+1} - 2L_i + L_{i-1}}{\Delta x^2}$$

L: Ligand level (Dpp or Wg)
$\Delta x$: Mesh size for PDE

Cell level

Cell center and contact lines →
← Dpp and Wg concentrations in each cell

● Cell center

$$\Delta[L]_i \approx \sum_{\substack{j\,:\,\text{cell j and}\\ \text{cell i contact}}} \frac{A_{ij}}{l_{ij}}\left([L]_j - [L]_i\right)$$

$A_{ij}$: Contact length between cell i and j
$l_{ij}$: Distance between cell centers of cell i and j

FIG. 4.1.10 Schematic diagram of coupling the mechanical submodel with a molecular signaling network. The coupling can be achieved at the tissue level (top row). Continuum partial differential equations (PDEs) on the domain provided by mechanical submodel are solved to update the chemical gradient at tissue level or, at the cell level (bottom row). Passive transport is applied to compute a continuum PDE model on the spatial domain, which consisting of individual cells. This cell network geometry is obtained from the mechanical submodel. (Source for cell simulation (top row): Nematbakhsh A, Sun W, Brodskiy PA, et al. Multi-scale computational study of the mechanical regulation of cell mitotic rounding in epithelia. *PLoS Comput Biol*. 2017; 13(5):e1005533.)

output of the Hh pathway leads to a change in the tension between cell bonds. To model this, a scaling factor was introduced into the line tension coefficient (described earlier), which influences the total energy function. Further, the scaling factor only depends on the ratio of the output of Hh signaling in two adjacent cells. This model was able to explain the experimental observations in the context of cell sorting at the anterior-posterior boundary in a *Drosophila* wing disc primordium. Similarly, Smith et al. incorporated chemical signaling into the vertex model to study the effect of Dpp on the growth of the wing disc. Here, they used an arbitrary Lagrangian-Eulerian formulation in conjunction with the finite element method (FEM) to solve the reaction-diffusion equations for morphogen concentration.[172] In this model, the polygons in the vertex model act as a mesh for the FEM to solve the concentration equations. Further, the node movements are governed by force laws. An ALE formulation was adopted to account for the domain movement while solving the concentration equations. Using this method, the authors investigated the role of Dpp in the wing disc by making the growth of the cell explicitly dependent on the morphogen concentration.

Additionally, mechanical models that are coupled with signaling are used to study the interplay between planar polarity proteins and cellular mechanics. For example, Salbreux et al. used a vertex model to explain the ordered packing of cone photoreceptors in the zebrafish retina by coupling mechanical terms and planar polarity proteins using a vertex model.[173] Using a differential equation model for planar polarity protein dynamics, the authors assumed that protein localization regulates cell mechanics through terms associated with cell tension, which in turn affects the tissue geometry. Further, the changes in tissue geometry affect the interactions between planar cellular polarity proteins within a cell or neighboring cells. As a result, the localization of the proteins is affected. After running simulations with different hypotheses, the authors concluded that the external force and progressive cell growth and division are essential for the observed packing behavior.

A variety of modeling platforms have been developed to combine both mechanical and chemical cues to study multicellular development. For example, Marin-Reira et al. have created a modeling platform called EmbryoMaker, which can simulate the basic behaviors of animal cells, extracellular signal-mediated interactions, and mechanical interactions between cells.[174] This platform is based on a subcellular element model for mesenchymal cells. For epithelial cells, it assumes that they are made of cylindrical SCEs, whereas standard SEM assumes them as spheres. Their modeling platform can specify gene regulatory networks that can regulate the mechanical activity of the cell behaviors and modes. Recently, Delile et al. created a cell-based simulation platform called Mecagen (based on SEM), which links gene regulatory network dynamics to individual cell behaviors that account for a wide range of morphogenetic events.[168] Using their platform, they

have tested the validity of the wide range of hypotheses about gene regulation by Wnt signaling; compartment formation through induction, epithelialization, and boundary sharpening; and epiboly in zebrafish embryos.

There are other modeling platforms such as Ingenue,[175] Virtual leaf,[176] CompuCell3D,[177] CellSys,[178] which allow coupling between microscale molecular process and macroscale mechanical properties and cell behaviors. These modeling platforms play a key role in developmental biology as they enable us to study large-scale integration of multiple processes at different scales from the molecular to the tissue level.

## 7. CASE STUDY: HYBRID BIOMECHANOCHEMICAL MODELS OF THE *DROSOPHILA* WING DISC

Growth regulation by mechanical factors plays a crucial role in several models explaining the mechanisms behind uniform growth in tissues such as the *Drosophila* wing disc.[179,180,180a] Many experimental observations indicate the presence of the differential proliferation rates, uneven stress, and pressure differences in the wing disc.[180] Multiple models have been developed to explore the interplay between the mechanics and signaling to investigate the mechanisms behind the uniform tissue growth during *Drosophila* imaginal disc growth.[68,89] In a model incorporating mechanical feedback control, Shraiman proposed that when the growth rate for a clone of cells is higher than the surrounding cells in a wing disc, the clone of cells experience mechanical stress.[75] This stress acts as feedback and reduces the growth rate of the cells resulting in a uniform growth throughout the tissue. This model was later extended to account for growth termination. In this model, cells at the lateral parts of the wing disc stop growing when they fall below a threshold concentration of the morphogen growth factor *Dpp*.[180] When cells within the lateral domain stop growing, the central parts experience mechanical stress. As this stress acts as a feedback, the cells in the central part stop growing, leading to growth termination. However, this model requires that the Dpp gradient does not scale with the tissue size. More recent analysis shows that the Dpp gradient scales with tissue growth.[45,181,182]

Later, Aegerter-Wilmsen et al. developed a continuum-level model to investigate the crosstalk between morphogens and mechanical forces. In this model, high levels of morphogen signaling at the center of the disc induce growth. Moreover, it assumes that growth is inhibited when compression exceeds a certain threshold. This model accounts for scaling of the morphogen Dpp with the size of the wing disc. Further, the model postulates that tangential stretching at lateral parts of the disc induces growth even though the morphogen levels are low. Also, it assumes that the wing pouch to be a two-dimensional elastic sheet with constant density and radially symmetric. This biochemical model with mechanical feedback predicted growth only in the center of the disc during initial growth. Further, it explained the position-dependent growth rates for clones with altered levels of Dpp morphogen signaling.[75]

Hufnagel et al. used a vertex model to simulate epithelial tissues while accounting for the cellular topology and geometry to study the role of mechanical factors.[180] In this model, the growth of the cells depends on morphogen concentration and compression. Cells grow when they sense the morphogen concentration exceeds a threshold. Growth is inhibited when they sense mechanical compression. Compression is calculated when the volume of a cell is close to a hypothetical target volume. Cells are chosen at random to divide with probability proportional to their growth rate. Based on these assumptions, their model predicts spatially uniform growth rates. At the center of the disc, growth terminates when the compressive stress is enough to oppose the growth induced by morphogens. At the lateral parts of the tissue, growth terminates when cells grow past the spatial distance where the local morphogen concentration is below a certain threshold. Together, mechanical feedback plays a key role in homogenizing growth. However, the above models were not able to explain the exact connection between mechanical stress and growth regulation.

To explain this relation, Aegerter-Wilmsen et al. included signaling pathways in their model. These signaling pathways explicitly connect the mechanical forces to growth regulation. In this model, a vertex model is coupled with a protein regulatory network to describe the crosstalk between mechanics and signaling in regulating the tissue size.[89] The protein regulatory network describes the complex interactions of proteins and cell-cell signaling (such as Notch, Dpp, and Wg). In their model, the mechanical compression in the cells modulates the activity of the multiple proteins in the regulatory network. Using this model, they were able to enumerate many experimental observations such as spatial uniform growth of a wing disc in the presence of a Dpp gradient. Computational models integrating mechanical control with secreted morphogens were able to explain the uniform growth of tissue. Even though the critical players during the mechanical

regulation of the growth were identified, the relation between the mechanical cues and signaling pathways requires further elucidation.

## 8. SUMMARY AND DISCUSSION OF FUTURE DIRECTIONS

Multiscale computational modeling coupled to novel imaging approaches and advanced quantitative analysis of experiments plays an increasingly important role in extending our understanding of the genetic and biophysical mechanisms of developing animal (and plant) organs. This chapter describes the multiple chemical and mechanical factors that regulate the growth of organs and how computational modeling is critical for unraveling the dynamic coupling between biochemical and mechanical cues. Multiscale computational models that couple cell mechanics and signal transduction submodels of protein and genetic networks are also increasingly used to analyze the interplay between biochemical and mechanical processes occurring during organ development, for studying cancer invasion and tissue regeneration.

As a case study, this chapter highlights how studies in the *Drosophila* wing imaginal disc have helped to reveal generally conserved genetic and biophysical mechanisms involved in regulating epithelial morphogenesis in a broad range of contexts. Multiscale models are gaining importance for their frequent usage in understanding systems-level biological phenomena. For example, modeling approaches can be used to predict new therapeutic targets by discerning key mechanisms of disease pathologies in tissues.

Computational modeling efforts to date are largely specific to a given biological context and are not readily generalizable. The choice of a computational modeling strategy depends on several factors such as computational cost; the level of resolution of the model; and the level of biological, mechanical, and geometric details included in the model. A key challenge in the field is how best to create computational modeling tools that are easily adaptable to new questions or employed by outside research groups. Such tools generally require dedicated software support and implementation on high-performance computer clusters. This chapter provides a description of several cell-based modeling approaches used in simulating morphogenesis.

Many biological problems in morphogenesis are three-dimensional (3D) in nature. For example, 3D models are needed to investigate the role of epithelial-ECM interactions during morphogenesis. However, high computational costs make this task very difficult. Integration of a large amount of spatiotemporal data increases the model complexity. This increase in complexity requires an efficient implementation of hybrid multiscale models.

With advances in experimental data acquisition and the development of image processing techniques including machine learning approaches, a key challenge for the modeling community is now to apply these advances in even more detailed and biologically calibrated multiscale simulation frameworks. In particular, multiscale modeling efforts must include sensitivity analysis to identify the essential parameters influencing the biological process. Parameter estimation for model building remains challenging, as many parameter values are difficult to obtain experimentally. Integration of advanced machine learning techniques with mechanistic modeling approaches are needed to better analyze the outcomes of model simulations and define causal relationships. Machine learning is well suited to predict the outcome from experimental inputs without formulating a specific causal mechanism. On the other hand, mechanistic modeling explicitly incorporates causal relationships, but it is not readily extendable to new contexts or parameter sets.

In principle, machine learning methods can be used in synergy with mechanistic models to overcome current limitations, which include the computational costs associated with handling multiple length and time scales.[183] One way of integrating machine learning methods with mechanistic models is to use surrogate models. Surrogate models are machine learning models obtained from the data produced from detailed mechanistic simulations. With surrogate models, a small number of detailed simulations can be run to provide input/output data that can be used to train an approximate machine learning-based model to be used for future predictions. Application of such approaches may prove useful to optimize computationally expensive multiscale simulations.[184]

Continued efforts in advancing the integration of multiscale mechanistic computational modeling with quantitative experimental analysis will continue leading the way for developing fundamental insights into the general principles of the genetics and mechanics of developmental systems.

## ACKNOWLEDGMENTS

VV and JZ were partially supported by NSF Award CBET-1553826, and JZ was supported in part by the NIH GrantR35GM124935. MA and WC were partially supported by the NSF Grant DMS-1762063 through the joint NSF DMS/NIH NIGMS Initiative to Support Research at the Interface of the Biological and Mathematical Sciences. MA was also partially supported by the NIH Grant UO1 HL116330.

## REFERENCES

1. Potter CJ, Xu T. Mechanisms of size control. *Curr Opin Genet Dev.* 2001;11(3):279−286. https://doi.org/10.1016/S0959-437X(00)00191-X.
2. Yu F-X, Zhao B, Guan K-L. Hippo pathway in organ size control, tissue homeostasis, and cancer. *Cell.* 2015;163(4): 811−828. https://doi.org/10.1016/j.cell.2015.10.044.
3. Kozma SC, Thomas G. Regulation of cell size in growth, development and human disease: PI3K, PKB and S6K. *Bioessays.* 2002;24(1):65−71. https://doi.org/10.1002/bies.10031.
4. Guillot C, Lecuit T. Mechanics of epithelial tissue homeostasis and morphogenesis. *Science.* 2013;340(6137): 1185−1189. https://doi.org/10.1126/science.1235249.
5. Martin AC, Kaschube M, Wieschaus EF. Pulsed contractions of an actin-myosin network drive apical constriction. *Nature.* 2009;457(7228):495−499. https://doi.org/10.1038/nature07522.
6. Lecuit T, Lenne P-F. Cell surface mechanics and the control of cell shape, tissue patterns and morphogenesis. *Nat Rev Mol Cell Biol.* 2007;8(8):633−644. https://doi.org/10.1038/nrm2222.
7. Lubarsky B, Krasnow MA. Tube morphogenesis: making and shaping biological tubes. *Cell.* 2003; 112(1):19−28.
8. Zartman JJ, Shvartsman SY. Unit operations of tissue development: epithelial folding. *Ann Rev Chem Biomol Eng.* 2010;1(1):231−246. https://doi.org/10.1146/annurev-chembioeng-073009-100919.
9. Saunders TE, Ingham PW. Open questions: how to get developmental biology into shape? *BMC Biol.* 2019; 17(1):17. https://doi.org/10.1186/s12915-019-0636-6.
10. Edgar BA. How flies get their size: genetics meets physiology. *Nat Rev Genet.* 2006;7(12):907. https://doi.org/10.1038/nrg1989.
11. Mirth CK, Riddiford LM. Size assessment and growth control: how adult size is determined in insects. *Bioessays.* 2007;29(4):344−355. https://doi.org/10.1002/bies.20552.
12. Vollmer J, Casares F, Iber D. Growth and size control during development. *Open Biol.* 2017;7(11). https://doi.org/10.1098/rsob.170190.
13. Twitty VC, Schwind JL. The growth of eyes and limbs transplanted heteroplastically between two species of Amblystoma. *J Exp Zool.* 1931;59(1):61−86. https://doi.org/10.1002/jez.1400590105.
14. Bryant PJ, Simpson P. Intrinsic and extrinsic control of growth in developing organs. *Q Rev Biol.* 1984;59(4): 387−415. https://doi.org/10.1086/414040.
15. Irvine KD, Harvey KF. Control of organ growth by patterning and hippo signaling in Drosophila. *Cold Spring Harb Perspect Biol.* 2015;7(6):a019224. https://doi.org/10.1101/cshperspect.a019224.

16. Hariharan IK. Organ size control: lessons from Drosophila. *Dev Cell.* 2015;34(3):255—265. https://doi.org/10.1016/j.devcel.2015.07.012.

17. Milán M, Campuzano S, García-Bellido A. Cell cycling and patterned cell proliferation in the wing primordium of Drosophila. *Proc Natl Acad Sci USA.* 1996;93(2): 640—645.

18. Worley MI, Setiawan L, Hariharan IK. Tie-dye: a combinatorial marking system to visualize and genetically manipulate clones during development in *Drosophila melanogaster. Development.* 2013;140(15):3275—3284. https://doi.org/10.1242/dev.096057.

19. Stocker H, Hafen E. Genetic control of cell size. *Curr Opin Genet Dev.* 2000;10(5):529—535. https://doi.org/10.1016/S0959-437X(00)00123-4.

20. Pan D. The hippo signaling pathway in development and cancer. *Dev Cell.* 2010;19(4):491—505. https://doi.org/10.1016/j.devcel.2010.09.011.

21. Halder G, Johnson RL. Hippo signaling: growth control and beyond. *Development.* 2011;138(1):9—22. https://doi.org/10.1242/dev.045500.

22. Zhao B, Tumaneng K, Guan K-L. The Hippo pathway in organ size control, tissue regeneration and stem cell self-renewal. *Nat Cell Biol.* 2011;13(8):877—883. https://doi.org/10.1038/ncb2303.

23. Yu F-X, Guan K-L. The Hippo pathway: regulators and regulations. *Genes Dev.* 2013;27(4):355—371. https://doi.org/10.1101/gad.210773.112.

24. Staley BK, Irvine KD. Warts and Yorkie mediate intestinal regeneration by influencing stem cell proliferation. *Curr Biol.* 2010;20(17):1580—1587. https://doi.org/10.1016/j.cub.2010.07.041.

25. Kf H, Cm P, Ik H. The Drosophila Mst ortholog, hippo, restricts growth and cell proliferation and promotes apoptosis. *Cell.* 2003;114(4):457—467.

26. Jia J, Zhang W, Wang B, Trinko R, Jiang J. The Drosophila Ste20 family kinase dMST functions as a tumor suppressor by restricting cell proliferation and promoting apoptosis. *Genes Dev.* 2003;17(20):2514—2519. https://doi.org/10.1101/gad.1134003.

27. Pantalacci S, Tapon N, Léopold P. The Salvador partner Hippo promotes apoptosis and cell-cycle exit in Drosophila. *Nat Cell Biol.* 2003;5(10):921—927. https://doi.org/10.1038/ncb1051.

28. Udan RS, Kango-Singh M, Nolo R, Tao C, Halder G. Hippo promotes proliferation arrest and apoptosis in the Salvador/Warts pathway. *Nat Cell Biol.* 2003;5(10): 914—920. https://doi.org/10.1038/ncb1050.

29. Wu S, Huang J, Dong J, Pan D. Hippo encodes a Ste-20 family protein kinase that restricts cell proliferation and promotes apoptosis in conjunction with salvador and warts. *Cell.* 2003;114(4):445—456.

30. Affolter M, Basler K. The Decapentaplegic morphogen gradient: from pattern formation to growth regulation. *Nat Rev Genet.* 2007;8(9):663—674. https://doi.org/10.1038/nrg2166.

31. Baker NE. Patterning signals and proliferation in Drosophila imaginal discs. *Curr Opin Genet Dev.* 2007; 17(4):287—293. https://doi.org/10.1016/j.gde.2007.05.005.

32. Wang S-H, Simcox A, Campbell G. Dual role for Drosophila epidermal growth factor receptor signaling in early wing disc development. *Genes Dev.* 2000; 14(18):2271—2276.

33. Zecca M, Struhl G. Control of growth and patterning of the Drosophila wing imaginal disc by EGFR-mediated signaling. *Development.* 2002;129(6):1369—1376.

34. Djiane A, Krejci A, Bernard F, Fexova S, Millen K, Bray SJ. Dissecting the mechanisms of Notch induced hyperplasia. *EMBO J.* 2013;32(1):60—71. https://doi.org/10.1038/emboj.2012.326.

35. Casso DJ, Biehs B, Kornberg TB. A novel interaction between hedgehog and Notch promotes proliferation at the anterior—posterior organizer of the Drosophila wing. *Genetics.* 2011;187(2):485—499. https://doi.org/10.1534/genetics.110.125138.

36. Pallavi SK, Shashidhara LS. Egfr/Ras pathway mediates interactions between peripodial and disc proper cells in Drosophila wing discs. *Development.* 2003;130(20): 4931—4941. https://doi.org/10.1242/dev.00719.

37. Zecca M, Basler K, Struhl G. Sequential organizing activities of engrailed, hedgehog and decapentaplegic in the Drosophila wing. *Development.* 1995;121(8): 2265—2278.

38. Restrepo S, Zartman JJ, Basler K. Coordination of patterning and growth by the morphogen DPP. *Curr Biol.* 2014;24(6):R245—R255. https://doi.org/10.1016/j.cub.2014.01.055.

39. Schwank G, Basler K. Regulation of organ growth by morphogen gradients. *Cold Spring Harb Perspect Biol.* 2010;2(1):a001669. https://doi.org/10.1101/cshperspect.a001669.

40. Wartlick O, Mumcu P, Jülicher F, Gonzalez-Gaitan M. Understanding morphogenetic growth control — lessons from flies. *Nat Rev Mol Cell Biol.* 2011;12(9):594—604. https://doi.org/10.1038/nrm3169.

41. Burke R, Basler K. Dpp receptors are autonomously required for cell proliferation in the entire developing Drosophila wing. *Development.* 1996;122(7): 2261—2269.

42. Capdevila J, Guerrero I. Targeted expression of the signaling molecule decapentaplegic induces pattern duplications and growth alterations in Drosophila wings. *EMBO J.* 1994;13(19):4459—4468. https://doi.org/10.1002/j.1460-2075.1994.tb06768.x.

43. Lecuit T, Brook WJ, Ng M, Calleja M, Sun H, Cohen SM. Two distinct mechanisms for long-range patterning by Decapentaplegic in the Drosophila wing. *Nature.* 1996; 381(6581):387. https://doi.org/10.1038/381387a0.

44. Posakony LG, Raftery LA, Gelbart WM. Wing formation in *Drosophila melanogaster* requires decapentaplegic gene function along the anterior-posterior compartment boundary. *Mech Dev.* 1990;33(1):69—82. https://doi.org/10.1016/0925-4773(90)90136-A.

45. Wartlick O, Mumcu P, Kicheva A, et al. Dynamics of Dpp signaling and proliferation control. *Science.* 2011;

331(6021):1154–1159. https://doi.org/10.1126/science.1200037.

46. Johnston LA, Gallant P. Control of growth and organ size in Drosophila. *Bioessays.* 2002;24(1):54–64. https://doi.org/10.1002/bies.10021.

47. Paluch EK, Nelson CM, Biais N, et al. Mechanotransduction: use the force(s). *BMC Biol.* 2015;13. https://doi.org/10.1186/s12915-015-0150-4.

48. Heisenberg C-P, Bellaïche Y. Forces in tissue morphogenesis and patterning. *Cell.* 2013;153(5):948–962. https://doi.org/10.1016/j.cell.2013.05.008.

49. Jaalouk DE, Lammerding J. Mechanotransduction gone awry. *Nat Rev Mol Cell Biol.* 2009;10(1):63–73. https://doi.org/10.1038/nrm2597.

50. Knöll R, Hoshijima M, Chien K. Cardiac mechanotransduction and implications for heart disease. *J Mol Med.* 2003;81(12):750–756. https://doi.org/10.1007/s00109-003-0488-x.

51. Shams H, Soheilypour M, Peyro M, Moussavi-Baygi R, Mofrad MRK. Looking "under the hood" of cellular mechanotransduction with computational tools: a systems biomechanics approach across multiple scales. *ACS Biomater Sci Eng.* 2017;3(11):2712–2726. https://doi.org/10.1021/acsbiomaterials.7b00117.

52. Cell Migration lab – cell adhesion. http://www.reading.ac.uk/cellmigration/adhesion.html.

53. Puckelwartz MJ, Kessler E, Zhang Y, et al. Disruption of nesprin-1 produces an Emery Dreifuss muscular dystrophy-like phenotype in mice. *Hum Mol Genet.* 2009;18(4):607–620. https://doi.org/10.1093/hmg/ddn386.

54. Vasquez CG, Martin AC. Force transmission in epithelial tissues. *Dev Dynam.* 2016;245(3):361–371. https://doi.org/10.1002/dvdy.24384.

55. Balaban NQ, Schwarz US, Riveline D, et al. Force and focal adhesion assembly: a close relationship studied using elastic micropatterned substrates. *Nat Cell Biol.* 2001;3(5):466. https://doi.org/10.1038/35074532.

56. Iyer KV, Pulford S, Mogilner A, Shivashankar GV. Mechanical activation of cells induces chromatin remodeling preceding MKL nuclear transport. *Biophys J.* 2012;103(7):1416–1428. https://doi.org/10.1016/j.bpj.2012.08.041.

57. Pioletti DP. Integration of mechanotransduction concepts in bone tissue engineering. *Comput Methods Biomech Biomed Eng.* 2013;16(10):1050–1055. https://doi.org/10.1080/10255842.2013.780602.

58. Santos LJ, Reis RL, Gomes ME. Harnessing magnetic-mechano actuation in regenerative medicine and tissue engineering. *Trends Biotechnol.* 2015;33(8):471–479. https://doi.org/10.1016/j.tibtech.2015.06.006.

59. Friedrich O, Schneidereit D, Nikolaev YA, et al. Adding dimension to cellular mechanotransduction: advances in biomedical engineering of multiaxial cell-stretch systems and their application to cardiovascular biomechanics and mechano-signaling. *Prog Biophys Mol Biol.* 2017;130:170–191. https://doi.org/10.1016/j.pbiomolbio.2017.06.011.

60. Resnick N, Yahav H, Shay-Salit A, et al. Fluid shear stress and the vascular endothelium: for better and for worse. *Prog Biophys Mol Biol.* 2003;81(3):177–199. https://doi.org/10.1016/S0079-6107(02)00052-4.

61. Brunette DM. Mechanical stretching increases the number of epithelial cells synthesizing DNA in culture. *J Cell Sci.* 1984;69(1):35–45.

62. Leung DY, Glagov S, Mathews MB. Cyclic stretching stimulates synthesis of matrix components by arterial smooth muscle cells in vitro. *Science.* 1976;191(4226):475–477. https://doi.org/10.1126/science.128820.

63. Montel F, Delarue M, Elgeti J, et al. Stress clamp experiments on multicellular tumor spheroids. *Phys Rev Lett.* 2011;107(18):188102. https://doi.org/10.1103/PhysRevLett.107.188102.

64. Dupont S, Morsut L, Aragona M, et al. Role of YAP/TAZ in mechanotransduction. *Nature.* 2011;474(7350):179–183. https://doi.org/10.1038/nature10137.

65. Wada K-I, Itoga K, Okano T, Yonemura S, Sasaki H. Hippo pathway regulation by cell morphology and stress fibers. *Development.* 2011;138(18):3907–3914. https://doi.org/10.1242/dev.070987.

66. Rauskolb C, Sun S, Sun G, Pan Y, Irvine KD. Cytoskeletal tension inhibits hippo signaling through an ajuba-warts complex. *Cell.* 2014;158(1):143–156. https://doi.org/10.1016/j.cell.2014.05.035.

67. Oh H, Irvine KD. Yorkie: the final destination of Hippo signaling. *Trends Cell Biol.* 2010;20(7):410–417. https://doi.org/10.1016/j.tcb.2010.04.005.

68. Pan Y, Heemskerk I, Ibar C, Shraiman BI, Irvine KD. Differential growth triggers mechanical feedback that elevates Hippo signaling. *Proc Natl Acad Sci USA.* 2016;113(45):E6974–E6983. https://doi.org/10.1073/pnas.1615012113.

69. Deng H, Wang W, Yu J, Zheng Y, Qing Y, Pan D. Spectrin regulates Hippo signaling by modulating cortical actomyosin activity. *eLife.* 2015;4:e06567. https://doi.org/10.7554/eLife.06567.

70. Fletcher GC, Elbediwy A, Khanal I, Ribeiro PS, Tapon N, Thompson BJ. The Spectrin cytoskeleton regulates the Hippo signalling pathway. *EMBO J.* 2015;34(7):940–954. https://doi.org/10.15252/embj.201489642.

71. Aragona M, Panciera T, Manfrin A, et al. A mechanical checkpoint controls multicellular growth through YAP/TAZ regulation by actin-processing factors. *Cell.* 2013;154(5):1047–1059. https://doi.org/10.1016/j.cell.2013.07.042.

72. Benham-Pyle BW, Pruitt BL, Nelson WJ. Cell adhesion. Mechanical strain induces E-cadherin-dependent Yap1 and β-catenin activation to drive cell cycle entry. *Science.* 2015;348(6238):1024–1027. https://doi.org/10.1126/science.aaa4559.

73. Zhao B, Wei X, Li W, et al. Inactivation of YAP oncoprotein by the Hippo pathway is involved in cell contact inhibition and tissue growth control. *Genes Dev.* 2007;21(21):2747–2761. https://doi.org/10.1101/gad.1602907.

74. Shraiman BI. Mechanical feedback as a possible regulator of tissue growth. *Proc Natl Acad Sci*. 2005;102(9): 3318–3323. https://doi.org/10.1073/pnas.0404782102.

75. Aegerter-Wilmsen T, Aegerter CM, Hafen E, Basler K. Model for the regulation of size in the wing imaginal disc of Drosophila. *Mech Dev*. 2007;124(4):318–326. https://doi.org/10.1016/j.mod.2006.12.005.

76. Brodland GW. How computational models can help unlock biological systems. *Semin Cell Dev Biol*. 2015;47(48): 62–73. https://doi.org/10.1016/j.semcdb.2015.07.001.

77. Turing AM. The chemical basis of morphogenesis. *Philos Trans R Soc Lond Ser B Biol Sci*. 1952;237(641):37–72.

78. Cartwright JHE. Labyrinthine turing pattern formation in the cerebral cortex. *J Theor Biol*. 2002;217(1):97–103. https://doi.org/10.1006/jtbi.2002.3012.

79. Gaffney EA, Monk NAM. Gene expression time delays and turing pattern formation systems. *Bull Math Biol*. 2006;68(1):99–130. https://doi.org/10.1007/s11538-006-9066-z.

80. Maini PK, Baker RE, Chuong C-M. The turing model comes of molecular age. *Science*. 2006;314(5804): 1397–1398. https://doi.org/10.1126/science.1136396.

81. Sun G-Q, Zhang G, Jin Z, Li L. Predator cannibalism can give rise to regular spatial pattern in a predator–prey system. *Nonlinear Dyn*. 2009;58(1–2):75–84. https://doi.org/10.1007/s11071-008-9462-z.

82. Reaction-Diffusion Model as a Framework for Understanding Biological Pattern Formation Science. http://science.sciencemag.org/content/329/5999/1616.

83. Ouyang Q, Swinney HL. Transition from a uniform state to hexagonal and striped Turing patterns. *Nature*. 1991; 352(6336):352610a0. https://doi.org/10.1038/352610a0.

84. Uriu K, Iwasa Y. Turing pattern formation with two kinds of cells and a diffusive chemical. *Bull Math Biol*. 2007; 69(8):2515–2536. https://doi.org/10.1007/s11538-007-9230-0.

85. Zhu J, Zhang Y-T, Alber MS, Newman SA. Bare bones pattern formation: a core regulatory network in varying geometries reproduces major features of vertebrate limb development and evolution. *PLoS One*. 2010;5(5): e10892. https://doi.org/10.1371/journal.pone.0010892.

86. Wolpert L. Positional information and the spatial pattern of cellular differentiation. *J Theor Biol*. 1969;25(1):1–47.

87. Briscoe J, Small S. Morphogen rules: design principles of gradient-mediated embryo patterning. *Development*. 2015;142(23):3996–4009.

88. Gurdon JB, Bourillot P-Y. Morphogen gradient interpretation. *Nature*. 2001;413(6858):797.

89. Aegerter-Wilmsen T, Heimlicher MB, Smith AC, et al. Integrating force-sensing and signaling pathways in a model for the regulation of wing imaginal disc size. *Development*. 2012;139(17):3221–3231. https://doi.org/10.1242/dev.082800.

90. Peskin CS. The immersed boundary method. *Acta Numer*. 2002;11:479–517. https://doi.org/10.1017/S0962492902000077.

91. Osher S, Sethian JA. Fronts propagating with curvature-dependent speed: algorithms based on Hamilton-Jacobi formulations. *J Comput Phys*. 1988;79(1):12–49. https://doi.org/10.1016/0021-9991(88)90002-2.

92. Chou C-S, Lo W-C, Gokoffski KK, et al. Spatial dynamics of multistage cell lineages in tissue stratification. *Biophys J*. 2010;99(10):3145–3154. https://doi.org/10.1016/j.bpj.2010.09.034.

93. Lo W-C, Chou C-S, Gokoffski KK, et al. Feedback regulation IN multistage cell lineages. *Math Biosci Eng*. 2009; 6(1):59–82. https://doi.org/10.3934/mbe.2009.6.59.

94. Zhang L, Lander AD, Nie Q. A reaction–diffusion mechanism influences cell lineage progression as a basis for formation, regeneration, and stability of intestinal crypts. *BMC Syst Biol*. 2012;6:93. https://doi.org/10.1186/1752-0509-6-93.

95. Du H, Nie Q, Holmes WR. The interplay between Wnt mediated expansion and negative regulation of growth promotes robust intestinal crypt structure and homeostasis. *PLoS Comput Biol*. 2015;11(8):e1004285. https://doi.org/10.1371/journal.pcbi.1004285.

96. Adam JA. A simplified mathematical model of tumor growth. *Math Biosci*. 1986;81(2):229–244. https://doi.org/10.1016/0025-5564(86)90119-7.

97. Cristini V, Lowengrub J, Nie Q. Nonlinear simulation of tumor growth. *J Math Biol*. 2003;46(3):191–224. https://doi.org/10.1007/s00285-002-0174-6.

98. Wise SM, Lowengrub JS, Frieboes HB, Cristini V. Three-dimensional multispecies nonlinear tumor growth—I: model and numerical method. *J Theor Biol*. 2008; 253(3):524–543. https://doi.org/10.1016/j.jtbi.2008.03.027.

99. Lander AD, Gokoffski KK, Wan FYM, Nie Q, Calof AL. Cell lineages and the logic of proliferative control. *PLoS Biol*. 2009;7(1):e1000015. https://doi.org/10.1371/journal.pbio.1000015.

100. Heldt FS, Lunstone R, Tyson JJ, Novák B. Dilution and titration of cell-cycle regulators may control cell size in budding yeast. *PLoS Comput Biol*. 2018;14(10): e1006548. https://doi.org/10.1371/journal.pcbi.1006548.

101. Chen W, Nie Q, Yi T-M, Chou C-S. Modelling of yeast mating reveals robustness strategies for cell-cell interactions. *PLoS Comput Biol*. 2016;12(7):e1004988. https://doi.org/10.1371/journal.pcbi.1004988.

102. Gou J, Lin L, Othmer HG. A model for the hippo pathway in the Drosophila wing disc. *Biophys J*. 2018;115(4): 737–747. https://doi.org/10.1016/j.bpj.2018.07.002.

103. Tyson JJ, Chen KC, Novak B. Sniffers, buzzers, toggles and blinkers: dynamics of regulatory and signaling pathways in the cell. *Curr Opin Cell Biol*. 2003;15(2):221–231. https://doi.org/10.1016/S0955-0674(03)00017-6.

104. Osborne JM, Fletcher AG, Pitt-Francis JM, Maini PK, Gavaghan DJ. Comparing individual-based approaches to modelling the self-organization of multicellular tissues. *PLoS Comput Biol*. 2017;13(2):e1005387. https://doi.org/10.1371/journal.pcbi.1005387.

105. Ahrens L, Tanaka S, Vonwil D, Christensen J, Iber D, Shastri VP. Generation of 3D soluble signal gradients in cell-laden hydrogels using passive diffusion. *Adv Biosyst.* 2019;3(1):1800237. https://doi.org/10.1002/adbi.201800237.

106. Fried P, Iber D. Dynamic scaling of morphogen gradients on growing domains. *Nat Commun.* 2014;5:5077. https://doi.org/10.1038/ncomms6077.

107. Unverdi SO, Tryggvason G. A front-tracking method for viscous, incompressible, multi-fluid flows. *J Comput Phys.* 1992;100(1):25–37. https://doi.org/10.1016/0021-9991(92)90307-K.

108. Glimm J, Grove JW, Li XL, Zhao N. Simple front tracking. *Contemp Math.* 1999;238(2):133–149.

109. Strychalski W, Copos CA, Lewis OL, Guy RD. A poroelastic immersed boundary method with applications to cell biology. *J Comput Phys.* 2015;282:77–97. https://doi.org/10.1016/j.jcp.2014.10.004.

110. Tanaka S. Simulation frameworks for morphogenetic problems. *Computation.* 2015;3(2):197–221. https://doi.org/10.3390/computation3020197.

111. Rejniak KA. An immersed boundary framework for modelling the growth of individual cells: an application to the early tumour development. *J Theor Biol.* 2007;247(1):186–204. https://doi.org/10.1016/j.jtbi.2007.02.019.

112. Fix GJ. Phase field models for free boundary problems. Free Boundary Problems: theory and Applications. *Pitman Res Notes Math Ser.* 1983;79.

113. Langer JS. Models of pattern formation in first-order phase transitions. *Dir Condens Matter Phys.* 1986;1:165–186. https://doi.org/10.1142/9789814415309_0005.

114. Wang Y, Luo Z, Zhang N, Kang Z. Topological shape optimization of microstructural metamaterials using a level set method. *Comput Mater Sci.* 2014;87:178–186. https://doi.org/10.1016/j.commatsci.2014.02.006.

115. CFD for the Design and Optimization of Slurry Bubble Column Reactors. https://doi.org/10.5772/intechopen.71361.

116. Macklin P, Lowengrub J. An improved geometry-aware curvature discretization for level set methods: application to tumor growth. *J Comput Phys.* 2006;215(2):392–401. https://doi.org/10.1016/j.jcp.2005.11.016.

117. Aras BS, Zhou YC, Dawes A, Chou C-S. The importance of mechanical constraints for proper polarization and psuedo-cleavage furrow generation in the early *Caenorhabditis elegans* embryo. *PLoS Comput Biol.* 2018;14(7):e1006294. https://doi.org/10.1371/journal.pcbi.1006294.

118. Brun CC, Lepore N, Pennec X, et al. A nonconservative Lagrangian framework for statistical fluid registration—SAFIRA. *IEEE Trans Med Imaging.* 2011;30(2):184–202. https://doi.org/10.1109/TMI.2010.2067451.

119. Ovadia J, Nie Q. Stem cell niche structure as an inherent cause of undulating epithelial morphologies. *Biophys J.* 2013;104(1):237–246. https://doi.org/10.1016/j.bpj.2012.11.3807.

120. Qiu Y, Chen W, Nie Q. Stochastic dynamics of cell lineage in tissue homeostasis. *Discrete Contin Dyn Syst.* 2019;22:1–24. https://doi.org/10.3934/dcdsb.2018339.

121. Iber D, Tanaka S, Fried P, Germann P, Menshykau D. Simulating tissue morphogenesis and signaling. *Methods Mol Biol.* 2015:323–338. https://doi.org/10.1007/978-1-4939-1164-6_21.

122. Gillespie DT. Exact stochastic simulation of coupled chemical reactions. *J Phys Chem.* 1977;81(25):2340–2361. https://doi.org/10.1021/j100540a008.

123. Erban R, Chapman J, Maini P. *A Practical Guide to Stochastic Simulations of Reaction-Diffusion Processes*; 2007. arXiv:0704.1908 [physics, q-bio] http://arxiv.org/abs/0704.1908.

124. Drawert B, Jacob B, Li Z, Yi T-M, Petzold L. A hybrid smoothed dissipative particle dynamics (SDPD) spatial stochastic simulation algorithm (sSSA) for advection–diffusion–reaction problems. *J Comput Phys.* 2019;378:1–17. https://doi.org/10.1016/j.jcp.2018.10.043.

125. Shinbrot T, Chun Y, Caicedo-Carvajal C, Foty R. Cellular morphogenesis in silico. *Biophys J.* 2009;97(4):958–967. https://doi.org/10.1016/j.bpj.2009.05.020.

126. Dicko M, Saramito P, Blanchard G, Lye C, Sanson B, Étienne J. Geometry can provide long-range mechanical guidance for embryogenesis. *PLoS Comput Biol.* 2017;13(3):e1005443.

127. Larson DE, Johnson RI, Swat M, Cordero JB, Glazier JA, Cagan RL. Computer simulation of cellular patterning within the Drosophila pupal eye. *PLoS Comput Biol.* 2010;6(7):e1000841.

128. Kursawe J, Brodskiy PA, Zartman JJ, Baker RE, Fletcher AG. Capabilities and limitations of tissue size control through passive mechanical forces. *PLoS Comput Biol.* 2015;11(12):e1004679. https://doi.org/10.1371/journal.pcbi.1004679.

129. Nematbakhsh A, Sun W, Brodskiy PA, et al. Multi-scale computational study of the mechanical regulation of cell mitotic rounding in epithelia. *PLoS Comput Biol.* 2017;13(5):e1005533.

130. Khalilgharibi N, Fouchard J, Asadipour N, et al. Stress relaxation in epithelial monolayers is controlled by actomyosin. *bioRxiv.* 2018:302158. https://doi.org/10.1101/302158.

131. Bambardekar K, Clément R, Blanc O, Chardès C, Lenne P-F. Direct laser manipulation reveals the mechanics of cell contacts in vivo. *Proc Natl Acad Sci.* 2015;112(5):1416–1421. https://doi.org/10.1073/pnas.1418732112.

132. Clément R, Dehapiot B, Collinet C, Lecuit T, Lenne P-F. Viscoelastic dissipation stabilizes cell shape changes during tissue morphogenesis. *Curr Biol.* 2017;27(20):3132–3142.e4. https://doi.org/10.1016/j.cub.2017.09.005.

133. Rodriguez EK, Hoger A, McCulloch AD. Stress-dependent finite growth in soft elastic tissues. *J Biomech.* 1994;27(4):455–467.

134. Taber LA. Biomechanics of cardiovascular development. *Annu Rev Biomed Eng.* 2001;3(1):1−25. https://doi.org/10.1146/annurev.bioeng.3.1.1.

135. Muñoz JJ, Barrett K, Miodownik M. A deformation gradient decomposition method for the analysis of the mechanics of morphogenesis. *J Biomech.* 2007;40(6):1372−1380. https://doi.org/10.1016/j.jbiomech.2006.05.006.

136. Ranft J, Basan M, Elgeti J, Joanny J-F, Prost J, Jülicher F. Fluidization of tissues by cell division and apoptosis. *Proc Natl Acad Sci.* 2010;107(49):20863−20868. https://doi.org/10.1073/pnas.1011086107.

137. Dillon R, Othmer HG. A mathematical model for outgrowth and spatial patterning of the vertebrate limb bud. *J Theor Biol.* 1999;197(3):295−330. https://doi.org/10.1006/jtbi.1998.0876.

138. Streichan SJ, Lefebvre MF, Noll N, Wieschaus EF, Shraiman BI. Global morphogenetic flow is accurately predicted by the spatial distribution of myosin motors. *eLife.* 2018;7. https://doi.org/10.7554/eLife.27454.

139. Blanchard GB, Fletcher AG, Schumacher LJ. The devil is in the mesoscale: mechanical and behavioural heterogeneity in collective cell movement. *Semin Cell Dev Biol.* 2018.

140. Tozluoğlu M, Duda M, Kirkland NJ, et al. Planar differential growth rates determine the position of folds in complex epithelia. *bioRxiv.* 2019:515528.

141. Voß-Böhme A, Starruß J, de Back W. Cellular Potts model. *Encycl Syst Biol.* 2013:386−390. https://doi.org/10.1007/978-1-4419-9863-7_298.

142. Swat MH, Hester SD, Balter AI, Heiland RW, Zaitlen BL, Glazier JA. Multicell simulations of development and disease using the CompuCell3D simulation environment. *Syst Biol.* 2009:361−428.

143. Palm MM, Merks RMH. Vascular networks due to dynamically arrested crystalline ordering of elongated cells. *Phys Rev.* 2013;87(1):012725. https://doi.org/10.1103/PhysRevE.87.012725.

144. Hester SD, Belmonte JM, Gens JS, Clendenon SG, Glazier JA. A multi-cell, multi-scale model of vertebrate segmentation and somite formation. *PLoS Comput Biol.* 2011;7(10):e1002155. https://doi.org/10.1371/journal.pcbi.1002155. San Francisco.

145. Magno R, Grieneisen VA, Marée AF. The biophysical nature of cells: potential cell behaviours revealed by analytical and computational studies of cell surface mechanics. *BMC Biophys.* 2015;8(1):8. https://doi.org/10.1186/s13628-015-0022-x.

146. Popławski NJ, Swat M, Scott Gens J, Glazier JA. Adhesion between cells, diffusion of growth factors, and elasticity of the AER produce the paddle shape of the chick limb. *Phys A Stat Mech Appl.* 2007;373:521−532. https://doi.org/10.1016/j.physa.2006.05.028.

147. Chaturvedi R, Huang C, Kazmierczak B, et al. On multiscale approaches to three-dimensional modelling of morphogenesis. *J R Soc Interface.* 2005;2(3):237−253. https://doi.org/10.1098/rsif.2005.0033.

148. Cerruti B, Puliafito A, Shewan AM, et al. Polarity, cell division, and out-of-equilibrium dynamics control the growth of epithelial structures. *J Cell Biol.* 2013;203(2):359. https://doi.org/10.1083/jcb.201305044.

149. Käfer J, Hayashi T, Marée AF, Carthew RW, Graner F. Cell adhesion and cortex contractility determine cell patterning in the Drosophilaretina. *Proc Natl Acad Sci.* 2007;104(47):18549−18554.

150. Voss-Böhme A. Multi-scale modeling in morphogenesis: a critical analysis of the cellular Potts model. *PLoS One.* 2012;7(9):e42852.

151. Thorne BC, Bailey AM, Peirce SM. Combining experiments with multi-cell agent-based modeling to study biological tissue patterning. *Briefings Bioinf.* 2007;8(4):245−257.

152. Fletcher AG, Cooper F, Baker RE. Mechanocellular models of epithelial morphogenesis. *Philos Trans R Soc B Biol Sci.* 2017;372(1720). https://doi.org/10.1098/rstb.2015.0519.

153. Nagai T, Honda H. A dynamic cell model for the formation of epithelial tissues. *Philos Mag B.* 2001;81(7):699−719. https://doi.org/10.1080/13642810108205772.

154. Odell GM, Oster G, Alberch P, Burnside B. The mechanical basis of morphogenesis: I. Epithelial folding and invagination. *Dev Biol.* 1981;85(2):446−462.

155. Mao Y, Tournier AL, Hoppe A, Kester L, Thompson BJ, Tapon N. Differential proliferation rates generate patterns of mechanical tension that orient tissue growth. *EMBO J.* 2013;32(21):2790−2803. https://doi.org/10.1038/emboj.2013.197.

156. Aliee M, Röper J-C, Landsberg KP, et al. Physical mechanisms shaping the Drosophila dorsoventral compartment boundary. *Curr Biol.* 2012;22(11):967−976.

157. Osterfield M, Du X, Schüpbach T, Wieschaus E, Shvartsman SY. Three-dimensional epithelial morphogenesis in the developing Drosophila egg. *Dev Cell.* 2013;24(4):400−410.

158. Barton DL, Henkes S, Weijer CJ, Sknepnek R. Active Vertex Model for cell-resolution description of epithelial tissue mechanics. *PLoS Comput Biol.* 2017;13(6):e1005569. https://doi.org/10.1371/journal.pcbi.1005569.

159. Staddon MF, Cavanaugh KE, Munro EM, Gardel ML, Banerjee S. *Mechanosensitive Junction Remodelling Promotes Robust Epithelial Morphogenesis.* 2019. https://doi.org/10.1101/648980.

160. Strickland D, Lin Y, Wagner E, et al. TULIPs: tunable, light-controlled interacting protein tags for cell biology. *Nat Methods.* 2012;9(4):379.

161. Fletcher AG, Osterfield M, Baker RE, Shvartsman SY. Vertex models of epithelial morphogenesis. *Biophys J.* 2014;106(11):2291−2304. https://doi.org/10.1016/j.bpj.2013.11.4498.

162. Newman TJ. Modeling multicellular structures using the subcellular element model. *Single Cell Based Model Biol Med.* 2007:221−239.

163. Sandersius SA, Newman TJ. Modeling cell rheology with the subcellular element model. *Phys Biol.* 2008;5(1):015002.

164. Gardiner BS, Wong KK, Joldes GR, et al. Discrete element framework for modelling extracellular matrix, deformable cells and subcellular components. *PLoS Comput Biol.* 2015;11(10):e1004544.

165. Christley S, Lee B, Dai X, Nie Q. Integrative multicellular biological modeling: a case study of 3D epidermal development using GPU algorithms. *BMC Syst Biol.* 2010;4(1):107.

166. Veldhuis JH, Mashburn D, Hutson MS, Brodland GW. Practical aspects of the cellular force inference toolkit (CellFIT). *Methods Cell Biol.* 2015;125:331–351. https://doi.org/10.1016/bs.mcb.2014.10.010.

167. Noll N, Streichan SJ, Shraiman BI. *Geometry of Epithelial Cells Provides a Robust Method for Image Based Inference of Stress within Tissues.* 2018. arXiv preprint arXiv:1812.04678.

168. Delile J, Herrmann M, Peyriéras N, Doursat R. A cell-based computational model of early embryogenesis coupling mechanical behaviour and gene regulation. *Nat Commun.* 2017;8:13929. https://doi.org/10.1038/ncomms13929.

169. Menshykau D, Blanc P, Unal E, Sapin V, Iber D. An interplay of geometry and signaling enables robust lung branching morphogenesis. *Development.* 2014;141(23):4526–4536. https://doi.org/10.1242/dev.116202.

170. Cilfone NA, Kirschner DE, Linderman JJ. Strategies for efficient numerical implementation of hybrid multi-scale Agent-based models to describe biological systems. *Cell Mol Bioeng.* 2015;8(1):119–136. https://doi.org/10.1007/s12195-014-0363-6.

171. Schilling S, Willecke M, Aegerter-Wilmsen T, Cirpka OA, Basler K, von Mering C. Cell-sorting at the a/P boundary in the Drosophila wing primordium: a computational model to consolidate observed non-local effects of Hh signaling. *PLoS Comput Biol.* 2011;7(4):e1002025. https://doi.org/10.1371/journal.pcbi.1002025.

172. Smith AM, Baker RE, Kay D, Maini PK. Incorporating chemical signalling factors into cell-based models of growing epithelial tissues. *J Math Biol.* 2012;65(3):441–463.

173. Salbreux G, Barthel LK, Raymond PA, Lubensky DK. Coupling mechanical deformations and planar cell polarity to create regular patterns in the zebrafish retina. *PLoS Comput Biol.* 2012;8(8):e1002618. https://doi.org/10.1371/journal.pcbi.1002618.

174. Marin-Riera M, Brun-Usan M, Zimm R, Välikangas T, Salazar-Ciudad I. Computational modeling of development by epithelia, mesenchyme and their interactions: a unified model. *Bioinformatics.* 2016;32(2):219–225. https://doi.org/10.1093/bioinformatics/btv527.

175. Meir E, Munro EM, Odell GM, Von Dassow G. Ingeneue: a versatile tool for reconstituting genetic networks, with examples from the segment polarity network. *J Exp Zool.* 2002;294(3):216–251. https://doi.org/10.1002/jez.10187.

176. Merks RMH, Guravage M, Inzé D, Beemster GTS. VirtualLeaf: an open-source framework for cell-based modeling of plant tissue growth and development. *Plant Physiol.* 2011;155(2):656–666. https://doi.org/10.1104/pp.110.167619.

177. Swat MH, Thomas GL, Belmonte JM, Shirinifard A, Hmeljak D, Glazier JA. Chapter 13 —multi-scale modeling of tissues using CompuCell3D. *Methods Cell Biol.* 2012;110:325–366. https://doi.org/10.1016/B978-0-12-388403-9.00013-8.

178. Hoehme S, Drasdo D. A cell-based simulation software for multi-cellular systems. *Bioinformatics.* 2010;26(20):2641–2642. https://doi.org/10.1093/bioinformatics/btq437.

179. Aegerter-Wilmsen T, Smith AC, Christen AJ, Aegerter CM, Hafen E, Basler K. Exploring the effects of mechanical feedback on epithelial topology. *Development.* 2010;137(3):499–506. https://doi.org/10.1242/dev.041731.

180. Hufnagel L, Teleman AA, Rouault H, Cohen SM, Shraiman BI. On the mechanism of wing size determination in fly development. *Proc Natl Acad Sci USA.* 2007;104(10):3835–3840. https://doi.org/10.1073/pnas.0607134104.

180a. Buchmann A, Alber M, Zartman JJ. Sizing it up: the mechanical feedback hypothesis of organ growth regulation. In: *Seminars in cell & developmental biology.* 35. Academic Press; 2014, (November):73–81.

181. Hamaratoglu F, de Lachapelle AM, Pyrowolakis G, Bergmann S, Affolter M. Dpp signaling activity requires pentagone to scale with tissue size in the growing Drosophila wing imaginal disc. *PLoS Biol.* 2011;9(10):e1001182. https://doi.org/10.1371/journal.pbio.1001182.

182. Ben-Zvi D, Pyrowolakis G, Barkai N, Shilo B-Z. Expansion-repression mechanism for scaling the Dpp activation gradient in Drosophila wing imaginal discs. *Curr Biol.* 2011;21(16):1391–1396. https://doi.org/10.1016/j.cub.2011.07.015.

183. Baker Ruth E, Peña J-M, Jayamohan J, Jérusalem A. Mechanistic models versus machine learning, a fight worth fighting for the biological community? *Biol Lett.* 2018;14(5):20170660. https://doi.org/10.1098/rsbl.2017.0660.

184. Xu C, Wang C, Ji F, Yuan X. Finite-element neural network-based solving 3-D differential equations in MFL. *IEEE Trans Magn.* 2012;48(12):4747–4756. https://doi.org/10.1109/TMAG.2012.2207732.

# Computational Morphogenesis of Embryonic Bone Development: Past, Present, and Future

MATTHEW E. DOLACK • CHANYOUNG LEE • YING RU • ARSALAN MARGHOUB •
JOAN T. RICHTSMEIER • ETHYLIN WANG JABS • MEHRAN MOAZEN •
DIEGO A. GARZÓN-ALVARADO • REUBEN H. KRAFT

## 1. INTRODUCTION

Computational prediction of bone formation and growth is a challenge many researchers across numerous fields hope to solve. If a robust model was developed, validated, and implemented into the healthcare system, improved prognoses with respect to developmental bone abnormalities (e.g., craniosynostosis and clubfoot), as well as improvements in planning reconstructive surgery, would result. Prognoses would improve in part by physicians' newfound ability to test potential osteogenic growth outcomes, which would then allow them to recommend the safest, most effective treatment plan. According to the Centers for Disease Control and Prevention, it is estimated that between 2004 and 2006, 1 in every 709 births include a developmental bone abnormality, specifically a musculoskeletal birth defect.[1] Skeletal anomalies are an important area of research and study because of their high incidence rate,[1] expensive treatment plans,[2] and contribution to increased rates of infant mortality.[2] Unfortunately, osteogenic formation and growth models with validated prediction capabilities do not currently exist.

Techniques for simulating bone development during embryogenesis can be split into two main categories: cellular biomechanics[3] and ligand-receptor-based signaling.[4,5] Predictive models that use cellular mechanics at their core typically build off of the differential adhesion hypothesis[6,7] and "mechanically integrated cell motile behavior."[3,8,9] Cellular mechanics-based models are not used as frequently as their ligand-receptor-based counterparts for simulating embryonic skeletal development because of limitations in patterning ability. Ligand-receptor-based models use coupled reaction-diffusion (RD) equations to predict osteogenic development.[4,10]

Research into bone-modeling proteins lends itself to RD-based embryonic bone modeling because it provides a framework for pattern formation that is linked to chemical interaction and evolution.[11–13] Cellular biomechanics and ligand-receptor-based signaling theory can be paired together to create more robust, comprehensive models.[4] Because RD systems are at the core of many osteogenic formation and growth techniques, they will be discussed in greater detail. Readers interested in cellular mechanics-based embryonic models are directed to Delile et al.[3] and Wyczalkowski et al.[6]

## 2. REACTION-DIFFUSION SYSTEMS

An RD equation can be classified as a semilinear parabolic partial differential equation composed of three terms:

$$\frac{\partial u}{\partial t} = R(u) + D\nabla^2 u \tag{4.2.1}$$

where $u$ is the molar concentration of a chemical substance, $t$ is time, $R(u)$ is local reactions, $D$ is mass diffusivity, and $\nabla^2$ is the Laplace operator.[14,15] For example, the Laplacian of the scalar concentration $u$ in three dimensions is defined as $\sum_{i=1}^{3} \frac{\partial^2 u}{\partial x_i^2}$. Molar concentration in an RD system changes temporally and spatially until a steady-state solution is reached. Mass diffusivity is commonly assumed to be Fickian,[16] and if all local reaction terms were assumed to be negligible the RD equation would simplify into Fick's second law of diffusion (i.e., the heat equation). A pair of RD equations is considered coupled when the local reaction term for one RD equation is dependent on the molar concentration variable from the other RD equation, such that

Mechanobiology. https://doi.org/10.1016/B978-0-12-817931-4.00011-X

$$\frac{\partial u}{\partial t} = R_1(u, v) + D_u \nabla^2 u \qquad (4.2.2)$$

$$\frac{\partial v}{\partial t} = R_2(u, v) + D_v \nabla^2 v \qquad (4.2.3)$$

where $v$ is the molar concentration of a different chemical substance. The local reaction term $R_i(u, v)$ describes chemical propagation and degradation. $R_1(u, v)$ and $R_2(u, v)$ in Eqs. 4.2.2 and 4.2.3, respectively, can be equal, but they do not have to be. Local reaction terms are selected based on the intended application of the RD system. The RD equation format described in Eq. 4.2.1 as well as the RD system format described in Eqs. 4.2.2 and 4.2.3 were theorized by Turing in 1952.[14] When using an RD system to predict osteogenic formation and growth, there are well-established local reaction terms that are documented to have morphologic significance.[14,17]

## 3. MORPHOGENIC REACTION-DIFFUSION SYSTEMS

Morphogenesis is a field of study in developmental biology that aims to understand how biological organisms develop their characteristic shape. An example of morphogenesis is how an accumulation of initially unstructured mesenchymal cells undergo differentiation and growth into the bones of the cranial vault.[4] Cellular organization and distribution during morphogenesis is widely considered a patterning-related effect.[17] Use of RD systems to predict naturally occurring nonhomogeneous chemical patterns was first documented in 1952 by Alan Turing.[14]

### 3.1. Turing Patterns

Turing published his seminal work on natural pattern formation entitled *The Chemical Basis of Morphogenesis* in 1952.[14] Turing theorized that "a system of chemical substances … reacting together and diffusing through a tissue, is adequate to account for the main phenomena of morphogenesis."[14] The term "Turing pattern" describes RD systems that are used to predict patterns that occur within nature (e.g., zebra stripes; cheetah spots). An RD system in the format of Eqs. 4.2.2 and 4.2.3, which were introduced in Section 2, has the capability of producing Turing patterns. Turing patterns arise when initial nearly homogeneous molar concentrations develop instabilities that are then amplified to generate a nonhomogeneous spatial pattern.[14,18] Instability can result from either statistical or random disturbances in the initial molar concentration.[14] A homogeneous Neumann boundary condition (i.e., zero flux) is used in RD systems because all domain boundaries must be impermeable to molar leakage.

The ability of an RD system to predict naturally occurring patterns has been previously demonstrated (Fig. 4.2.1). Within the dashed boxes of Fig. 4.2.1 the upper images show the patterning of a young *Pomacanthus* angelfish and the lower images show the respective Turing pattern predictions. All pattern predictions within the dashed boxes are from the same simulation, with differences resulting from time-step evolution.[19] The patterns in Fig. 4.2.1 were created using a two-equation coupled RD system.[19]

Turing patterns are only created through the solution of coupled RD equations. Fig. 4.2.2 reveals that Turing patterns can be vastly different from one another depending on local reaction terms and their associated parameters. All stable Turing patterns form a nonhomogeneous pattern that will reach steady state.[20] Turing patterns were further investigated with respect to local reaction terms by Alfred Gierer and Hans Meinhardt.[17,21]

### 3.2. Common Morphogenic Reaction-Diffusion Systems

Advancements made by Gierer and Meinhardt along with the established significance of Turing patterns provided the groundwork for extending RD systems to embryonic patterning.[17,20] Embryonic patterning, also referred to as morphogenic patterning, is created using coupled RD systems commonly composed of only two RD equations. When Turing first theorized RD system pattern formation, he referred to all molar concentration terms as "morphogens."[14] Morphogens can be loosely defined as chemical substances capable of governing morphogenic pattern development. Later, in 1972, when Gierer and Meinhardt studied morphogenic RD systems, they reclassified the molar concentration terms as activator, inhibitor, or substrate.[17] Using the RD system described in Eqs. 4.2.2 and 4.2.3 as an example, $u$ could be activator and $v$ could be inhibitor. In a pattern-producing system, the activator has short-range self-enhancement effects and both inhibitor and substrate have long-range activator inhibition effects.[22] Activator-substrate and activator-inhibitor models are two commonly used RD systems in developmental biology.[22–24]

#### 3.2.1. Activator-substrate models

In activator-substrate systems, short-range activator self-enhancement causes localized substrate depletion.[22] Inhibition effects result from the complete depletion of substrate, which has the effect of halting further activator production. In an RD system, the rate of diffusion

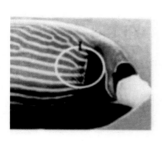
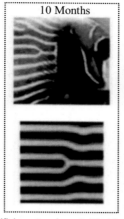
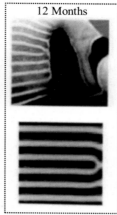
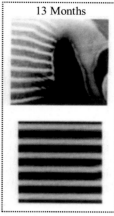

FIG. 4.2.1 *Pomacanthus* angelfish patterning at 10, 12, and 13 months of age. Each biological angelfish pattern is also computationally replicated with a reaction-diffusion-based Turing pattern. (From Kondo S, Asai R. A reaction–diffusion wave on the skin of the marine angelfish Pomacanthus. *Nature.* 1995; with permission.)

for each chemical substance must not be equal.[22] More specifically, the rate of inhibitor diffusion must be greater than the rate of activator diffusion. This ensures the eventual cessation of activator diffusion within the system[22] and the emergence of a pattern. A commonly used activator-substrate system is

$$\frac{\partial a}{\partial t} = \alpha_a - \beta_a a + \gamma_a a^2 s + D_a \nabla^2 a \qquad (4.2.4)$$

$$\frac{\partial s}{\partial t} = \alpha_s - \gamma_s a^2 s - \beta_s s + D_s \nabla^2 s \qquad (4.2.5)$$

where $\alpha_a$ and $\alpha_s$ are activator and substrate production rates, respectively, and $\beta_a$, $\beta_s$, $\gamma_a$, and $\gamma_s$ are reaction coefficients.[17] Subscripts $a$ and $s$ identify activator and substrate coefficients, respectively. An activator-substrate system is described schematically in Fig. 4.2.3A: ① represents nonlinear self-enhancement of activator, which shares a proportionality to the substrate concentration[25] and ② and ③ represent the substrate inhibition and activator production, respectively.[25] Fig. 4.2.3B is a one-dimensional (1D) plot of an activator-substrate system

with the activator concentration in blue and the substrate concentration in green. Because activator consumes substrate, the levels of activator and substrate concentration are out of phase with one another.[26]

### 3.2.2. Activator-inhibitor models

In activator-inhibitor systems, a positive feedback loop and a negative feedback loop are coupled together.[27] The self-promotion of activator, which is indicated by ① in Fig. 4.2.4A, is the positive feedback loop. ② and ③ in Fig. 4.2.4A form the negative feedback loop where activator production promotes inhibitor production, which in turn promotes activator degradation. Fig. 4.2.4B is a 1D plot of an activator-inhibitor system with the activator concentration in blue and the inhibitor concentration in orange. Inspection of Fig. 4.2.4B reveals that activator and inhibitor concentrations are in phase with one another.[26] Furthermore, inhibitor concentration can never exceed activator concentration because of the self-enhancement properties of activator and the system constraint that inhibitor must diffuse faster

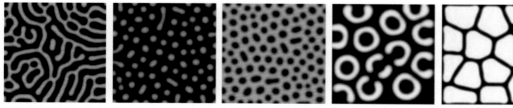

FIG. 4.2.2 Various types of Turing patterns. (From Kondo S, Miura T. Reaction-diffusion model as a framework for understanding biological pattern formation. *Science.* 2010; with permission.)

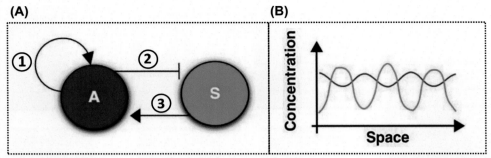

FIG. 4.2.3 Activator-substrate system interaction **(A)** and one-dimensional domain plot **(B)**. (From Marcon L, Sharpe J. Turing patterns in development: What about the horse part? *Current Opinion in Genetics & Development*. 2012; with permission.)

than activator.[26] The system constraint that inhibitor is allowed to diffuse faster than activator is purely mathematical.[26] Failure to satisfy this constraint will result in a nonpattern producing activator-inhibitor system.[26]

The chemical networks responsible for cellular self-organization during embryogenesis are considered replicable by the Gierer-Meinhardt model.[21,28] The Gierer-Meinhardt model is an activator-inhibitor system composed of two coupled RD equations and is defined as follows:

$$\frac{\partial a}{\partial t} = \alpha_a - \beta_a a + \gamma_a \frac{a^2}{h} + D_a \nabla^2 a \qquad (4.2.6)$$

$$\frac{\partial h}{\partial t} = \alpha_h - \beta_h h + \gamma_h a^2 + D_h \nabla^2 h \qquad (4.2.7)$$

where $\beta_a, \beta_h, \gamma_a,$ and $\gamma_h$ are reaction coefficients.[17] Subscripts $a$ and $h$ identify activator and inhibitor coefficients, respectively. Each RD equation in the Gierer-Meinhardt model has three local reaction terms. $\alpha_a$ and $\alpha_h$ represent activator and inhibitor production, respectively.[21] $\beta_a a$ and $\beta_h h$ represent activator and inhibitor degradation, respectively.[21] $\gamma_a \frac{a^2}{h}$ and $\gamma_h a^2$ represent nonlinear activator production, with the former including a $\frac{1}{h}$ inhibition term.[21] When an RD system is used for predicting embryonic bone development the generated patterns can reflect the initial position of mineralization, or centers of ossification.[29,30] Two theories for predicting embryonic bone development are discussed in Sections 5 and 6. Before discussing these theories, however, the spatiotemporal factors affecting RD patterning must be discussed.

## 4. SPATIOTEMPORAL FACTORS AFFECTING PATTERNING

RD patterning exhibits extreme sensitivity to domain mesh, domain geometry, and initial molar concentration. Initial condition changes and mesh alterations can produce patterning ranging from functionally identical to substantially different. Similar outcomes can result from changes to model domain. Fig. 4.2.2 showed how slight changes to reaction coefficients

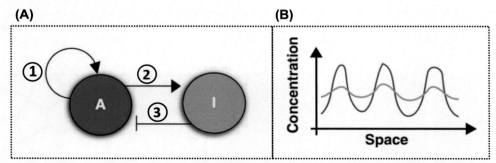

FIG. 4.2.4 Activator-inhibitor system interaction **(A)** and one-dimensional domain plot **(B)**. (From Marcon L, Sharpe J. Turing patterns in development: What about the horse part? *Current Opinion in Genetics & Development*. 2012; with permission.)

and diffusion rates can produce significant pattern variation.[28] When multiple RD variables (e.g., mesh, initial concentration, and domain) are adjusted simultaneously, the effects on pattern variation can be compounded. By way of new results and findings, we present a study on the behavior of RD systems where the primary interest is model reproducibility.

## 4.1. Mesh Dependency and Initial Molar Concentration Sensitivity

RD patterning has a unique dependence on domain mesh and initial molar concentration. Fundamentally, this arises from the spatial dependence of the Laplace operator in Euclidean space, which is an operator found in every RD equation. A 268-element unstructured triangular mesh is shown in Fig. 4.2.5A, and a 225-element structured quadrilateral mesh is shown in Fig. 4.2.5B. In the center of Fig. 4.2.5B is a red transparent square that represents a prescribed region of elevated activator concentration. This prescribed region is representative of an RD system initial condition. If the region of elevated activator concentration found in Fig. 4.2.5B was mapped to Fig. 4.2.5A, the initially square prescribed region would be distorted according to the shape of proximal mesh elements. A qualitative mapping of how the region distorts is indicated by the blue region (Fig. 4.2.5A) relative to the original, undistorted prescribed region (purple). If an RD system were to use the meshing schemes discussed in Fig. 4.2.5, as well as their respective prescribed regions of initial

activator concentration, the systems would return a different patterning. More specifically, both meshes would return patterning with the same geometric objects (i.e., spikes, stripes, or swirls) but in domain locations different from one another. In the case of Fig. 4.2.5, patterning will be spatiotemporally different because more activator is present in Fig. 4.2.5A than in Fig. 4.2.5B. Additionally, *distribution differences in the prescribed initial condition will result in dissimilar amplification of initial activator concentration fluctuations.*[18] In conclusion, it is theorized that activator concentration differences along with dissimilar amplification pathways cause geometrically similar, yet spatiotemporally different, patterns to form. Even if the nodal molar concentrations in Fig. 4.2.5 were assumed to be variable for each mesh element, the mapping would still be imperfect and solution inconsistencies would result. Fig. 4.2.5 and its analysis are a theory for how initial molar concentration and domain mesh can yield mesh-dependent patterning in an RD system. Examples of mesh-dependent patterning behavior are presented in the remainder of this section.

The qualitative effects of initial activator concentration and domain mesh on RD patterning described in Fig. 4.2.5 are demonstrated quantitatively in Fig. 4.2.6 that uses the Gierer-Meinhardt RD model. The same unstructured mesh is used for Fig. 4.2.6A and B. Additionally, Fig. 4.2.6C and D use the same structured mesh. Statistics for these two meshes can be found in Table A1 of Appendix A. The four domains in Fig. 4.2.6

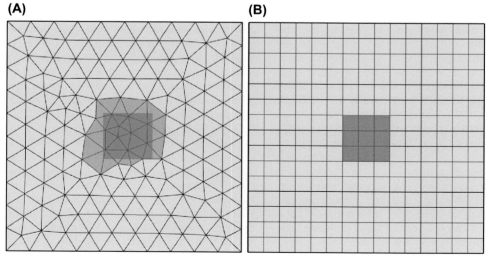

**(A)**  **(B)**

FIG. 4.2.5 Mapping an initial molar concentration between two different meshes. The red square in **(B)** identifies a prescribed region of elevated molar concentration. The blue region in **(A)** is the mapping of the red square found in **(B)**.

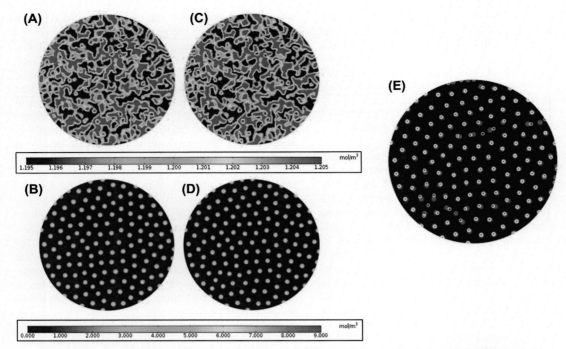

FIG. 4.2.6 Reaction-diffusion sensitivity analysis with respect to domain mesh and initial activator concentration. **(A and C)** The initial random distribution of activator applied over the unstructured and structured meshes, respectively. **(B and D)** Gierer-Meinhardt-based patterning generated using **(A)** and **(C)**, respectively. **(E)** is **(B)** superimposed with **(D)**.

describe activator distribution, with Fig. 4.2.6A and C detailing the prescribed initial condition and Fig. 4.2.6B and D detailing the returned Gierer-Meinhardt pattern. The random activator distributions shown in Fig. 4.2.6A and C were created using the same mapping (i.e., they have the same distribution). All parameters and coefficients used to resolve the Gierer-Meinhardt patterns contained in Fig. 4.2.6 can be found in Table A2 of Appendix A. Additionally, the stability analysis used to ensure patterning in Fig. 4.2.6 is discussed in Appendix A. The COMSOL simulations detailed in Fig. 4.2.6 are available on GitHub (https://github.com/PSUCompBio/ReactionDiffusionStudies). Inspection of Fig. 4.2.6B and D reveals patterning in the form of localized regions of elevated concentration, or hot spots, that appear to be similar with respect to the diameter and number produced. However, when Fig. 4.2.6B and D is superimposed (Fig. 4.2.6E), spatial differences in hot-spot locations become evident. The positions of some hot spots correspond almost perfectly with one another, whereas others are considerably disparate. Using the structural similarity (SSIM) index function in MATLAB the similarity of Fig. 4.2.6D with respect to Fig. 4.2.6B was calculated to

be 93.9%. Fig. 4.2.6E elucidates how sensitive RD models are to the prescribed initial molar concentration and domain mesh. More specifically, RD models that are identical to one another except for their domain mesh can produce patterns that, *although similar, are spatially different*. The two meshes used to generate Fig. 4.2.6B and D are extremely fine and more than adequate to achieve convergence to a single solution if it were possible. Potential ramifications of these findings include the inability to replicate patterning from one RD study to another unless the exact same domain mesh is used.

The mesh dependency detailed in Fig. 4.2.6 can potentially be explained as either phenotypic variation or individual variation depending on the application of the RD system. Phenotypic variation in the context of RD systems typically describes differences in physical morphology (e.g., craniosynostosis and polydactylism).[4,31,32] Individual variation in the context of RD systems describes differences in naturally occurring patterns. From a biologist's perspective, mesh-dependent behavior in RD systems is not important because only the produced pattern is significant. This idea, however, is troubling for a computationalist because solutions would not be reproducible. Furthermore, it complicates modeling situations where the time and place of pattern emergence is more important than the produced pattern. Two examples of individual variation that align with the mesh-dependent patterning provided in Fig. 4.2.6 are shown in Fig. 4.2.7. Two different cheetah forearms are shown in Fig. 4.2.7A and the underside of two different Hula painted frogs are shown in Fig. 4.2.7B. No two animal coats or skins are identical.[33] The cheetah forearms in Fig. 4.2.7A contain circled regions of interest for easier comparison. The frog undersides in Fig. 4.2.7B have distinctly noncircular spots indicated with yellow arrows and strings of spots indicated with blue contours.

Some embryonic bone growth models use RD systems that include non-RD equations.[10,24,34] RD systems of this form, however, still exhibit mesh-dependent patterning behavior. Any equation that cannot satisfy Eq. 4.2.1, which was previously introduced, is a non-RD equation when present in an RD system. An example of such a system is as follows:

$$\frac{\partial a}{\partial t} = H(o_s - o) \left[ \alpha_a + \alpha_o o - \beta_a a + \gamma_a \frac{a^2}{h} + D_a \nabla^2 a \right] \quad (4.2.8)$$

$$\frac{\partial h}{\partial t} = H(o_s - o) \left[ \alpha_h - \beta_h h + \gamma_h a^2 + D_h \nabla^2 h \right] \quad (4.2.9)$$

$$\frac{\partial o}{\partial t} = \eta H(o_s - o) H \left( \left[ \frac{a^2}{h} \right] - a_T \right) \quad (4.2.10)$$

where $o$ is molar concentration, $H(\dots)$ are Heaviside step functions, $\alpha_o$ is a reaction coefficient that promotes molar $o$ production, $\eta$ is a reaction multiplier, and both

$a_T$ and $o_s$ represent molar concentration thresholds. The remaining parameters in Eqs. 4.2.8 and 4.2.9 were previously described. The Heaviside step function $H(o_s - o)$ has the discontinuous form of $H(o_s \gg o) = 1$ and $H(o_s \ll o) = 0$. Additionally, the Heaviside step function $H\left( \left[ \frac{a^2}{h} \right] - a_T \right)$ has the discontinuous form of $H(a^2/h \gg a_T) = 1$ and $H(a^2/h \ll a_T) = 0$. Equations 4.2.8–4.2.10 describe an RD system that was adapted from Lee's doctoral thesis.[4] Fig. 4.2.8A and C contain activator distributions based on Eq. 4.2.8 using an unstructured and a structured mesh, respectively. The distribution of molar $o$ concentration from Eq. 4.2.10 is shown in Fig. 4.2.8B and D using an unstructured and a structured mesh, respectively. The meshes used in Fig. 4.2.8 are identical to the meshes used in Fig. 4.2.6. All distributions shown in Fig. 4.2.8A−D were generated using the same initial concentration mapping and the parameters contained in Table A3 of Appendix A. The COMSOL simulations detailed in Fig. 4.2.8 are available on GitHub (https://github.com/PSUCompBio/ReactionDiffusionStudies). Inspection of the activator distributions in Fig. 4.2.8 with respect to those in Fig. 4.2.6 suggests that RD systems composed of one or more non-RD equations can result in patterning that is uniquely different from well-established Turing patterns. Outlines of prominent regions within the unstructured domains of Fig. 4.2.8A and B were superimposed with the structured domains of Fig. 4.2.8C and D, respectively. The superimposed results for activator distribution are shown in Fig. 4.2.8E, and the superimposed results for molar $o$ distribution

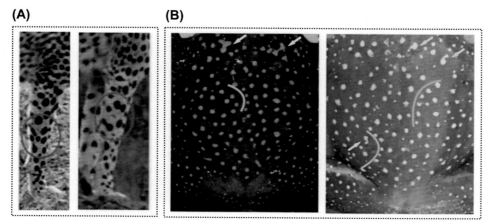

**(A)**   **(B)**

FIG. 4.2.7 Individual variation within a species compared using the **(A)** cheetah forearm and **(B)** Hula painted frog. (From Chelysheva EV. A new approach to cheetah identification. Cat News. 2004 and Perl RGB, Geffen E, Malka Y et al. Population genetic analysis of the recently rediscovered Hula painted frog (Latonia nigriventer) reveals high genetic diversity and low inbreeding. *Sci Rep.* 2018; with permission.)

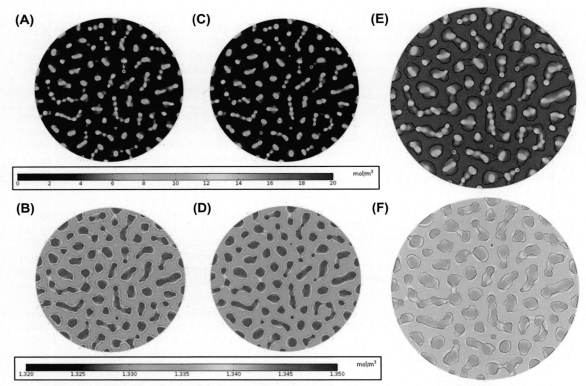

FIG. 4.2.8 **(A–D)** Domain distributions for a three-equation reaction-diffusion (RD) system composed of two RD equations and one non-RD equation. **(A and C)** Activator distributions generated using an unstructured and a structured mesh, respectively. **(B and D)** Molar *o* distributions generated using an unstructured and a structured mesh, respectively. **(E and F)** The superimposed distributions for activator and molar *o* concentration, respectively.

are shown in Fig. 4.2.8F. A significant difference between the superimposed results of Fig. 4.2.6E and F is that in Fig. 4.2.8F, there are no pattern outliers. Fig. 4.2.8E and F yielded SSIM index percentages of 96.5% and 85.9%, respectively. *Therefore mesh-dependent patterning behavior occurs even in RD systems that are not composed entirely of RD equations.*

These differences, however, can perhaps be classified as individual phenotypic variation. Phenotypic variation in RD systems is manifested through altered morphology. Examples of altered morphology produced via genetic manipulation are shown in Fig. 4.2.9 using the paw of *Mus musculus*.[32] Each paw in Fig. 4.2.9 contains five digits of similar, yet different, morphology. Assuming the distributions of molar *o* concentration in Fig. 4.2.8B and D represent newly developed embryonic bone, definite parallels can be drawn to the paw specimens contained in Fig. 4.2.9. More specifically, mesh-dependent patterning that

results in nonidentical embryonic bone development can be described, according to Fig. 4.2.9, as individual phenotypic variation. In addition to phenotypic variation caused by a genetic variant, it can result from natural (i.e., normal) development. Naturally occurring phenotypic variation is the reason identical twins, and even clones, are not exactly alike. Therefore the embryonic bone models in Fig. 4.2.8B and D are likely representative of naturally occurring individual phenotypic variation too. Whether or not differences in the mesh correspond with individual differences (e.g., the distribution of undifferentiated mesenchymal cells) is an open question. However, when trying to understand how a mutation changes the activator-inhibitor partnership, a mesh-invariant system becomes necessary.

## 4.2. Domain Geometry

Another RD parameter that affects patterning is domain geometry. In RD systems, even simple changes in

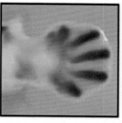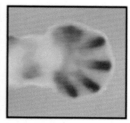

FIG. 4.2.9 Individual phenotypic variation in the paw of *Mus musculus*. (From Sheth R, Marcon L, Félix Bastida M et al. Hox genes regulate digit patterning by controlling the wavelength of a Turing-type mechanism. *Science*. 2012; with permission.)

domain geometry can have an impact on patterning. Fig. 4.2.10 contains three meshed domains that are slightly different in shape. Each domain is patterned using the Schnakenberg RD model. The Schnakenberg model is a coupled two-equation RD system.[35] In Fig. 4.2.10, the upper images represent the meshed domain, while the lower images are the resulting Schnakenberg pattern. Identical parameter values and initial conditions were used to generate all the patterning in Fig. 4.2.10. The interested reader is directed to Zhu et al.[36] for a comprehensive breakdown of the simulation used to generate Fig. 4.2.10. All domains in Fig. 4.2.10 are of similar dimensions, with Fig. 4.2.10A—C representing the simple shapes of a square, circle, and ellipse, respectively. If the domains in Fig. 4.2.10 were enlarged in a manner that preserved their overall shape, there would be no effect on the hot-spot patterning produced.[37] Furthermore, the patterning would be similar, yet spatiotemporally different. If the domains in Fig. 4.2.10 were decreased past a critical dimension, there would be a patterning shift from hot spots to stripes.[36,38] Inspection of Fig. 4.2.10 *reveals that simple geometry variations can produce patterning that is dissimilar in shape (i.e., circular vs. oblong hot spots) and spatially different.*[36,38,39] Because the meshes used in Fig. 4.2.10A—C are not identical, the resulting patterning differences can be partially attributed to mesh dependency. These findings complicate the use of RD-based patterning solutions when applied to complex domain geometries.

## 4.3. Nonrandom Initial Concentration Perturbations

Nonrandom initial concentration perturbations are another RD parameter capable of affecting patterning. RD systems can produce shaped patterns if prompted with nonrandom initial concentration perturbations. Pattern shaping describes the alignment of small-scale shapes into large-scale patterns. A nonrandom initial

concentration perturbation is shown in Fig. 4.2.11A using four localized hot spots of activator that are 0.5% greater than the background activator concentration. Fig. 4.2.11B contains the patterned domain that was generated using the activator perturbation in Fig. 4.2.11A. The patterning in Fig. 4.2.11B was generated using the Gierer-Meinhardt model. All associated model parameters are outlined in Table A2 of Appendix A apart from the initial activator distribution. The COMSOL simulation detailed in Fig. 4.2.11 is available on GitHub (https://github.com/PSUCompBio/ReactionDiffusionStudies). The shaped patterning present in Fig. 4.2.11B is partially highlighted in Fig. 4.2.11C using dashed red circles. Using the circular pattern indicators in Fig. 4.2.11C the presence of four semisymmetric quadrants becomes apparent. Additionally, inspection of Fig. 4.2.11B reveals that activator hot spots align themselves in concentric circles that emanate from each initial concentration perturbation. This finding is significant from a computationalist's perspective because it suggests that RD systems can be prompted to generate large-scale patterns. From a biologist's perspective, this finding is significant because nonrandom initial concentration perturbations are more likely to occur over a real biological domain than a completely random distribution. In the next section an RD system developed for modeling embryonic cranial vault growth is presented.

## 5. A MODEL FOR CRANIAL VAULT GROWTH

"The cranial vault develops and grows through the mechanism of intramembranous ossification."[40] During intramembranous ossification, sheets of mesenchymal stem cells (MSCs) undergo differentiation into osteoblasts that then cluster into condensations and form centers of ossification.[4,31,40] Ossification centers secrete an unmineralized organic matrix called osteoid that is "primarily composed of type I collagen fibers

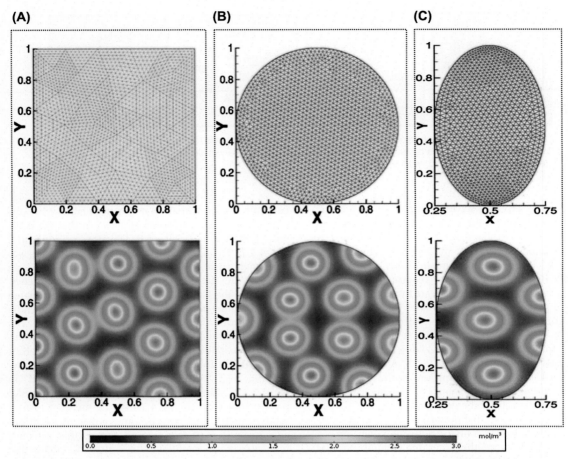

FIG. 4.2.10 Pattern sensitivity with respect to changing domain geometry explored using **(A)** square, **(B)** circle, and **(C)** ellipse domains. (From Zhu J, Zhang Y-T, Newman SA et al. Application of discontinuous Galerkin methods for reaction-diffusion systems in developmental biology. *J Sci Comput*. 2009; with permission.)

and serves as a template for inorganic hydroxyapatite crystals."[40] Osteoblasts trapped within the expanding matrix produced by the ossification centers become osteocytes.[40] "Once sufficient bone matrix is produced to form a small island of bone, additional osteoblasts are recruited to the surface, where there is continued" bone production.[40] As the bones of the cranial vault grow and approximate, one another sutures form between them.[41] When sutures are patent, they function as growth sites and allow new bone to form at the edges of the approximating bone fronts.[41] A suture can only function as an intermembranous growth site if sufficient osteoblasts are available at the bone front and all cells within the suture exist in an undifferentiated state.[41] Craniosynostosis, a condition in which sutures

close prematurely, manifests itself as mineralization of the cells of the suture and a cessation of growth at the edges of approximating bone fronts.[31] A theory for intramembranous cranial vault ossification using an activator-inhibitor RD system is shown in Fig. 4.2.12. When osteoblasts cluster into ossification centers, there is a release of activator.[4,34] Activator secretion signals MSCs to differentiate into osteoblasts.[4,34] Because inhibitor degrades long-range MSC differentiation, the formation of nonossified sutures results.[4] Bone modeling research suggests that activator and inhibitor in Fig. 4.2.12 represent bone morphogenetic proteins (BMPs) and Noggin proteins, respectively.[4,11,23,24] BMP is a morphogenetic bone growth protein that when secreted induces MSC differentiation into

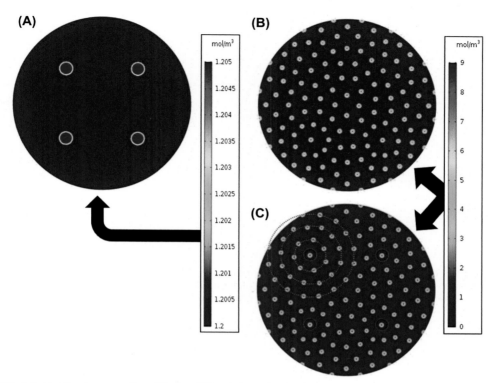

FIG. 4.2.11 Shaping a reaction-diffusion (RD) solution using nonrandom initial concentration perturbations. **(A)** A nonrandom initial concentration perturbation and **(B)** the resulting RD pattern. The shaped patterning in **(B)** is identified using dashed red circles in **(C)**.

osteoblasts.[11,12] BMP secretion has the added effect of signaling the release of Noggin.[11] Noggin reduces the effectiveness of BMP-induced MSC differentiation.[12] Additionally, newly differentiated osteoblasts exhibit the self-enhancement effect of BMP secretion.[42] A significant assumption of the activator (BMP)-inhibitor (Noggin) model in Fig. 4.2.12 is that the compounded effects of Noggin cause suture formation to occur.[4] The previously introduced RD system outlined in Eqs. 4.2.8–4.2.10 was developed based on the ossification system described in Fig. 4.2.12.

Using the framework introduced in Fig. 4.2.12, a schematic for intramembranous cranial vault ossification was developed. The developed schematic is shown in Fig. 4.2.13 and builds off an activator-inhibitor RD model. Fig. 4.2.13 is split into two biological response levels: molecular and cellular. Activator-inhibitor-related processes take place at the molecular level, and ossification-related processes take place at the cellular level. At the molecular level in Fig. 4.2.13, activator simultaneously promotes the production of itself (①) and the inhibitor (②). The inhibitor then inhibits

activator production (③). Through a series of chemical pathways the activator induces MSC differentiation into osteoblasts (④).[4] Differentiated osteoblasts then begin expressing activator (⑤), which promotes adjacent undifferentiated MSCs to differentiate.[4] Lastly, inhibitor at the molecular level has the cellular level effect of inhibiting osteoblast differentiation (⑥). Using Fig. 4.2.13, a three-equation RD system was developed. The developed RD system was introduced in Section 4.1 as an example using Eqs. 4.2.8–4.2.10. The only clarification needed with respect to Eqs. 4.2.8–4.2.10 is that the molar $o$ concentration in Eq. 4.2.10 represents osteoblast concentration. Osteoblasts have an approximate diameter of 20 μm.[43] Therefore each mesh element in the meshed domains documented in Table A1 of Appendix A represents approximately 120 osteoblasts.

## 6. SIMULATING CRANIAL VAULT GROWTH WITH IMBEDDED MECHANICAL STRAIN

Recent research suggests that mechanical stimuli are an important component of bone development and

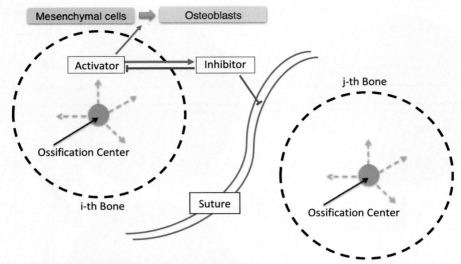

FIG. 4.2.12 A theorized framework for intramembranous cranial vault ossification based on an activator-inhibitor reaction-diffusion system. (From Lee C. A computational analysis of bone formation in the cranial vault using a reaction-diffusion-strain model. 2018; with permission.)

remodeling.[44–47] During embryonic development of the cranial vault the underlying brain grows rapidly.[48,49] Rapid growth of the underlying brain suggests that the flat bones of the developing cranial vault may be under nonnegligible intracranial pressure, or applied mechanical strain.[50–52] Therefore mechanical strain must be incorporated into Eqs. 4.2.8–4.2.10 if a comprehensive predictive model for cranial vault ossification is to be developed. Before adjusting

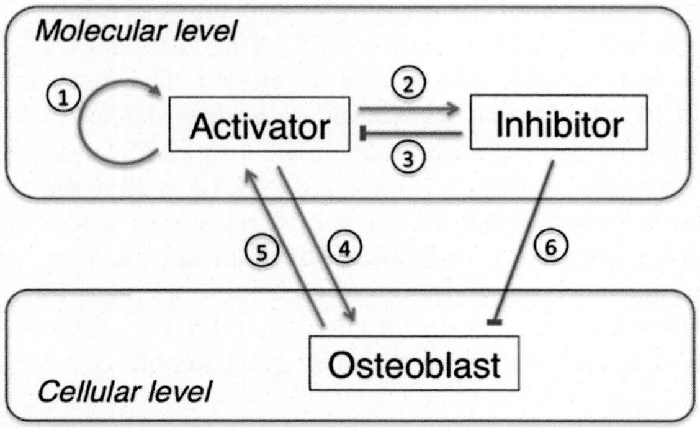

FIG. 4.2.13 A theorized schematic for intramembranous cranial vault ossification that builds off an activator-inhibitor reaction-diffusion model. (From Lee C. A computational analysis of bone formation in the cranial vault using a reaction-diffusion-strain model. 2018; with permission.)

Eqs. 4.2.8–4.2.10, a revised version of the schematic in Fig. 4.2.13 was developed to include mechanical strain. The revised schematic is provided in Fig. 4.2.14. Because strain effects are imparted from the growing brain on the developing neurocranium, a tissue-level response was added to Fig. 4.2.14. Tensile strain at the tissue level in Fig. 4.2.14 promotes the production of the activator (⑦ and ⑧).[4] Additionally, the accumulated volumetric strain greater than a threshold value promotes osteoblast production (⑨).[4] Using Fig. 4.2.14 as well as Eqs. 4.2.8–4.2.10, the following RD system with imbedded mechanical strain was developed:

$$\frac{\partial a}{\partial t} + V \cdot \nabla a = H(o_s - o)\left[ (\acute{E}_v + \acute{E}_0)(\alpha_a + \alpha_o o) - \beta_a a \right.$$
$$\left. + \gamma_a \frac{a^2}{h} + D_a \nabla^2 a \right] \qquad (4.2.11)$$

$$\frac{\partial h}{\partial t} + V \cdot \nabla h = H(o_s - o)\left[ \alpha_h - \beta_h h + \gamma_h a^2 + D_h \nabla^2 h \right] \qquad (4.2.12)$$

$$\frac{\partial o}{\partial t} + V \cdot \nabla o = \eta H(o_s - o) \, H\left( \left[ \frac{a^2}{h} \right] - a_T \right) H(E_v - E_T) \qquad (4.2.13)$$

where $\acute{E}_v$, $E_v$, $\acute{E}_0$, and $E_T$ are all volumetric measures of strain rate, accumulated strain, strain rate threshold, and accumulated strain threshold, respectively. Advection effects in Eqs 4.2.11–4.2.13 are represented by $V \cdot \nabla a$, $V \cdot \nabla h$, and $V \cdot \nabla o$, respectively, where $V$ represents brain growth velocity and $\nabla$ represents the spatial gradient

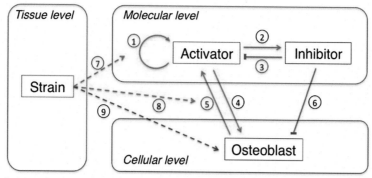

FIG. 4.2.14 A revised schematic for intramembranous cranial vault ossification that builds off an activator-inhibitor reaction-diffusion model and incorporates mechanical strain. (From Lee C. A computational analysis of bone formation in the cranial vault using a reaction-diffusion-strain model. 2018; with permission.)

of activator, inhibitor, or osteoblast. The remaining parameters in Eqs. 4.2.11—4.2.13 were previously described. The Heaviside step function $H(E_v - E_T)$ in Eq. 4.2.13 has the discontinuous form of $H(E_v \gg E_T) = 1$ and $H(E_v \ll E_T) = 0$. Eqs. 4.2.11—4.2.13 were first used to predict cranial vault development in the mouse. The first step in this process was determining the strain field applied to the mouse neurocranium.

Using magnetic resonance microscopic (MRM) scans of embryonic mice brains as well as micro computed tomographic (μCT) scans of embryonic mice neurocrania, a series of computational cranial domains at developmental intervals were developed. Each domain volume is composed of three surfaces: the internal surface coincides with the underlying brain, the external surface coincides with the differentiating MSCs, and the inner layer connects the internal and external surfaces.[4] A normal mapping of domain displacements at developmental intervals was then used to calculate the strain field induced by the growing brain. The process of domain development and strain field extraction is outlined in Fig. 4.2.15. MRM scans of the mice brain and μCT scans of the mice neurocrania are shown in Fig. 4.2.15A for embryonic day 15.5 (E15.5) and E17.5 as well as the day of birth (postnatal day 0, or P0). The light blue volumes in Fig. 4.2.15A represent the mice brain and are anatomically oriented within their respective μCT neurocranium rendering. A superior view of the extracted E15.5, E17.5, and P0 computational domains is shown in Fig. 4.2.15B. Also shown in Fig. 4.2.15B is the computational domain for developmental day E13.5, which is assumed to be an ellipse.[4] The domains in Fig. 4.2.15B were then inlaid with one another as seen in Fig. 4.2.15C and a normal mapping was performed to determine their

respective displacement fields. The resulting displacement fields were then used to calculate the Green-Lagrange strain for use in Eqs. 4.2.11 and 4.2.13. Interpolation of the displacement field allows strain values between the four developmental milestones to be determined. After determining the applied strain field and the starting computational domain, the RD system was run.

With the induced strain field from the growing brain calculated and the elliptic computational domain of E13.5 established, the RD system described in Eqs. 4.2.11—4.2.13 was run. The system was evaluated using the open-source numerical solver OpenFOAM. A structural mechanics solver was paired with the finite volume method to evaluate the RD system. Those interested in a more detailed explanation of the RD system and its solution method are directed to Lee[4] and the full source code located on GitHub (https://github.com/PSUCompBio/skull-growth-modeling). Fig. 4.2.16 compares experimentally observed cranial vault growth (lower image in each dotted rectangle) in *Mus musculus* against model predictions (upper image in each dotted rectangle) for developmental days E14.5 through P0. The predicted regions of cranial vault growth in Fig. 4.2.16 represent regions of osteoblast concentration that are greater than a threshold saturation. All parameters used to generate the cranial vault predictions in Fig. 4.2.16 are outlined in Table B1 of Appendix B Inspection of Fig. 4.2.16 reveals model predictions that align well with experimental observations for E16.5, E17.5, and P0 except for the development of a nonsingular interparietal bone. Additionally, the model prematurely predicts frontal and parietal bone growth in E14.5 and E15, respectively. However, the predicted order of

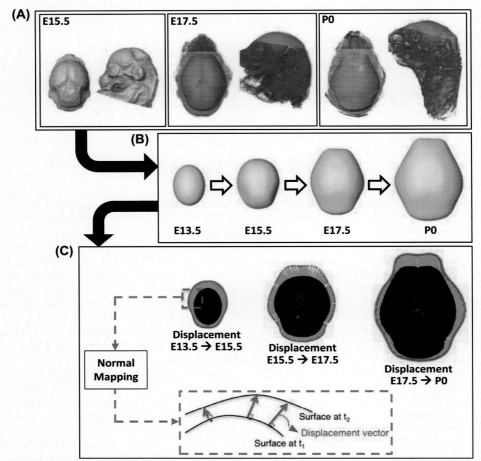

FIG. 4.2.15 Computational domain development and strain field extraction for the cranial vault of *Mus musculus*. (From Lee C. A computational analysis of bone formation in the cranial vault using a reaction-diffusion-strain model. 2018; with permission.)

emergence of the parietal and interparietal bones matches experimental observations.

Growth predictions were validated by linking model parameters to genetic mutations and replicating the resulting developmental abnormalities in the cranial vault. Two mutation-based validation studies were conducted using the RD model. The findings for both studies are outlined in Fig. 4.2.17. Fig. 4.2.17A shows the predictive ability of the model to fuse the coronal suture by reducing the osteoblast production coefficient, $\alpha_o$. In the study by Clendenning and Mortlock,[53] it was determined that the growth factor *Gdf6*, a member of the BMP family, has a similar effect on the coronal suture when made genetically inoperable. Genes made inoperable are commonly referred to as knocked

out. When *Gdf6* functions normally (wild type), the coronal suture remains patent (Fig. 4.2.17A, upper left), but when *Gdf6* is knocked out (*Gdf6*−/−), bone forms and the coronal suture closes prematurely (Fig. 4.2.17A, upper right).[53] Model predictions detailing a similar effect are documented in the lower images of Fig. 4.2.17A through the reduction of $\alpha_o$. In this case of craniosynostosis, *Gdf6* operates as $\alpha_o$ in the computational model. Yu et al.[54] conducted experiments that showed that Axin2 is associated with premature closure of the metopic suture. Fig. 4.2.17B shows the influence of Axin2 on metopic suture formation, revealing that when Axin2 is knocked out, the metopic suture closes. Because Axin2 inhibits MSC differentiation into osteoblasts,[55] all model parameters with inhibitory roles

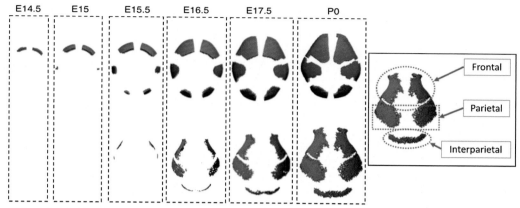

FIG. 4.2.16 Experimental bone growth observations (lower image in each dotted rectangle) compared against model predictions (upper image in each dotted rectangle) at different developmental stages. (From Lee C. A computational analysis of bone formation in the cranial vault using a reaction-diffusion-strain model. 2018; with permission.)

were investigated with respect to metopic suture effects. The investigation revealed that either increasing inhibitor degradation, $\beta_h$, or decreasing inhibitor interaction, $\gamma_h$, closed the metopic suture by fusing both frontal bones. Along with a reference prediction of cranial development the effects of increasing $\beta_h$ and decreasing $\gamma_h$ on metopic suture formation are shown in the lower images of Fig. 4.2.17B. Based on these findings, we interpret $\beta_h$ and $\gamma_h$ as representative

of the function of Axin2 in Eq. 4.2.12. In the next section, recent advancements in postnatal cranial vault modeling are discussed.

## 7. ADVANCEMENTS IN POSTNATAL CRANIAL VAULT MODELING

Marghoub et al.[51,56] suggest that early postnatal cranial vault development, specifically suture formation, can

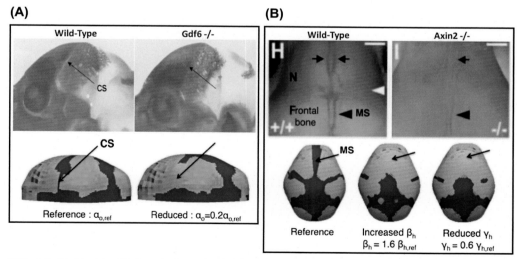

FIG. 4.2.17 Model validation studies using **(A)** *Gdf6*-induced coronal suture fusion and **(B)** Axin2-induced metopic suture fusion. CS, *Coronal suture*; MS, *Metopic suture*. (From Lee C. A computational analysis of bone formation in the cranial vault using a reaction-diffusion-strain model. 2018 and Yu H-MI, Jerchow B, Sheu T-J, et al. The role of Axin2 in calvarial morphogenesis and craniosynostosis.*Development*. 2005 and Clendenning DE, Mortlock DP. The BMP ligand Gdf6 prevents differentiation of coronal suture mesenchyme in early cranial development. *PLOS ONE*. 2012; with permission.)

be modeled using a purely hydrostatic strain-based tissue-differentiation algorithm. The hydrostatic strain field in the model by Marghoub et al.' is calculated via an expansion analysis of underlying mice brain growth.[51,56] Furthermore, the strain field is the sole variable that drives neurocranium growth and suture formation.[51,56] The mechanobiological differentiation algorithm of Marghoub et al. was adapted from Carter et al.[57] and computes by increasing the elastic modulus of suture elements that are proximal to nearby calvarial bone and below a threshold level of hydrostatic strain.[51,56] The elastic modulus is increased in increments of 250 MPa, a value determined through the extrapolation of mouse calvarial properties between P10 and P20 for each period of simulated growth.[51,58] Fig. 4.2.18 compares in silico suture growth and cessation against an ex vivo wild-type mouse at P3, P7, and P10. Inspection of Fig. 4.2.18 reveals "a gradual reduction in suture sizes across all … sutures from P3 to P10."[51] Additionally, the in silico results "capture the overall pattern of bone formation across the skull."[51] Although the model of Marghoub et al.' yields promising results, it is not without its limitations. For example, using solely the level of hydrostatic strain to propagate growth assumes that all biological and nonbiological mechanisms, specifically those documented to have a role in cranial development, are negligible.[51]

Garzón-Alvarado[59] and Garzón-Alvarado et al.[24] proposed a biochemical model for approximating the complex mechanisms that regulate membranous bone growth in the neurocranium.[60] Burgos-Flórez et al.[60] later adapted this biochemical model to simulate "the formation, maintenance and interdigitation of cranial sutures during human prenatal development and infancy."[60] The work of Burgos-Flórez et al.' is novel because it "describes how sutures form and maintain their phenotypical characteristics … [when] complex biochemical regulatory mechanisms"[60] are dictating neurocranium growth. Using their RD-based biochemical model, Burgos-Flórez et al. were able to "determine the time and location of suture formation during prenatal development and the emergence of the interdigitated patterns seen in sutures during infancy."[60] Additionally, they determined that "suture fate is dependent on the ability of suture cells to respond to biochemical signals coming from the developing flat bones"[60] of the neurocranium. This finding suggests that premature suture fusion (i.e., craniosynostosis) may result from atypical osteoinhibitory sensing ability.[60] Burgos-Flórez et al. also showed "that interdigitated suture morphologies are the result of local variations in the concentration of biochemical factors along opposing bone fronts, which conjointly regulate bone formation and resorption events at the sutures."[60] Fig. 4.2.19 shows a time evolution of suture interdigitation. Interdigitation

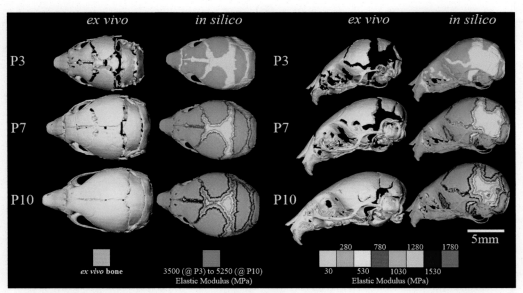

FIG. 4.2.18 Computationally predicted postnatal cranial vault growth compared to experimentally observed wild-type development at P3, P7, and P10. (From Marghoub A, Libby J, Babbs C et al. Characterizing and modeling bone formation during mouse calvarial development. *Phys Rev Lett.* 2019; with permission.)

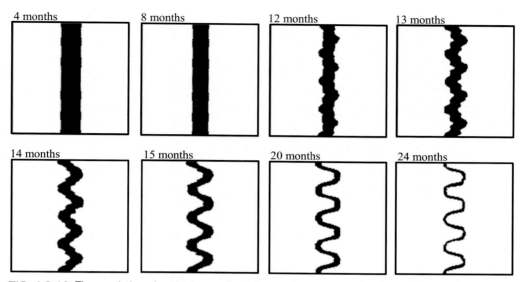

FIG. 4.2.19 Time evolution of sagittal suture interdigitation from 4 months to 2 years. (From Burgos-Flórez FJ, Gavilán-Alfonso ME, Garzón-Alvarado DA. Flat bones and sutures formation in the human cranial vault during prenatal development and infancy: A computational model. *Journal of Theoretical Biology*. 2016; with permission.)

begins at approximately 12 months and is followed by continuous narrowing until cessation at 24 months. In Fig. 4.2.20, a morphologic comparison of suture interdigitation and fusion between 'the computationally predicted calvaria by Burgos-Flórez et al. and an adult calvaria is shown.

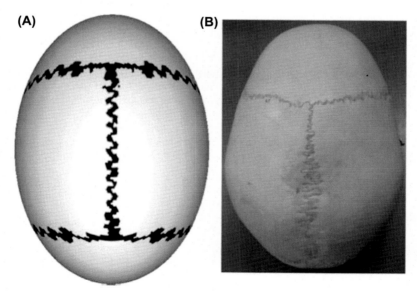

FIG. 4.2.20 Morphologic comparison of suture interdigitation and fusion between **(A)** computationally predicted calvaria and **(B)** an adult calvaria. (From Burgos-Flórez FJ, Gavilán-Alfonso ME, Garzón-Alvarado DA. Flat bones and sutures formation in the human cranial vault during prenatal development and infancy: A computational model. *Journal of Theoretical Biology*. 2016; with permission.)

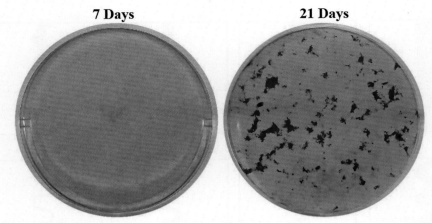

FIG. 4.2.21 Unstrained osteoblast cultures at 7 and 21 days. (Courtesy of Ying Ru at Icahn School of Medicine at Mount Sinai.)

## 8. FUTURE WORK AND CONCLUSION

Our intention is to move forward with the without-strain bone growth model given by Eqs. 4.2.8–4.2.10 and the with-strain bone growth model given by Eqs. 4.2.11–4.2.13 through further refinement and validation. Model validation will be done primarily through morphologic comparison of in vitro osteogenic differentiation assays and in vivo bone growth experiments. Two in vitro osteogenic assays at different stages of differentiation are shown in Fig. 4.2.21. Both assays were cultured using a differentiation protocol adapted from Lee et al.[61] The assays have undergone alizarin staining to identify regions of osteoblast mineralization (Fig. 4.2.21). At 7 and 21 days, there is approximately 0.0% and 8.3% osteoblast mineralization, respectively, in a 3.48-cm-diameter petri dish. The assays in Fig. 4.2.21 are useful for model comparison and validation because many of the complexities associated with in vivo growth (e.g., three-dimensionality, changing size and shape of the domain, and changing boundary conditions with growth) are not present. Model validation studies will also be conducted using osteogenic assays at longer growth durations and greater mineralization percentages than those in Fig. 4.2.21. Strain effects can also be incorporated into the osteogenic assay using the Flexcell culture straining system. A Flexcell culture plate is depicted in Fig. 4.2.22 for a test specimen before and during strain loading. Test specimen strain results from vacuum pressure applied circumferentially to the deformable membrane identified in Fig. 4.2.22. An investigation into potential factors affecting mesh dependence in RD systems is also planned.

Another RD concept that will be further explored is whether mesh-dependent RD systems can be stabilized to produce mesh-independent results. Preliminary findings suggest that RD stabilization terms exist. When stabilization terms are incorporated into an RD system, it allows mesh-independent results to be produced by a mesh-dependent system. We theorize that mechanical strain acts as a stabilization term. RD pattern stabilization is an important focus

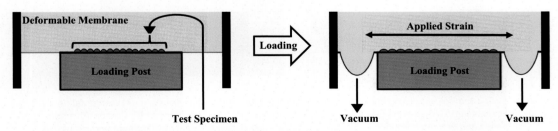

FIG. 4.2.22 A Flexcell culture plate depicting a test specimen before and during strain loading.

because it enables solution reproducibility and allows for control of changing phenotypes. Embryonic ossification is a complex biological process. Even so, RD-based modeling of embryonic ossification has advanced considerably in the past decade, with researchers across various fields contributing promising findings. As our understanding of embryonic ossification improves and the capabilities of computational modeling advance, the development of a validated osteogenic growth model draws closer.

## APPENDIX A

Section 4.1 detailed reaction-diffusion (RD) sensitivity with respect to initial molar concentration and domain mesh. Two meshes, one unstructured and the other structured, were used in the analysis. Table A1 details both the meshes used. The Gierer-Meinhardt RD model used in Sections 4.1 and 4.3 was run using COMSOL Multiphysics. All parameters and coefficients used to simulate the Gierer-Meinhardt RD system are detailed in Table A2.

RD systems need to be properly initiated to ensure nonhomogeneous pattern formation.[62] Using the stability analysis documented by Koch and Meinhardt,[62] equations for steady-state molar concentration and a

parameter inequality constraint were derived. Steady-state molar concentration equations were derived by setting both the Laplacian and temporal rate of change terms equal to zero in each RD equation and then solving the resulting system of equations.[4,34] When this process is applied to the Gierer-Meinhardt model the following equations result:

$$a_0 = \frac{\beta_h \gamma_a}{\beta_a \gamma_h} + \frac{\alpha_a}{\beta_a} \qquad \text{(A1)}$$

$$h_0 = \frac{\gamma_h a_0^2}{\beta_h} \qquad \text{(A2)}$$

where $a_0$ is the steady-state activator concentration and $h_0$ is the steady-state inhibitor concentration.[4] If the steady-state activator and inhibitor concentrations were then perturbed the system solution would become:

$$a = a_0 + \delta_{a_0} e^{\omega t} \cos(2\pi kx) \qquad \text{(A3)}$$

$$h = h_0 + \delta_{h_0} e^{\omega t} \cos(2\pi kx) \qquad \text{(A4)}$$

where $\delta_{i_0}$ is the steady-state perturbation amplitude, $\omega$ is the perturbation frequency, and $k$ is the wave number.[62] Patterning in an RD system occurs when the real part of complex $\omega$ is positive.[34] Manipulation of the Gierer-Meinhardt model along with Eqs. A1–A4 result in the following inequality constraint:

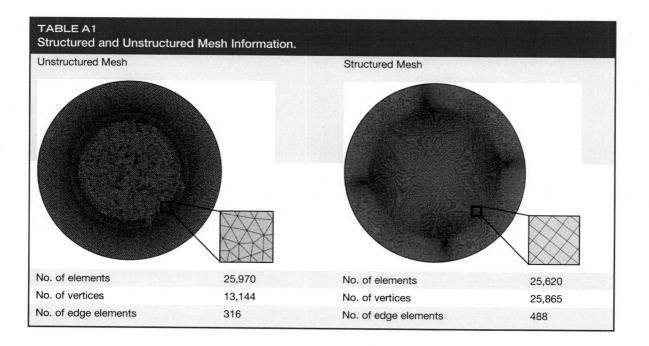

**TABLE A1**
**Structured and Unstructured Mesh Information.**

| Unstructured Mesh | | Structured Mesh | |
|---|---|---|---|
| No. of elements | 25,970 | No. of elements | 25,620 |
| No. of vertices | 13,144 | No. of vertices | 25,865 |
| No. of edge elements | 316 | No. of edge elements | 488 |

**TABLE A2**
**The Gierer-Meinhardt Model Parameters and Coefficients.**

| Simulated Time | 1.037e07 (s) |
|---|---|
| $\Delta t$ | 3.111e03 (s) |
| $a_0$ | 1.20 ($\pm$0.5%) (mol/m$^3$) |
| $h_0$ | 1.44 (mol/m$^3$) |
| $\alpha_a$ | 1.00e-05 (mol$/[\text{m}^3 * \text{s}]$) |
| $\alpha_h$ | 1.00e-09 (mol$/[\text{m}^3 * \text{s}]$) |
| $\beta_a$ | 5.00e-05 (1/s) |
| $\beta_h$ | 1.00e-04 (1/s) |
| $\gamma_a$ | 5.00e-05 (1/s) |
| $\gamma_h$ | 1.00e-04 (m$^3$/mol$*$s) |
| $D_a$ | 1.20e-12 (m$^2$/s) |
| $D_h$ | 1.20e-10 (m$^2$/s) |

**TABLE A3**
**Parameters and Coefficients Used in Three-Equation Reaction-Diffusion System.**

| Simulated Time | 1.037e07 (s) |
|---|---|
| $\Delta t$ | 3.111e03 (s) |
| $a_0$ | 1.20 ($\pm$0.5%) (mol/m$^3$) |
| $h_0$ | 1.44 (mol/m$^3$) |
| $o_0$ | 1.33 (mol/m$^3$) |
| $\alpha_a$ | 1.00e-08 (mol$/[\text{m}^3 * \text{s}]$) |
| $\alpha_h$ | 1.00e-09 (mol$/[\text{m}^3 * \text{s}]$) |
| $\alpha_o$ | 1.00e-07 (1/s) |
| $\beta_a$ | 5.00e-05 (1/s) |
| $\beta_h$ | 1.00e-04 (1/s) |
| $\gamma_a$ | 5.00e-05 (1/s) |
| $\gamma_h$ | 1.00e-04 (m$^3$/mol$*$s) |
| $D_a$ | 2.20e-12 (m$^2$/s) |
| $D_h$ | 2.20e-10 (m$^2$/s) |
| $\eta$ | 4.00e-07 (mol$/[\text{m}^3 * \text{s}]$) |
| $a_T$ | 3.50 (mol/m$^3$) |
| $o_s$ | 1.30 (mol/m$^3$) |

$$\frac{2\beta_h\gamma_a}{\beta_h\gamma_a + \alpha_a\gamma_h} - 1 \le \frac{\beta_h}{\beta_a} \le \frac{D_h}{D_a}\left[\sqrt{\frac{2\beta_h\gamma_a}{\beta_h\gamma_a + \alpha_a\gamma_h}} - 1\right] \quad \text{(A5)}$$

which must be satisfied for patterning to occur.[4] The parameters outlined in Table A2 satisfy Eq. (A5), yielding $0.667 \le 2.00 \le 29.1$ when evaluated. Section 4.1 also discussed a coupled RD system composed of a single non-RD equation and two RD equations. All parameters and coefficients used to simulate this three-equation RD system in COMSOL Multiphysics are listed in Table A3.

## APPENDIX B

All parameters used in Eqs. 4.2.11−4.2.13 to predict the growth outcomes shown in Fig. 4.2.16 of Section 6 are outlined in Table B1.

**TABLE B1**
**Parameters Used in Cranial Vault Simulation.**

| Simulated Time | 1.685e06 (s) |
|---|---|
| $\Delta t$ | 1.728e02 (s) |
| $a_0$ | 1.20e-3 ($\pm$0.5%) (kg/m$^3$) |
| $h_0$ | 1.44e-3 (kg/m$^3$) |
| $o_0$ | 0.00 (kg/m$^3$) |
| $\alpha_a$ | 7.00e-03 (kg/m$^3$) |
| $\alpha_h$ | 2.00e-09 (kg$/[\text{m}^3 * \text{s}]$) |
| $\alpha_o$ | 4.00e02 (1/s) |
| $\beta_a$ | 1.00e-04 (1/s) |
| $\beta_h$ | 2.00e-04 (1/s) |
| $\gamma_a$ | 1.00e-04 (1/s) |
| $\gamma_h$ | 2.5e-01 (m$^3$/[kg$*$s]) |
| $D_a$ | 5.00e-13 (m$^2$/s) |
| $D_h$ | 1.50e-10 (m$^2$/s) |
| $\eta$ | 1.00e-07 (kg$/[\text{m}^3 * \text{s}]$) |
| $a_T$ | 2.00e-03 (kg/m$^3$) |
| $o_s$ | 1.00e-03 (kg/m$^3$) |
| $E_T$ | 5.00e-01 (unitless) |
| $\acute{E}_0$ | 3.00e-06 (1/s) |

From Lee C. A computational analysis of bone formation in the cranial vault using a reaction-diffusion-strain model. 2018; with permission.

## ACKNOWLEDGMENTS

This work was supported in part through instrumentation funded by the National Science Foundation through grant OCI0821527 and by the National Institutes of Health grants R01 DE022988 and P01HD078233. The content is solely the responsibility of the authors and does not necessarily represent the official views of the National Institutes of Health. This work was also supported by the Royal Academy of Engineering (Grant No.10216/119 to M.M.).

## REFERENCES

1. Parker SE, Mai CT, Canfield MA, et al. Updated national birth prevalence estimates for selected birth defects in the United States, 2004–2006. *Birth Defects Res A Clin Mol Teratol*. 2010;88(12):1008–1016. https://doi.org/10.1002/bdra.20735.

2. Trainor PA, Richtsmeier JT. Facing up to the challenges of advancing craniofacial research. *Am J Med Genet*. 2015; 167(7):1451–1454. https://doi.org/10.1002/ajmg.a.37065.

3. Delile J, Herrmann M, Peyriéras N, Doursat R. A cell-based computational model of early embryogenesis coupling mechanical behaviour and gene regulation. *Nat Commun*. 2017;8. https://doi.org/10.1038/ncomms13929.

4. Lee C. *A Computational Analysis of Bone Formation in the Cranial Vault Using a Reaction-Diffusion-Strain Model*; May 2018. https://etda.libraries.psu.edu/catalog/14979cxl503.

5. Wittwer LD, Croce R, Aland S, Iber D. *Simulating Organogenesis in COMSOL: Phase-Field Based Simulations of Embryonic Lung Branching Morphogenesis*. Vol. 7. 2016.

6. Wyczalkowski MA, Chen Z, Filas BA, Varner VD, Taber LA. Computational models for mechanics of morphogenesis. *Birth Defects Res C Embryo Today*. 2012;96(2):132–152. https://doi.org/10.1002/bdrc.21013.

7. Steinberg MS. Differential adhesion in morphogenesis: a modern view. *Curr Opin Genet Dev*. 2007;17(4): 281–286. https://doi.org/10.1016/j.gde.2007.05.002.

8. Keller R. Physical biology returns to morphogenesis. *Science*. 2012;338(6104):201–203. https://doi.org/10.1126/science.1230718.

9. Ingber DE. Mechanical control of tissue morphogenesis during embryological development. *Int J Dev Biol*. 2003; 50(2–3):255–266. https://doi.org/10.1387/ijdb.052044di.

10. Raspopovic J, Marcon L, Russo L, Sharpe J. Digit patterning is controlled by a Bmp-Sox9-Wnt Turing network modulated by morphogen gradients. *Science*. 2014;345(6196): 566–570. https://doi.org/10.1126/science.1252960.

11. Zhu W, Kim J, Cheng C, et al. Noggin regulation of bone morphogenetic protein (BMP) 2/7 heterodimer activity in vitro. *Bone*. 2006;39(1):61–71. https://doi.org/10.1016/j.bone.2005.12.018.

12. Chen D, Zhao M, Mundy GR. Bone morphogenetic proteins. *Growth Factors*. 2004;22(4):233–241. https://doi.org/10.1080/08977190412331279890.

13. Tezuka K, Wada Y, Takahashi A, Kikuchi M. Computer-simulated bone architecture in a simple bone-remodeling model based on a reaction-diffusion system. *J Bone Miner Metab*. 2005;23(1):1–7. https://doi.org/10.1007/s00774-004-0533-z.

14. Turing AM. The chemical basis of morphogenesis. *Philos Trans R Soc Lond Ser B Biol Sci*. 1952;237(641):37–72. https://doi.org/10.1098/rstb.1952.0012.

15. Dlamini PG, Khumalo M. On the computation of blow-up solutions for semilinear ODEs and parabolic PDEs. *Math Probl Eng*. 2012;2012:15. https://doi.org/10.1155/2012/162034.

16. Haugh JM. Analysis of reaction-diffusion systems with anomalous subdiffusion. *Biophys J*. 2009;97(2):435–442. https://doi.org/10.1016/j.bpj.2009.05.014.

17. Gierer A, Meinhardt H. A theory of biological pattern formation. *Kybernetik*. 1972;12(1):30–39. https://doi.org/10.1007/BF00289234.

18. Siregar P, Julen N, Hufnagl P, Mutter G. A general framework dedicated to computational morphogenesis part I — constitutive equations. *Biosystems*. 2018;173:298–313. https://doi.org/10.1016/j.biosystems.2018.07.003.

19. Kondo S, Asai R. A reaction-diffusion wave on the skin of the marine angelfish Pomacanthus. *Nature*. 1995; 376(6543):765–768. https://doi.org/10.1038/376765a0.

20. Newman SA, Comper WD. 'Generic' physical mechanisms of morphogenesis and pattern formation. *Development*. 1990;110(1):1–17.

21. Meinhardt H, Gierer A. Applications of a theory of biological pattern formation based on lateral inhibition. *J Cell Sci*. 1974;15(2):321.

22. Meinhardt H. Models of biological pattern formation: from elementary steps to the organization of embryonic axes. *Curr Top Dev Biol*. 2008;81:1–63. https://doi.org/10.1016/S0070-2153(07)81001-5. Elsevier.

23. Lee C, Richtsmeier JT, Kraft RH. A multiscale computational model for the growth of the cranial vault in craniosynostosis. *Int Mech Eng Congress Expo*. 2014;2014. https://doi.org/10.1115/IMECE2014-38728.

24. Garzón-Alvarado DA, González A, Gutiérrez ML. Growth of the flat bones of the membranous neurocranium: a computational model. *Comput Methods Progr Biomed*. 2013;112(3):655–664. https://doi.org/10.1016/j.cmpb.2013.07.027.

25. Meinhardt H. Turing's theory of morphogenesis of 1952 and the subsequent discovery of the crucial role of local self-enhancement and long-range inhibition. *Interface Focus*. 2012;2(4):407–416. https://doi.org/10.1098/rsfs.2011.0097.

26. Green JBA, Sharpe J. Positional information and reaction-diffusion: two big ideas in developmental biology combine. *Development*. 2015;142(7):1203–1211. https://doi.org/10.1242/dev.114991.

27. Cao Y, Wang H, Ouyang Q, Tu Y. The free energy cost of accurate biochemical oscillations. *Nat Phys*. 2015;11(9): 772–778. https://doi.org/10.1038/nphys3412.

28. Kondo S, Miura T. Reaction-diffusion model as a framework for understanding biological pattern formation. *Science*. 2010;329(5999):1616–1620. https://doi.org/10.1126/science.1179047.

29. Garzón-Alvarado DA, García-Aznar JM, Doblaré M. Appearance and location of secondary ossification centres may be explained by a reaction-diffusion mechanism. *Comput Biol Med*. 2009;39(6):554–561. https://doi.org/10.1016/j.compbiomed.2009.03.012.

30. López-Vaca OR, Garzón-Alvarado DA. Spongiosa primary development: a biochemical hypothesis by Turing patterns formations. *Comput Math Methods Med*. 2012;2012:748302. https://doi.org/10.1155/2012/748302.

31. OpenStax College. *Anatomy & Physiology*. Rice University Press; 2013. https://openstax.org/details/books/anatomy-and-physiology.

32. Sheth R, Marcon L, Félix Bastida M, et al. Hox genes regulate digit patterning by controlling the wavelength of a Turing-type mechanism. *Science*. 2012;338(6113):1476–1480. https://doi.org/10.1126/science.1226804.

33. Cabrera A. Subspecific and individual variation in the Burchell zebras. *J Mammal*. 1936;17(2):89–112. https://doi.org/10.2307/1374181.

34. Lee C, Richtsmeier JT, Kraft RH. A computational analysis of bone formation in the cranial vault in the mouse. *Front Bioeng Biotechnol*. 2015;3. https://doi.org/10.3389/fbioe.2015.00024.

35. Beentjes CHL. *Pattern Formation Analysis in the Schnakenberg Model*. Oxford, UK: University of Oxford; 2005:18.

36. Zhu J, Zhang Y-T, Newman SA, Alber M. Application of discontinuous Galerkin methods for reaction-diffusion systems in developmental biology. *J Sci Comput*. 2009;40(1):391–418. https://doi.org/10.1007/s10915-008-9218-4.

37. Maini PK, Woolley TE, Gaffney EA, Baker RE. Turing's theory of developmental pattern formation. In: *The Once and Future Turing: Computing the World*. 2016. https://doi.org/10.1017/CBO9780511863196.014.

38. Arrarás A, Gaspar FJ, Portero L, Rodrigo C. Domain decomposition multigrid methods for nonlinear reaction-diffusion problems. *Commun Nonlinear Sci Numer Simul*. 2015;20(3):699–710. https://doi.org/10.1016/j.cnsns.2014.06.044.

39. Zhang R, Yu X, Zhu J, Loula AFD. Direct discontinuous Galerkin method for nonlinear reaction-diffusion systems in pattern formation. *Appl Math Model*. 2014;38(5):1612–1621. https://doi.org/10.1016/j.apm.2013.09.008.

40. Burr DB, Allen MR. *Basic and Applied Bone Biology*. 1st ed. New York: Academic Press; 2014. https://www.sciencedirect.com/book/9780124160156/basic-and-applied-bone-biology.

41. Opperman LA. Cranial sutures as intramembranous bone growth sites. *Dev Dynam*. 2000;219(4):472–485. https://doi.org/10.1002/1097-0177(2000)9999:9999<::AID-DVDY1073>3.0.CO;2-F.

42. Lai T-H, Fong Y-C, Fu W-M, Yang R-S, Tang C-H. Osteoblasts-derived BMP-2 enhances the motility of prostate cancer cells via activation of integrins. *Prostate*. 2008;68(12):1341–1353. https://doi.org/10.1002/pros.20799.

43. Pompe W, Rödel G, Weiss H-J, Mertig M. *Bio-Nanomaterials: Designing Materials Inspired by Nature*. 1st ed. Weinheim, Germany: Wiley-VCH; 2013.

44. Li R, Liang L, Dou Y, et al. Mechanical strain regulates osteogenic and adipogenic differentiation of bone marrow mesenchymal stem cells. *Biomed Res Int*. 2015;2015:10. https://doi.org/10.1155/2015/873251.

45. Palomares KTS, Gleason RE, Mason ZD, et al. Mechanical stimulation alters tissue differentiation and molecular expression during bone healing. *J Orthop Res*. 2009;27(9):1123–1132. https://doi.org/10.1002/jor.20863.

46. Sumanasinghe RD, Bernacki SH, Loboa EG. Osteogenic differentiation of human mesenchymal stem cells in collagen matrices: effect of uniaxial cyclic tensile strain on bone morphogenetic protein (BMP-2) mRNA expression. *Tissue Eng*. 2006;12(12):3459–3465. https://doi.org/10.1089/ten.2006.12.3459.

47. Maul TM, Chew DW, Nieponice A, Vorp DA. Mechanical stimuli differentially control stem cell behavior: morphology, proliferation, and differentiation. *Biomech Model Mechanobiol*. 2011;10(6):939–953. https://doi.org/10.1007/s10237-010-0285-8.

48. Jin S-W, Sim K-B, Kim S-D. Development and growth of the normal cranial vault: an embryologic review. *J Korean Neurosurg Soc*. 2016;59(3):192–196. https://doi.org/10.3340/jkns.2016.59.3.192.

49. Ting M-C, Wu NL, Roybal PG, et al. EphA4 as an effector of Twist1 in the guidance of osteogenic precursor cells during calvarial bone growth and in craniosynostosis. *Development*. 2009;136(5):855–864. https://doi.org/10.1242/dev.028605.

50. Malde O, Libby J, Moazen M. An overview of modelling craniosynostosis using the finite element method. *Mol Syndromol*. 2019;10(1–2):74–82. https://doi.org/10.1159/000490833.

51. Marghoub A, Libby J, Babbs C, Ventikos Y, Fagan MJ, Moazen M. Characterizing and modeling bone formation during mouse calvarial development. *Phys Rev Lett*. 2019;122(4). https://doi.org/10.1103/PhysRevLett.122.048103.

52. Weickenmeier J, Fischer C, Carter D, Kuhl E, Goriely A. Dimensional, geometrical, and physical constraints in skull growth. *Phys Rev Lett*. 2017;118(24). https://doi.org/10.1103/PhysRevLett.118.248101.

53. Clendenning DE, Mortlock DP. The BMP ligand Gdf6 prevents differentiation of coronal suture mesenchyme in early cranial development. *PLoS One*. 2012;7(5):e36789. https://doi.org/10.1371/journal.pone.0036789.

54. Yu H-MI, Jerchow B, Sheu T-J, et al. The role of Axin2 in calvarial morphogenesis and craniosynostosis. *Development*. 2005;132(8):1995–2005. https://doi.org/10.1242/dev.01786.

55. Yan Y, Tang D, Chen M, et al. Axin2 controls bone remodeling through the β-catenin-BMP signaling pathway in adult mice. *J Cell Sci*. 2009;122(19):3566–3578. https://doi.org/10.1242/jcs.051904.

56. Marghoub A, Libby J, Babbs C, Pauws E, Fagan MJ, Moazen M. Predicting calvarial growth in normal and craniosynostotic mice using a computational approach. *J Anat.* 2018;232(3):440–448. https://doi.org/10.1111/joa.12764.
57. Carter DR, Blenman PR, Beaupré GS. Correlations between mechanical stress history and tissue differentiation in initial fracture healing. *J Orthop Res.* 1988;6(5):736–748. https://doi.org/10.1002/jor.1100060517.
58. Moazen M, Peskett E, Babbs C, Pauws E, Fagan MJ. Mechanical properties of calvarial bones in a mouse model for craniosynostosis. *PLoS One.* 2015;10(5):e0125757. https://doi.org/10.1371/journal.pone.0125757.
59. Garzón-Alvarado DA. A hypothesis on the formation of the primary ossification centers in the membranous neurocranium: a mathematical and computational model. *J Theor Biol.* 2013;317:366–376. https://doi.org/10.1016/j.jtbi.2012.09.015.
60. Burgos-Flórez FJ, Gavilán-Alfonso ME, Garzón-Alvarado DA. Flat bones and sutures formation in the human cranial vault during prenatal development and infancy: a computational model. *J Theor Biol.* 2016;393:127–144. https://doi.org/10.1016/j.jtbi.2016.01.006.
61. Lee D-F, Su J, Kim HS, et al. Modeling familial cancer with induced pluripotent stem cells. *Cell.* 2015;161(2):240–254. https://doi.org/10.1016/j.cell.2015.02.045.
62. Koch AJ, Meinhardt H. Biological pattern formation: from basic mechanisms to complex structures. *Rev Mod Phys.* 1994;66(4):1481–1507. https://doi.org/10.1103/RevModPhys.66.1481.

# CHAPTER 5

# Future Prospects and Challenges

GLEN L. NIEBUR, PHD

## 1 INTRODUCTION

Mechanobiology is now a firmly established interdisciplinary field in biology, biomedical engineering, and biophysics. Hypotheses exploring links between physical stimuli and biology are now a routine part of studies of cell and tissue function in both healthy and diseased states. Although early studies focused on the musculoskeletal and cardiovascular systems, some role for mechanobiology is now recognized in most tissues, even those that are not routinely subjected to external loading. In these tissues, the contractile forces of the cell cytoskeleton itself, which are generated by most eukaryotic cells, depend on the extracellular matrix (ECM) mechanical properties, and changes in these forces affect intracellular processes through the focal adhesion complex, cytoskeletal remodeling, and altered forces on the nuclear membrane. These processes regulate gene transcription and ultimately regulate the cell activity and fate.[1]

With continually evolving technology, scientists and engineers are now able to study mechanobiology at both smaller and larger scales than previously possible. At the same time, improved quantitative tools in molecular biology, such as RNA-sequencing, in situ hybridization, immunohistochemistry, super-resolution microscopy, and artificial intelligence-based image analysis, have made it possible to more directly measure both the mechanical stimuli acting on cells and the response at both the RNA and protein levels. These technologies will continue to drive mechanobiological research in the next decades.

The goal of this chapter is to highlight some research areas that are expected to make major impacts on the field of mechanobiology in the next decade. Certainly, many other research avenues offer high potential as well, and it is likely that as-yet-unknown techniques

will have a disruptive and revolutionary impact on the field. However, the goal is to highlight the present state of the art.

## 2 CELL AND CYTOSKELETAL MECHANICS

The mechanical behavior of the cell is at the heart of mechanobiology and as such is a logical starting point to discuss future innovation. Novel technologies for quantifying cell mechanical properties are essential to future progress in the field. In addition to describing the relationships between force and deformation in cell populations, mechanical measurements of cells have proven useful for phenotyping cells and detecting normal or abnormal function.[2,3]

Early cell mechanics measurements employed micropipette aspiration to measure membrane tensions and were based on a biophysical analysis of the simple experiment (c.f.[4]). However, these models fail to capture the cell mechanics when attached to substrates or their ability to generate force. Newer approaches, such as atomic force microscopy,[5,6] traction force microscopy,[7,8] or micropillar arrays,[9-11] can provide improved approaches to study cells subjected to inhibitors or promoters of mechanobiological pathways. The Förster resonance energy transfer (FRET)-based force sensors discussed earlier in this book will provide even greater resolution to probe the mechanics of molecules within a cell and their response to external mechanical stimuli.

Many experiments have demonstrated that cell behavior depends on the mechanical characteristics of the culture environment.[12-15] Cells in three dimensions and in heterogeneous cell populations often present different morphologies and, therefore, different mechanical properties than their counterparts in two

Mechanobiology. https://doi.org/10.1016/B978-0-12-817931-4.00012-1

dimensions. Recent studies have begun to address this important problem, which should provide more accurate and detailed descriptions of cell mechanics within tissues.[16,17] These techniques leverage new imaging modalities to directly manipulate cells or tissue with the light beam from the microscope.

## 3 GENE EDITING

The advent of Crispr/Cas9 gene editing is revolutionizing biology, and mechanobiology has been no exception. While genetic engineering has been possible for decades, the Crispr/Cas9 system is more accurate and easier to use than previous techniques. As such, genome editing will become available to a growing number of laboratories. This may allow, for example, more rapid development of FRET-based force sensors in cells, more accurate lineage tracing of cells that are implicated in mechanobiology, alteration of mechanobiological pathway genes in specific cell lineages, and improved understanding of the progression of tissue-engineered constructs.[18] The technology has already been widely applied in the development of zebra fish and mouse models for orthopedics.[19]

Gene editing is likely to play a growing role in mechanobiology through engineered tissue- and organ-mimicking culture systems. Tissue-engineered experimental systems can be seeded with genetically engineered cells to investigate the role of individual genes and their dependent pathways on mechanobiological responses of cells to injury.[20]

## 4 GENETICALLY MODIFIED ORGANISMS

The development of Cre-Lox recombination in both cell lines and animals has had one of the greatest impacts on mechanobiology research. Tissue-specific knockout or constitutive activation of genes can be accomplished by using Cre promotors specific to a cell lineage that is specific to the tissue of interest. With these models, it is possible to study the role of a single gene in mechanobiological processes with minimal cross talk to other tissues, thereby eliminating confounding systemic effects. Similarly, tamoxifen- or tetracycline-driven knockout or knockin allows researchers to study the effects of mechanoregulatory genes proximal to mechanical interventions or at other specific time points in the animal's lifespan by spiking drinking water or feed with the drug at the desired time points. With these technologies, genetic modifications that might normally be embryonically or early postnatally lethal can now be studied. For example, the WNT pathway has been implicated in many mechanobiological processes. However, WNT is essential to normal development in most organisms, and global knockout of the gene results in abnormal development, birth defects, or embryonic death[21]. However, using dental metalloproteinase (DMP-1), a gene that is uniquely expressed in osteocytes, as a promoter for Cre recombination, the LRP5 coreceptor for WNT was deleted in mechanosensing osteocytes, but not in the bone-forming osteoblasts or their earlier precursors[22] to test the specific role of WNT signaling in bone mechanotransduction. Alternatively, WNT can be constitutively activated in osteocytes, which demonstrates the bone formation is hyperactivated[23]. Similarly, tamoxifen-driven deletion or constitutive expression of the YAP/Taz transcription factors, which are mechanically regulated and play a role in the HIPPO pathway regulation, allowed normal skeletal development and growth of the animal, followed by deletion of the gene during mechanical interventions[24,25].

Although Cre/Lox recombination has been employed in mechanobiology research for more than a decade, it will continue to be a workhorse method in the future. Although tissue and organ-on-a-chip models will continue to advance, it is likely that the small animal model will be a critical step to human translation for the foreseeable future.

## 5 IN VIVO LOADING

When using animal models to study mechanobiology, it is necessary to devise a means to control or alter the mechanical environment of the cells or tissue. Several well-established animal models exist for bone,[26−29] tendon,[30] and blood vessels.[31,32] These can either be hyperloaded or unloaded conditions that allow the effects of mechanics to be observed over time. A major recent advance in this field has been the application of intravital microscopy to observe mechanobiological signaling by secondary messenger molecules during loading.[32,33] With these techniques, it is possible to directly observe the response of cells embedded in a tissue to the altered mechanical environment. Undoubtedly, new reporter molecules and more powerful imaging techniques will allow more detailed measurements of other molecular trafficking within the cell.

In vivo loading models have also benefitted from enhanced biomechanical modeling and imaging technologies. High-resolution computed tomography, magnetic resonance imaging (MRI), and ultrasonography provide a means to assess changes in tissue composition and geometry longitudinally.[34−37] When coupled with computational modeling, these techniques provide a unique means to assess the mechanobiological response of normal or healing tissues.

# 6 THREE-DIMENSIONAL TISSUE CULTURE AND ORGANS-ON-A-CHIP

Cells live in three-dimensional niches in vivo, and recapitulating those niches in experiments is essential to fully understand their behavior. An important recent advance in tissue engineering has been the development of tissue-engineered and organ culture models. A major National Institutes of Health initiative to create a "human-body-on-a-chip" for use in disease modeling and drug testing led to the development of new techniques to engineer tissues subjected to unique chemical and mechanical environments. Application of these techniques to mechanobiology is advantageous because precise control and measurements of the loads is possible.[38,39] Organ culture—harvested tissue explants or whole bones cultured in bioreactors—can play a similarly powerful role by enabling controlled loading to quantify changes in tissue composition, tissue growth, or gene expression.[40–43] These are particularly advantageous to study the effects of pathologic loads on tissues, as altered gene and protein expression can be mapped to a range of altered mechanical environments to develop and test relevant mechanobiological hypotheses.[44–46] By using tissues from genetically altered animals, specific pathways can be probed.

Tissue-engineered constructs cultured in bioreactors open the possibility to more specifically control the cell phenotype or genotype to create specific disease models.[38,39] In combination with additive manufacturing methods, complex cell-seeded geometries can be created to mimic tissues.[47] In combination with other techniques such as gene editing[20] and advanced imaging techniques, tissue-engineered constructs open the door to rigorously determine the cellular mechanisms underlying mechanobiological phenomena. However, a major challenge to address is the degree to which the tissue-engineered construct recapitulates the real tissue.

# 7 OMICS

Omics technology will greatly enhance our ability to study mechanobiology. Researchers were previously constrained to studies of a few gene transcripts by quantitative polymerase chain reaction or a few proteins by Western blots or enzyme-linked immunosorbent assay. However, RNA-sequencing, proteomics, and metabolomics now provide a means to explore the response of cells to mechanical loading across the genome. In conjunction with novel animal models and tissue-engineered experimental systems, omics approaches can provide detailed insight into the multiple interacting pathways that are activated by mechanical loading.[31,48] The advent of single-cell RNA-sequencing has made these methods even more powerful. By uniquely tagging each RNA fragment in a cell, the variability of the transcriptome within a cell population can be quantified, and differing cell populations within a sample can be identified to understand the response on a systemic level.[49,50] Although only a few years old, single-cell RNA-sequencing is now available at most major research institutions. These capabilities will allow mechanobiological research to progress on complex tissues where multiple cell lineages interact to achieve normal function.

# 8 IMAGING AND IMAGE ANALYSIS

Improved methods of imaging have driven the field of biology forward throughout history. Today, researchers have the enormous power to identify and quantify not only physical structures but also the presence of specific molecules including individual RNA segments within the nucleus of a cell. Spinning disc confocal, light sheet, and super-resolution microscopic techniques can now produce enormous three-dimensional image data sets with in situ hybridization or immunohistochemical staining.[51] For hard tissues, serial imaging using focused ion beam milling and scanning electron microscopy has been used to obtain detailed images of the lacunocanalicular system.[52]

Manual processing of large image data sets would be impossible. However, automated segmentation and pattern recognition techniques will provide a means to interrogate the data and identify regions of interest to researchers.[53,54] Novel methods of cataloging features in the data using graph and network theory will further allow relationships between features to be identified automatically, providing a powerful tool for exploratory research and hypothesis development.

Imaging and advanced image processing methods can also be used to generate detailed geometric models for computational mechanics. This is especially powerful when the models can be combined with measurements of cell signaling via reporter molecules,[33,55] local cell populations,[56] or mechanical behavior.[57] Multimodal image registration provides a means to merge data sets with structural and molecular data, such as X-ray computed tomography, single-photon emission computed tomography, positron emission tomography, and MRI, with microscopic images that include spatial immunohistochemical, histomorphometric, and in situ hybridization data. Methods that exploit multimodality imaging and correlate it with computational models of tissue mechanics will continue to be important. In particular, registering and interpolating data at differing resolutions or with missing data will be essential to merge experimental and computational results to better understand mechanobiological experiments.

## 9 COMPUTATIONAL MODELS

Computational models should continue to improve our insight into mechanobiological processes. Accurate calculations of tissue mechanics during physical activities are essential to understand how changes in ECM properties, tissue and organ geometry, and applied loads can affect the mechanical environment experienced by cells. Adaptive remodeling algorithms have been developed for skeletal and cardiovascular tissues. Recently, these have been enhanced to capture the specific activities of cells, including intercellular signaling and matrix production. As seen in the last two chapters, computational models in mechanobiology play a different role than in traditional engineering. Rather than providing a design tool to assess the performance of the system, mechanobiological models provide an important tool to develop hypotheses regarding the interactions of mechanics and biochemistry during development, where mechanical loads and changing material properties can influence the diffusion of molecules to affect morphogenesis.

Although not discussed in this book, organ- and tissue-scale models also provide powerful tools to investigate the relationships between mechanics and ECM remodeling, growth, and development, with applications in skin,[58] brain,[59] bone,[60–63] tendon,[64–66] and blood vessels.[67–69] Such models have been used to make connections between mechanical cues and altered morphology by predicting difference in outcomes associated with loaded and unloaded limbs.

Integrating models at multiple scales and with differing underlying physical and chemical principles is an essential goal. Coupling phenomenological models of tissue mechanics at the macroscopic scale to the local chemical and biochemical environment[70–73] is necessary to understand the processes within the cell that drive mechanobiological responses in both healthy and diseased states. With modern graphics processing unit solvers, it is now possible to solve these coupled equations across length scales. An example of this includes the models of thrombus formation under the influence of biochemical factors

and blood flow, which provide a means to hypothesize how cell adhesion competes with fluid shear.[74]

Moving forward, such coupled models incorporating both mechanical and biochemical signaling will be one of the most powerful approaches to interrogate mechanobiological processes because they will allow integration of information from the two domains. One important approach may be Bayesian inference, which provides a robust means to map between experimental observations and theoretical models to estimate parameters.[8,75] The advantage of these approaches is that parameters can be estimated with confidence intervals on a population, thereby allowing a direct way to apply the results to experiments. One important role of Bayesian inference will be the integration of data at different scales or densities. For example, current methods allow imaging of protein and gene expression in cells on a relatively small number of two-dimensional planes in a tissue. In contrast, the stress or strain field may be calculated from a finite element model with effectively continuous data in three dimensions.

## 10 DATA MINING, ARTIFICIAL INTELLIGENCE, AND BIOINFORMATICS

One of the key technologic challenges in mechanobiology is the explosion of experimental methods and the quantity of data that can be collected from them. For example, RNA-sequencing can produce millions of reads of RNA fragments that must be assembled into transcripts. A single data set can easily exceed 1 GB for omics-based investigations. Manually processing this data is impossible. Instead, researchers depend on high-speed computers and networks to perform bioinformatic processing and pattern recognition. Similarly, the imaging methods described produce enormous amounts of data that include multiple mechanical and biochemical signals. In many mechanobiological studies, multiple data-intensive experimental techniques may be employed, with the goal of correlating the results in time and space.

As an example of the potential, a recent clinical trial combined coronary computed tomography angiography–based modeling with proteomics in order to quantify the risk factors for coronary artery disease (CAD). A deep learning algorithm was used to identify patterns in proteins that were associated with plaques detected in the angiographic images. The researchers were able to identify 35 genes of 358 tested that were associated with CAD while compensating for interaction effects. In orthopedics, statistical shape matching, essentially a data mining technique, has been used to identify geometric and density features of bones that are associated with fracture risk.[76–79]

These big data approaches offer the potential to make connections and interpret extremely dense data sets from multiple sciences. Patterns in the data can further be used to develop hypotheses to identify mechanisms and potential drug targets.

## 11  SUMMARY

Mechanobiological research has driven the development of new experimental approaches in biology, which have been enabled by new technologies. Interdisciplinary collaborations of biologists, physicists, chemists, and engineers have been essential to these advances. The ability to identify critical questions and seek out mathematicians, scientists, and engineers with the knowledge to develop new experimental and computational approaches to answer them has become, and will continue to be, the norm rather than the exception in mechanobiology. Teams that are able to integrate across scales to quantify cellular level responses and integrate them across a tissue or organ are essential to progress in the field.[48]

Perhaps the greatest advance in mechanobiology is that it is now a mainstream concept in biology, biophysics, and bioengineering. Questions about the mechanical environment of cells and how it contributes to their function are now an integral part of many studies of biological functions of cells. The importance of the cellular microenvironment or niche is now widely recognized, and the cellular microenvironment includes not only the neighboring cell phenotypes and biochemical environment but also the mechanical environment. Recognizing the importance of a cell's physical surroundings and understanding the resulting effects on cell function will open new avenues to understand aging, health, and disease and their relationship to changes in tissue and organ composition.

## REFERENCES

1. Mammoto A, Mammoto T, Ingber DE. Mechanosensitive mechanisms in transcriptional regulation. *J Cell Sci.* 2012;125:3061–3073.
2. Chen J, Zheng Y, Tan Q, et al. Classification of cell types using a microfluidic device for mechanical and electrical measurement on single cells. *Lab Chip.* 2011;11:3174.
3. Qi D, Hoelzle DJ, Rowat AC. Probing single cells using flow in microfluidic devices. *Eur Phys J Spec Top.* 2012; 204:85–101.
4. Jacobs CR, Huang H, Kwon RY. *Introduction to Cell Mechanics and Mechanobiology.* Garland Science; 2012.
5. Darling EM, Zauscher S, Guilak F. Viscoelastic properties of zonal articular chondrocytes measured by atomic force microscopy. *Osteoarthr Cartil.* 2006;14:571–579.
6. Darling EM, Zauscher S, Block JA, Guilak F. A thin-layer model for viscoelastic, stress-relaxation testing of cells using atomic force microscopy: Do cell properties reflect metastatic potential? *Biophys J.* 2007;92:1784–1791.
7. Burton K, Taylor DL. Traction forces of cytokinesis measured with optically modified elastic substrata. *Nature.* 1997;385:450–454.
8. Huang Y, Schell C, Huber TB, et al. Traction force microscopy with optimized regularization and automated bayesian parameter selection for comparing cells. *Sci Rep.* 2019;9:539.
9. Tan JL, Tien J, Pirone DM, Gray DS, Bhadriraju K, Chen CS. Cells lying on a bed of microneedles: an approach to isolate mechanical force. *Proc Natl Acad Sci USA.* 2003; 100:1484–1489.
10. Beussman KM, Rodriguez ML, Leonard A, Taparia N, Thompson CR, Sniadecki NJ. Micropost arrays for measuring stem cell-derived cardiomyocyte contractility. *Methods.* 2016;94:43–50.
11. Ribeiro AJ, Denisin AK, Wilson RE, Pruitt BL. For whom the cells pull: hydrogel and micropost devices for measuring traction forces. *Methods.* 2016;94:51–64.
12. Engler AJ, Sen S, Sweeney HL, Discher DE. Matrix elasticity directs stem cell lineage specification. *Cell.* 2006;126: 677–689.
13. Holle AW, Engler AJ. Cell rheology: stressed-out stem cells. *Nat Mater.* 2010;9:4–6.
14. Reilly GC, Engler AJ. Intrinsic extracellular matrix properties regulate stem cell differentiation. *J Biomech.* 2010;43: 55–62.
15. Tse JR, Engler AJ. Stiffness gradients mimicking in vivo tissue variation regulate mesenchymal stem cell fate. *PLoS One.* 2011;6:e15978.
16. Moo EK, Sibole SC, Han SK, Herzog W. Three-dimensional micro-scale strain mapping in living biological soft tissues. *Acta Biomater.* 2018;70:260–269.
17. Dagro A, Rajbhandari L, Orrego S, Kang SH, Venkatesan A, Ramesh KT. Quantifying the local mechanical properties of cells in a fibrous three-dimensional microenvironment. *Biophys J.* 2019;117:817–828.
18. Adkar SS, Wu CL, Willard VP, et al. Step-wise chondrogenesis of human induced pluripotent stem cells and purification via a reporter allele generated by crispr-cas9 genome editing. *Stem Cells.* 2019;37:65–76.
19. Wu N, Liu B, Du H, et al. The progress of crispr/cas9-mediated gene editing in generating mouse/zebrafish models of human skeletal diseases. *Comput Struct Biotechnol J.* 2019;17:954–962.
20. Acun A, Zorlutuna P. Crispr/cas9 edited induced pluripotent stem cell-based vascular tissues to model aging and disease-dependent impairment. *Tissue Eng A.* 2019;25: 759–772.

21. Aoki K, Taketo MM. Tissue-specific transgenic, conditional knockout and knock-in mice of genes in the canonical wnt signaling pathway. *Methods Mol Biol.* 2008;468:307—331.

22. Zhao L, Shim JW, Dodge TR, Robling AG, Yokota H. Inactivation of lrp5 in osteocytes reduces young's modulus and responsiveness to the mechanical loading. *Bone.* 2013;54: 35—43.

23. Osorio J. Bone. Osteocyte-specific activation of the canonical wnt-beta catenin pathway stimulates bone formation. *Nat Rev Endocrinol.* 2015;11:192.

24. McDermott AM, Herberg S, Mason DE, Collins JM, Pearson HB, Dawahare JH, Tang R, Patwa AN, Grinstaff MW, Kelly DJ, Alsberg E, Boerckel JD. Recapitulating bone development through engineered mesenchymal condensations and mechanical cues for tissue regeneration. *Sci Transl Med.* 2019;11.

25. Kegelman CD, Mason DE, Dawahare JH, Horan DJ, Vigil GD, Howard SS, Robling AG, Bellido TM, Boerckel JD. Skeletal cell yap and taz combinatorially promote bone development. *Faseb Journal.* 2018;32: 2706—2721.

26. Grimston SK, Goldberg DB, Watkins M, Brodt MD, Silva MJ, Civitelli R. Connexin43 deficiency reduces the sensitivity of cortical bone to the effects of muscle paralysis. *J Bone Miner Res.* 2011;26:2151—2160.

27. Lynch ME, Main RP, Xu Q, et al. Tibial compression is anabolic in the adult mouse skeleton despite reduced responsiveness with aging. *Bone.* 2011;49:439—446.

28. Bivi N, Pacheco-Costa R, Brun LR, et al. Absence of cx43 selectively from osteocytes enhances responsiveness to mechanical force in mice. *J Orthop Res.* 2013;31:1075—1081.

29. Main RP, Lynch ME, van der Meulen MC. Load-induced changes in bone stiffness and cancellous and cortical bone mass following tibial compression diminish with age in female mice. *J Exp Biol.* 2014;217:1775—1783.

30. Thampatty BP, Wang JH. Mechanobiology of young and aging tendons: in vivo studies with treadmill running. *J Orthop Res.* 2018;36:557—565.

31. Simmons RD, Kumar S, Jo H. The role of endothelial mechanosensitive genes in atherosclerosis and omics approaches. *Arch Biochem Biophys.* 2016;591:111—131.

32. Hilscher MB, Sehrawat T, Arab JP, et al. Mechanical stretch increases expression of cxcl1 in liver sinusoidal endothelial cells to recruit neutrophils, generate sinusoidal microthrombi, and promote portal hypertension. *Gastroenterology.* 2019;157, 193-209 e199.

33. Lewis KJ, Frikha-Benayed D, Louie J, et al. Osteocyte calcium signals encode strain magnitude and loading frequency in vivo. *Proc Natl Acad Sci USA.* 2017;114: 11775—11780.

34. Boerckel JD, Uhrig Ba, Willett NJ, Huebsch N, Guldberg RE. Mechanical regulation of vascular growth and tissue regeneration in vivo. *Proc Natl Acad Sci USA.* 2011.

35. Birkhold AI, Razi H, Weinkamer R, Duda GN, Checa S, Willie BM. Monitoring in vivo (re)modeling: a computational approach using 4D microCT data to quantify bone surface movements. *Bone.* 2015;75:210—221.

36. Hoerth RM, Baum D, Knotel D, et al. Registering 2d and 3d imaging data of bone during healing. *Connect Tissue Res.* 2015;56:133—143.

37. Birkhold AI, Razi H, Duda GN, Checa S, Willie BM. Tomography-based quantification of regional differences in cortical bone surface remodeling and mechanoresponse. *Calcif Tissue Int.* 2017;100:255—270.

38. Lozito TP, Alexander PG, Lin H, Gottardi R, Cheng AW, Tuan RS. Three-dimensional osteochondral microtissue to model pathogenesis of osteoarthritis. *Stem Cell Res Ther.* 2013;4(Suppl 1):S6.

39. Occhetta P, Mainardi A, Votta E, et al. Hyperphysiological compression of articular cartilage induces an osteoarthritic phenotype in a cartilage-on-a-chip model. *Nat Biomed Eng.* 2019;3:545—557.

40. Vivanco J, Garcia S, Ploeg HL, Alvarez G, Cullen D, Smith EL. Apparent elastic modulus of ex vivo trabecular bovine bone increases with dynamic loading. *Proc Inst Mech Eng H J Eng Med.* 2013;227:904—912.

41. Birmingham E, Niebur GL, McNamara LM, McHugh PE. An experimental and computational investigation of bone formation in mechanically loaded trabecular bone explants. *Ann Biomed Eng.* 2016;44:1191—1203.

42. Meyer LA, Johnson MG, Cullen DM, et al. Combined exposure to big endothelin-1 and mechanical loading in bovine sternal cores promotes osteogenesis. *Bone.* 2016; 85:115—122.

43. Hemmatian H, Jalali R, Semeins CM, et al. Mechanical loading differentially affects osteocytes in fibulae from lactating mice compared to osteocytes in virgin mice: possible role for lacuna size. *Calcif Tissue Int.* 2018;103: 675—685.

44. Sun L, Chandra S, Sucosky P. Ex vivo evidence for the contribution of hemodynamic shear stress abnormalities to the early pathogenesis of calcific bicuspid aortic valve disease. *PLoS One.* 2012;7:e48843.

45. Atkins SK, Moore AN, Sucosky P. Bicuspid aortic valve hemodynamics does not promote remodeling in porcine aortic wall concavity. *World J Cardiol.* 2016;8:89—97.

46. Liu J, Cornelius K, Graham M, et al. Design and computational validation of a novel bioreactor for conditioning vascular tissue to time-varying multidirectional fluid shear stress. *Cardiovasc Eng Technol.* 2019;10:531—542.

47. Zhang YS, Arneri A, Bersini S, et al. Bioprinting 3d microfibrous scaffolds for engineering endothelialized myocardium and heart-on-a-chip. *Biomaterials.* 2016;110:45—59.

48. Scheuren A, Wehrle E, Flohr F, Muller R. Bone mechanobiology in mice: toward single-cell in vivo mechanomics. *Biomechanics Model Mechanobiol.* 2017;16:2017—2034.

49. Macosko EZ, Basu A, Satija R, et al. Highly parallel genome-wide expression profiling of individual cells using nanoliter droplets. *Cell.* 2015;161:1202—1214.

50. Barron M, Zhang S, Li J. A sparse differential clustering algorithm for tracing cell type changes via single-cell RNA-sequencing data. *Nucleic Acids Res.* 2018;46:e14.

51. Zhang Y, Nichols EL, Zellmer AM, et al. Generating intravital super-resolution movies with conventional

microscopy reveals actin dynamics that construct pioneer axons. *Development.* 2019;146.

52. Schneider P, Meier M, Wepf R, Muller R. Serial FIB/SEM imaging for quantitative 3d assessment of the osteocyte lacuno-canalicular network. *Bone.* 2011;49:304−311.

53. Guldner IH, Yang L, Cowdrick KR, et al. An integrative platform for three-dimensional quantitative analysis of spatially heterogeneous metastasis landscapes. *Sci Rep.* 2016;6.

54. Repp F, Kollmannsberger P, Roschger A, et al. Coalignment of osteocyte canaliculi and collagen fibers in human osteonal bone. *J Struct Biol.* 2017;199:177−186.

55. Verbruggen SW, Vaughan TJ, McNamara LM. Fluid flow in the osteocyte mechanical environment: a fluid-structure interaction approach. *Biomechanics Model Mechanobiol.* 2014;13:85−97.

56. Cresswell EN, Goff MG, Nguyen TM, Lee WX, Hernandez CJ. Spatial relationships between bone formation and mechanical stress within cancellous bone. *J Biomech.* 2016;49:222−228.

57. Turnbull TL, Baumann AP, Roeder RK. Fatigue microcracks that initiate fracture are located near elevated intracortical porosity but not elevated mineralization. *J Biomech.* 2014; 47:3135−3142.

58. Zöllner AM, Holland MA, Honda KS, Gosain AK, Kuhl E. Growth on demand: reviewing the mechanobiology of stretched skin. *J Mech Behav Biomed Mater.* 2013;28:495−509.

59. Holland MA, Miller KE, Kuhl E. Emerging brain morphologies from axonal elongation. *Ann Biomed Eng.* 2015;43:1640−1653.

60. Carter DR, Van Der Meulen MC, Beaupre GS. Mechanical factors in bone growth and development. *Bone.* 1996;18:5S−10S.

61. van der Meulen MCH, Prendergast PJ. Mechanics in skeletal development, adaptation and disease. *Philos Trans R Soc Lond A.* 2000;358:565−578.

62. Nowlan NC, Murphy P, Prendergast PJ. A dynamic pattern of mechanical stimulation promotes ossification in avian embryonic long bones. *J Biomech.* 2008;41:249−258.

63. Verbruggen SW, Kainz B, Shelmerdine SC, et al. Altered biomechanical stimulation of the developing hip joint in presence of hip dysplasia risk factors. *J Biomech.* 2018;78:1−9.

64. Wren TA, Beaupre GS, Carter DR. Tendon and ligament adaptation to exercise, immobilization, and remobilization. *J Rehabil Res Dev.* 2000;37:217−224.

65. Wren TA, Beaupre GS, Carter DR. Mechanobiology of tendon adaptation to compressive loading through fibrocartilaginous metaplasia. *J Rehabil Res Dev.* 2000;37:135−143.

66. Thompson MS, Bajuri MN, Khayyeri H, Isaksson H. Mechanobiological modelling of tendons: review and future opportunities. *Proc Inst Mech Eng H.* 2017;231:369−377.

67. Taber LA, Humphrey JD. Stress-modulated growth, residual stress, and vascular heterogeneity. *J Biomech Eng.* 2001;123:528−535.

68. Valentin A, Humphrey JD, Holzapfel GA. A finite element-based constrained mixture implementation for arterial growth, remodeling, and adaptation: theory and numerical verification. *Int J Numer Method Biomed Eng.* 2013;29:822−849.

69. Stalhand J, Holzapfel GA. Length adaptation of smooth muscle contractile filaments in response to sustained activation. *J Theor Biol.* 2016;397:13−21.

70. Bailon-Plaza A, van der Meulen MCH. A mathematical framework to study the effects of growth factor influences on fracture healing. *J Theor Biol.* 2001;212:191−209.

71. Bailon-Plaza A, van der Meulen MCH. Beneficial effects of moderate, early loading and adverse effects of delayed or excessive loading on bone healing. *J Biomech.* 2003;36:1069−1077.

72. Checa S, Prendergast PJ. A mechanobiological model for tissue differentiation that includes angiogenesis: a lattice-based modeling approach. *Ann Biomed Eng.* 2009;37:129−145.

73. Khayyeri H, Checa S, Tagil M, Prendergast PJ. Corroboration of mechanobiological simulations of tissue differentiation in an in vivo bone chamber using a lattice-modeling approach. *J Orthop Res.* 2009;27:1659−1666.

74. Xu Z, Chen N, Kamocka MM, Rosen ED, Alber M. A multiscale model of thrombus development. *J R Soc Interface.* 2008;5:705−722.

75. Hines KE. A primer on bayesian inference for biophysical systems. *Biophys J.* 2015;108:2103−2113.

76. Bredbenner TL, Eliason TD, Potter RS, Mason RL, Havill LM, Nicolella DP. Statistical shape modeling describes variation in tibia and femur surface geometry between control and incidence groups from the osteoarthritis initiative database. *J Biomech.* 2010;43:1780−1786.

77. Nicolella DP, Bredbenner TL. Development of a parametric finite element model of the proximal femur using statistical shape and density modelling. *Comput Methods Biomech Biomed Eng.* 2012;15:101−110.

78. Bredbenner TL, Eliason TD, Francis WL, McFarland JM, Merkle AC, Nicolella DP. Development and validation of a statistical shape modeling-based finite element model of the cervical spine under low-level multiple direction loading conditions. *Front Bioeng Biotechnol.* 2014;2:58.

79. Bredbenner TL, Mason RL, Havill LM, Orwoll ES, Nicolella DP, S. Osteoporotic Fractures in Men. Fracture risk predictions based on statistical shape and density modeling of the proximal femur. *J Bone Miner Res.* 2014; 29:2090−2100.

# Glossary

Adherens junction: Adherens junctions are protein complexes that form cell-cell junctions. Transmembrane proteins called cadherins form bonds with one another. These junctions are very important in endothelial tissues, where they help cells to bind together to form barriers.

Cadherin: Calcium-dependent adhesion. A cell adhesion molecule that forms adherens junctions.

Cardiomyocyte: A cardiac cell that is capable of contracting or beating to generate a force.

Cardiofibroblast: A cardiac cell that does not contract or beat.

Cellular Potts Model (CPM): A computational model in which a rectangular lattice is used to compute the action of a group of cells. Each cell occupies a number of points on the lattice. The cells occupying each lattice region are updated iteratively according to an energy function. These models are commonly used to simulate cell sorting and developmental biology.

Chondrocyte: Chondrocytes are cells embedded in the cartilage. They form the cartilage matrix and maintain it.

Connexin: Connexins are cell-cell junctions that form a channel between cells to allow molecules to be transported. Connexins can regulate molecules that can cross through the channel. Connexins (cx-43) are essential to osteocyte function, allowing cells to communicate.

Cilium (plural cilia): A cilium is an organelle that takes the form of a protuberance from the cell body. Some cilia are motile, allowing cells to propel themselves through fluid, and others are nonmotile and act as mechano- and chemosensors. Cilia are critical structures on the hair cells of the ear that deflect in response to pressure changes to open calcium channels. The primary cilium extends from the cell centriole and is made up of microtubules. It can extend or contract in response to mechanical loads in bone and tendon and is believed to play a role in both mechanical and chemical sensing.

Elastic: An elastic material is one that deforms when forces are applied to it, and it returns to its original shape after the loads are removed.

Endothelium: The layer of cells that line the lumen of blood or lymphatic vessels.

Epithelium: The layer of cells that line the outside of organs or tissues to form a barrier.

Extracellular matrix (ECM): Proteins, proteoglycans, and other molecules that make up a tissue or organ and lie outside the cell.

Fibroblast: A cell that secretes extracellular matrix molecules such as collagen.

Finite element model (FEM): A finite element model is a computational model that solves a continuous system of partial differential equations in space (and possibly time) by approximating the global solution as a piecewise continuous set of solutions. Each part of the piecewise continuous solution is defined on one small but finitely sized region in space (and/or time). The global solution contains errors because the set of piecewise continuous approximations may have discontinuous derivatives at the boundaries between elements. If the derivatives are continuous across all the boundaries, the solution is exact. The magnitude of the error decreases as the size of the elements decreases, and the piecewise function approximates the smooth continuous function.

Focal adhesion: A cell structure composed of a cluster of integrins that forms an attachment of the cell to the extracellular matrix.

Functional adaptation: Deposition or resorption of extracellular matrix, changes in cell populations, or other biological actions that alter the structure, properties, or composition of a tissue in response to environmental factors.

Integrin: Integrins are transmembrane molecules that allow a cell to bind to a surface. Clusters of integrins are called focal adhesions. The integrin is attached to the cytoskeleton in the cytoplasm. The extracellular domain of the integrin attaches to proteins or peptides in the extracellular matrix. Tension on the integrin can cause conformal changes to the intracellular domain of the molecule, causing downstream signaling.

Ligament: A soft, elastic, fibrous tissue composed primarily of collagen and elastin that joins bones together, especially at articular joints.

Mechanobiology: A transdisciplinary subfield of biology, physics, and engineering that deals with biological processes that are controlled or influenced by mechanical factors, especially those that deform or constrain the shape of the cell.

Mechanostat theory: A theory developed in the 1960s by Harold Frost and colleagues that likened the response of bone to mechanical loads to a thermostat. Bone remodeling is in equilibrium for some small region around a set point and responds with either formation or resorption when the mechanical signal falls outside this region. This is analogous to a thermostat in which the heating system starts only if the temperature falls below the set temperature by some fixed amount and the cooling system starts when the temperature exceeds the setting by a fixed amount.

*Modulus:* A material constant that represents the change in stress for a unit change in strain. In linearly elastic materials the modulus is constant. In nonlinearly elastic materials, the modulus is a function of the present strain and is defined as the slope (or derivative) of the stress—strain curve. The *shear* modulus relates the stress parallel to a surface, i.e., the shear stress, to the resulting shearing or distorting strain. The *bulk* modulus relates the hydrostatic pressure to the change in volume. *Young's* modulus relates the tensile stress to the elongation.

*Osteocyte:* An osteocyte is a terminally differentiated osteoblast embedded in the mineralized bone matrix. Osteocytes form a network to communicate with cells on the bone surface and the vasculature.

*Osteoblasts:* Mesenchymal lineage cells that form osteoid, which is later mineralized to form bone.

*Osteoclasts:* Multinucleated hematopoietic lineage cells that resorb bone. Osteoclasts are a type of macrophage.

*Pericyte:* A cell that wraps around the endothelial cells that make up the blood or lymphatic vessels. It is an important part of the blood brain barrier.

*Progressive open-angle glaucoma (POAG):* A progressive neurodegenerative disease that leads to loss of vision.

*Retinal ganglion cell (RGC):* A cell that transmits visual information to the brain.

*Tendon:* A soft, elastic, fibrous, tissue composed primarily of collagen and elastin that joins muscle to bone.

*Tenocyte:* A cell embedded in tendon that forms and maintains the extracellular matrix.

*Wolff's law:* "Every change in the ... function of bone ... is followed by certain definite changes in ... internal architecture and external conformation in accordance with mathematical laws."

# Index

*Note*: Page numbers followed by "t" indicate tables, "f" indicate figures and "b" indicate boxes.

Printed in the United States
By Bookmasters